# Physiology and Biochemistry of Plant–Pathogen Interactions

# Physiology and Biochemistry of Plant–Pathogen Interactions

**I. J. Misaghi**
*University of Arizona*
*Tucson, Arizona*

PLENUM PRESS • NEW YORK AND LONDON

Library of Congress Cataloging in Publication Data

Misaghi, I. J., Date–
Physiology and biochemistry of plant–pathogen interactions.

Bibliography: p.
Includes index.
1. Plant diseases. 2. Plant diseases—Physiological aspects. 3. Host–parasite relationships. 4. Host plants. 5. Micro-organisms, Phytopathogenic. 6. Botanical chemistry. I. Title.
SB731.M54 1982 581.2′3 82-18594
ISBN 0-306-41059-1



Plenum Press is a Division of
Plenum Publishing Corporation
233 Spring Street, New York, N.Y. 10013



Printed in the United States of America

**To Trish and Mark**

# Preface

There has been a significant surge of interest in the study of the physiology and biochemistry of plant host–parasite interactions in recent years, as evidenced by the number of research papers currently being published on the subject. The increased interest is probably based on the evidence that effective management of many plant diseases is, for the most part, contingent upon a clear understanding of the nature of host–parasite interactions. This intensified research effort calls for a greater number of books, such as this one, designed to compile, synthesize, and evaluate widely scattered pieces of information on this subject.

The study of host–parasite interactions concerns the struggle between plants and pathogens, which has been incessant throughout their coevolution. Such interactions are often highly complex. Pathogens have developed sophisticated offensive systems to parasitize plants, while plants have evolved diversified defensive strategies to ward off potential pathogens. In certain cases, the outcome of a specific host–parasite interaction seems to depend upon the presence or efficacy of the plant's defense system. A plant may become diseased when a parasite manages to invade it, unhindered by preexisting defense systems and/or without eliciting the plant's induced resistance response(s). Absence of disease may reflect the inability of the invading pathogen to overcome the plant's defense system(s).

This book, based on my lecture notes, is an attempt to bring together much of the literature on the physiology and biochemistry of plant–parasite interactions for advanced students, teachers, and researchers of the physiology and biochemistry of healthy and diseased plants. It is a brief, up-to-date, extensively referenced account of the current state of thinking in this unique and exciting area. Much of the book concerns the impressive amount of knowledge on host–parasite interactions accumulated during the seventies and early eighties. However, to provide the reader with an objective view, some early literature has also been included. Formulation of sound concepts requires a thorough knowledge of the nature of the phenomenon being considered. Therefore, whenever possible I have pointed out uncertainties and emphasized the speculative nature of certain suggestions.

Emphasis has been placed on organization and categorization of the treated subjects. Each chapter is divided into a number of sections and subsections. The emphasis on categorization is spurred by my own frequent frustration in having to reread some articles on plant disease physiology in their entirety to find a specific point. In addition to facilitating information search, categorization helps emphasize the interrelationships of different parts of a subject and aids comprehension. The topics covered are arranged to follow the sequence of events during host–parasite interactions. Beginning with the penetration and infection processes, the book proceeds into discussions of pathogen-produced metabolites, covers pathogen-produced alterations in plant structure and function, considers the nature of plant disease, and finally deals with the newly emerging concept of surface–surface interaction (recognition) between plant and pathogen.

I would like to express my sincere appreciation to Professor R. G. Grogan for instilling enthusiasm in me for plant pathological studies during my graduate work at the University of California. I am indebted to the following individuals for their critical review of portions of the manuscript and for their invaluable suggestions: F. B. Abeles, J. R. Aist, P. Albersheim, S.M. Alcorn, R. M. Allen, D. F. Bateman, A. A. Bell, P. J. Cotty, J. E. DeVay, J. M. Duniway, D. G. Gilchrist, R. N. Goodman, J. G. Hancock, Jr., J. P. Helgeson, R. B. Hine, J. G. Horsfall, R. G. Jensen, N. T. Keen, T. Kosuge, K.Matsuda, D. J. Merlo, R. E. Mitchell, M. S. Mount, J.W. O'Leary, M. W. Olsen, S. S. Patil, D. J. Samborski, R. P. Scheffer, L. Sequeira, M. E. Stanghellini, L. J. Stowell, I. Uritani, H. D. Van Etten, P. von Bretzel, H. E. Wheeler, and O. C. Yoder.

I am grateful to M. R. Nelson for his encouragement and support, to Linda Bower for typing the manuscript, to Alison Habel for her artistic illustrations, and to Carol Knapp and Claude Donndelinger for their help with the bibliography. This book could not have been written without the enthusiastic support, encouragement, and editorial skills of my wife Trish.

# Contents

# Chapter 1

# Definition



## I. INTRODUCTION

To facilitate scientific communication, certain phenomena should be defined. Unfortunately, however, a term or a definition deemed appropriate by one can be considered inadequate, confusing, or even misleading by another. This is probably because the phenomena are perceived differently by different individuals. Appropriate, universally accepted terms and definitions will ultimately be adopted when the real nature of the phenomena being defined is clearly understood. The terms and definitions appearing in this book and elsewhere should, therefore, not be considered accurate and final. The dynamic and progressive nature of plant pathology requires periodic reexamination of our views regarding different phenomena and concepts.

## II. DEFINITION OF A FEW TERMS

### A. Plant Disease

No definition for plant *disease* is universally accepted. According to Horsfall and Dimond (1959) ''Disease is a malfunctioning process that is caused by continuous irritation.''* Wheeler (1975) considers plant disease to include

*Reproduced with permission from Horsfall, J. G., and Dimond, A. E., eds. 1959. *Plant Pathology: An Advanced Treatise,* Vol. 1. Academic Press, New York.

"all malfunctions which result in unsatisfactory plant performance or which reduce a plant's ability to survive and maintain its ecological niche."* In a recent review Bateman (1978) has eloquently traced the development of the concept of disease since Julius Kühn (1858). He has shown how the concept has evolved and taken on added dimensions.

According to Bateman (1978) disease "is the injurious alteration of one or more ordered processes of energy utilization in a living system, caused by the continued irritation of primary causal factor or factors."† The above definition underlines Bateman's view that "Our theory and terms in pathology should be related to our concept of disease and our concept of disease needs to be compatible with other biological concepts."†

## B. Symptoms

*Symptoms* are generally viewed as characteristic alterations in form and function in a diseased plant.

## C. Pathogenicity and Virulence

*Pathogenicity,* the capacity of a pathogen to cause disease, is sometimes equated with *virulence*. Miles (1955) considers pathogenicity as an attribute of a species and virulence as the ability of a particular member of the species to cause disease under defined conditions. Nelson *et al.* (1970), on the other hand, do not consider pathogenicity as an attribute of a species. They define pathogenicity as the absolute disease-inducing ability of a pathogen and virulence as the relative level of the disease induced. Daly (1976c) has pointed out that since pathogenicity and virulence are determined by the outcome of host–pathogen interaction, they cannot be determined without the involvement of the host. According to Wood (1967), pathogenicity becomes virtually synonymous with virulence if it is qualified by terms such as *high* and *low*.

## D. Pathogenesis

To paraphrase Wheeler (1975), *pathogenesis* is the sequence of events that occur during the course of the disease.

*Reproduced with permission from Wheeler, M. 1975. *Plant Pathogenesis*. Springer-Verlag, New York.

†Reproduced with permission from Horsfall, J. G., and Cowling, E. B., eds. 1978. *Plant Disease: An Advanced Treatise,* Vol. 3. Academic Press, New York.

## E. Tolerance

*Tolerance* is generally viewed as the ability of plants to survive attacks by pathogens without appreciable reduction in yield (Schafer, 1971; Nelson, 1973; Rubin and Artsikhovskaya, 1963; Mussell, 1980).

According to Mussell (1980) tolerance with respect to biotic pathogens is the ability of a plant cell, whole plant, or field to perform at an acceptable level while providing a limited habitat for the pathogens.

## F. Compatibility and Incompatibility

*Compatibility* results when a plant and a pathogen are able to coexist at least during the early stages of disease development. The inability of a plant and a pathogen to coexist results in *incompatibility* and the absence of disease.

## G. Hypersensitive Reaction

*Hypersensitive reaction* is defined as an extreme cellular response to infection which may lead to a high degree of disease resistance. Hypersensitive reaction results in the sudden death of a limited number of host cells surrounding infection sites. The death of the cells is believed by some to restrict the growth and development of the pathogen (Chapter 12).

## H. Biotrophs

The term *biotrophs,* adopted by Luttrell (1974) as a substitute for the term *obligate parasites,* refers to organisms which can obtain food in nature only from the living cells of the host. No consideration is given to the ease with which they may be cultured in any phase of their life cycle.

*Chapter 2*

# Attraction to and Penetration of Plants by Pathogens



## I. INTRODUCTION

To infect plants, pathogens must manage to penetrate plants' physical barriers, including the cuticle and cell walls. Almost all pathogenic bacteria and a few fungi enter plant tissue passively through natural openings or wounds, while most of the pathogenic fungi have evolved sophisticated means for active penetration.

Progress in the study of the nature of attraction of pathogens to plant surfaces and their penetration has been slow, as evidenced by the appearance of only a few reports on the subject during the past few years. The literature on the subject has been reviewed by Wheeler (1975), Goodman (1976), Aist (1976a), Ingram *et al.* (1976), and Dodman (1979). This chapter covers certain aspects of attraction and penetration with emphasis on more recent studies.

## II. ATTRACTION OF PATHOGENS TO PLANTS

Although most pathogens contact plants passively with the aid of such agents as wind and water, chemotactic and electrotactic responses have also been observed in certain pathogens (Wheeler, 1975; Goodman, 1976; Dodman, 1979).

## A. Chemotaxis

Zoospores of certain members of Phycomycetes accumulate around the stomates (Royle and Thomas, 1973; Gerlach *et al.*, 1976) and roots of plants (Hickman, 1970; Ho and Zentmyer, 1977; Irwin, 1976; Malajczuk and McComb, 1977). Chemotactic responses of zoospores observed in these and a few other studies were nonspecific. However, Chi and Sabo (1978) reported that zoospores of *Phytophthora megasperma* exhibited specific chemotactic responses to primary roots of 2-day-old susceptible alfalfa seedlings while weak or no chemotaxis was observed in the presence of the primary roots of resistant alfalfa seedlings.

Nonspecific chemotactic attraction of nematodes (Webster, 1969; Rohde, 1972; Endo, 1975) and *Pseudomonas lachrymans* (Chet *et al.*, 1973) to roots has also been reported. Chet *et al.* (1973) studied chemotactic responses of *P. lachrymans* to certain chemicals and to leaf and root extracts as well as to guttation fluids of both compatible and incompatible plant species, using capillary assay techniques. While chemotactic attraction of the bacterium varied depending on the type of the attractants used, chemotaxis to plant species was found to be nonspecific. Raymundo and Ries (1980) found that chemotactic response of *Erwinia amylovora* was specific with respect to the attractant. The bacterium was attracted to apple nectar, to aspartate, and to several organic acids, but not to any of the simple sugars tested. The response was pH- and temperature-dependent.

Mulrean and Schroth (1979) suggested that *P. phaseolicola* is chemotactic and directed motility plays an important role in locating infection courts by the bacterium. The chemotactic responses of *P. phaseolicola* were further characterized by Mulrean (1980). Results of his studies showed that the nature and the magnitude of chemotaxis is influenced by the type and the quantity of the chemicals tested. Chemotaxis was virtually diminished in the presence of high concentrations of attractants and was not confined to the chemicals which can be used as food sources by the bacterium. For example, two amino acids which could not be utilized by the bacterium were found to be chemoattractants; two others served as chemorepellents.

Judging from the results of the study of the behavior of the bacterium on bean leaf surfaces, Mulrean (1980) suggested that chemotaxis may aid the bacterium to locate and occupy wound sites on bean leaves. On the average about twice as many motile bacterial cells were recovered from wounded areas on leaves than from nonwounded areas. Such a difference was not observed with nonmotile bacteria. Moreover, based on the results of the capillary assay it was concluded that *P. phaseolicola* was capable of detecting concentration gradients of leaf leachates established through the stomates and entering into leaves. With a few exceptions infiltration of bean leaves with certain chemoattractants prior to inoculation resulted in increased infectivity. On the other hand, infectivity of a motile isolate of the bacterium was reduced in leaves infiltrated with L-threonine, which served as a chemorepellent. Judging from the results of this study Mulrean (1980) suggested that survival and penetration of *P. phaseolicola* depend to some extent on chemotaxis.

The genetic, biochemical, and biophysical aspects of amino acid- and sugar-mediated chemotaxins have been studied extensively in enteric bacteria (Mesibov *et al,* 1973; Hazelbauer and Parkinson, 1977; Clancy *et al.,* 1981). Chemotactic response in bacteria is considered to be regulated by certain specific receptor proteins which are thought to be associated with the plasma membrane or the periplasmic layer. Changes in the levels of attractants or repellents are detected by these chemoreceptors. The signals received by these receptors are transmitted to flagella, causing them to respond appropriately.

Studies designed to elucidate the nature of chemotaxis of pathogens should continue mainly because a better knowledge of this unique phenomenon may facilitate development of methods to prevent attraction of pathogens to their hosts. For example, it might be possible to reduce the severity of certain diseases by the use of chemorepellents and other chemicals which interfere with chemotaxis (Mulrean, 1980).

When searching for chemoattractants and chemorepellents, highly purified chemicals should be used. This is because the presence of a trace amount of impurity in the substance which is being tested may lead to deceptive results (Mesibov and Adler, 1972).

## B. Electrotaxis

Khew and Zentmyer (1974) have studied electrotactic attraction of zoospores to plant roots. Current intensities within the range around plant roots (0.3–0.6 μA) were found to be sufficient to initiate electrotaxis for zoospores of seven species of *Phytophthora*.

## C. Motility

In some cases virulence of bacteria has been correlated with their motility. Huang (1974) reported that virulent strains of *Erwinia amylovora* moved much more rapidly than did avirulent ones. He found that the virulent bacteria developed normal-size flagella *in vitro* after 10 hr while the avirulent ones had fewer and shorter flagella. However, the difference between the two groups disappeared after 24 hr. The significance of motility in the invasion of bean leaves by *Pseudomonas phaseolicola* was studied by Panopoulos and Schroth (1974) using motile and nonmotile mutants of the bacterium. Infectivity of the bacterium on bean leaves was found to be correlated with flagellar motility. The motile strains were able to form up to 12 times as many lesions as nonmotile strains. There was no correlation between motility and systemic invasion. Contrary to the above two cases, motility of *Pseudomonas solanacearum* was not correlated with virulence (Kelman and Hruschka, 1973).

Bacterial motility is considered by some to have survival value under conditions when oxygen and nutrients are in limited supply (Smith and Doetsch, 1968). Smith and Doetsch (1968) and Kelman and Hruschka (1973) showed that motile strains of *P. fluorescens* and *P. solanacearum* multiplied selectively under certain conditions.

The influence of certain factors on the motility of *Erwinia amylovora* has been studied by Raymundo and Ries (1981). The motility of the bacterium did not require an exogenous energy source. Oxygen was required only in the absence of an energy source which could be utilized anaerobically. Motility was enhanced by the chelating agent EDTA and was exhibited only at pH 6–9.

## III. PENETRATION

Pathogens enter plants either passively, through natural openings and wounds, or actively by the aid of enzymes or mechanical forces.

### A. Passive Penetration

Certain pathogenic microorganisms penetrate plants through natural openings such as stomates, lenticels, leaf traces, trichomes, and hydathodes (Crosse, 1956; Layne, 1967; Matthee and Daines, 1969; Goodman, 1976). Natural openings and wounds are considered to be the main avenues for the penetration of bacteria into plants. Certain bacteria may even penetrate closed stomates possibly because their architectural features allow enough opening for bacterial ingress. The physical features of grass stomates, for example, were thought to allow enough apparent opening at the ends of stomatal apertures for bacterial entry even when they were closed across the center (Gitaitis *et al.*, 1981). Gitaitis *et al.* (1981) showed that both closed and open stomates in sweet corn leaves served as points of entry for *Pseudomonas alboprecipitans*. In the above host–parasite system abscisic-acid-induced partial closure of stomates prior to inoculation with the bacterium did not cause a significant reduction in disease incidence.

Many pathogens enter plants through wounds. Exposure of xylem elements following clipping of apple shoots facilitated entry of *Erwinia amylovora* into xylem elements and establishment of systemic infection (Crosse *et al.*, 1972). In the case of *Agrobacterium tumefaciens,* wounding not only provides attachment sites for the bacterium but also conditions cells to receive bacterial plasmid DNA (Chapter 11).

The infection of certain fungi is also aided by wounds. *Botryosphaeria* canker developed more frequently on cut-wounded and cold-injured apple stems than on unwounded stems following inoculation with *Botryosphaeria dothidea* (Brown and Hendrix, 1981). The fungus invaded wounded tissue readily and entered xylem vessels.

Viruses are thought to enter plants through wounds (Herridge and Schlegel, 1962; Matthews, 1970), particularly those caused by vectors (Hewitt *et al.*, 1958; Campbell and Grogan, 1963; Slykhuis, 1969). However, seemingly uninjured protoplasts isolated from higher plants can easily be infected with different viruses (Takebe and Otsuki, 1969; Shalla and Petersen, 1973; Takebe, 1975; Okuno *et al.*, 1977). Ectodesmata also have been implicated as points of entry for plant viruses, but the evidence is not thus far conclusive. Autoradiographical studies of epidermal strips of plants infected with $^{14}C$-labeled tobacco mosaic virus (TMV) showed accumulation of radioactivity around ectodesmata, particularly around those adjacent to leaf hair bases (Brants, 1965). The difference in susceptibility to TMV of two tobacco varieties also has been correlated with the number of ectodesmata (Thomas and Fulton, 1968).

The mechanism by which viruses enter subcellular organelles such as nuclei and chloroplasts is not known. The presence of TMV in nuclei has been attributed to entrapment of virus particles during cell division (Esau 1968). According to DeZoeten and Gaard (1969), spherical plant viruses can readily enter nuclei through nuclear pores aided by cytoplasmic diffusion pressure.

Certain indirect lines of evidence suggest that viruses may move from one cell to another through plasmodesmata. Zaitlin *et al.* (1981) compared the number of plasmodesmata in plants infected with strain L and strain LS-1 of TMV. The latter strain carries a temperature-sensitive lesion which prevents its cell-to-cell movement in the leaf tissue at the restrictive temperature for the virus. Results of electron microscopic examination showed that at the restrictive temperature fewer plasmodesmata were present in LS-1-infected plants compared to L-infected and noninfected plants. However, at the permissive temperature for the virus, an equal number of plasmodesmata were observed in noninfected plants and in plants infected with either strain. The size and the structure of plasmodesmata were the same in plants infected with either strain under all conditions. Evidence provided by Nishiguchi *et al.* (1978) and Zaitlin *et al.* (1981) showed that the lesion in LS-1 strain is due to a defect in viral movement, not viral replication. Both strains replicated equally well at the restrictive temperature in leaf mesophyll protoplasts *in vitro*. Moreover, *in vitro* translation products of both strains were identical with those reported for other TMV strains.

Results of *in vitro* peptide mapping of translation products (30 K, 130 K, and 165 K polypeptides) showed that only the 30 K polypeptide was slightly different in the two strains, suggesting a possible role for this polypeptide in cell-to-cell movement of the virus in the leaf. The evidence for the involvement of plasmodesmata in cell-to-cell movement of tobacco mosaic virus has also been provided by Wieringa–Brants (1981). In cowpea and tobacco plants infected with TMV and tobacco necrosis virus, the time required for the passage of viruses from epidermis into mesophyll cells was reduced in plants kept in dark compared to those kept in light. The reduction in epidermal passage time in dark-treated plants coincided with a substantial increase in the number of plasmodesmata in superficial cells of these plants. While results of this study may suggest a role for plasmodesmata in virus movement in leaf cells, the evidence is indirect and inconclusive. In addition to increasing the number of plasmodesmata, dark-treatment may cause other physical and chemical alterations in leaf tissue that facilitate entry of virus particles into plants.

Long-distance transport of viruses takes place mostly in phloem. However, movement through xylem has also been reported for a few viruses (DeZoeten, 1976). In phloem transport plasmodesmata may serve as points of entry and exit for viruses. Whether viruses move from one cell to another in the form of naked nucleic acid or as a complete virus particle is not known. The literature on virus transport has been reviewed by Schneider (1965) and DeZoeten (1976).

## B. Active Penetration

### *1. Penetration of the Cell Wall*

Certain fungi have evolved sophisticated means of active penetration into the host cells. Penetration of motile fungi begins with encystment of zoospores. With some fungi, penetration is accomplished with the aid of haustoria or infection pegs developed directly from the encysted zoospores. In others, such as *Pythium aphanidermatum,* penetration is more elaborate. It includes formation of a germ tube from the encysted zoospore and development of the germ tube into a cushion-shaped structure, the appressorium, from which a hypha penetrates the host (Kraft *et al.*, 1967). Appressoria are also formed during penetration of certain nonmotile fungus spores.

One of the most highly evolved methods of penetration is that of the members of Plasmodiophorales. A detailed description of the penetration of *Polymyxa betae* has been provided by Keskin and Fuchs (1969) and of *Plasmodiophora brassicae* by Aist and Williams (1971) and Williams *et al.* (1973). The sequence of events, which is basically the same in both fungi, starts with the attachment of zoospores to the root hairs. The encysted zoospores form penetration structures

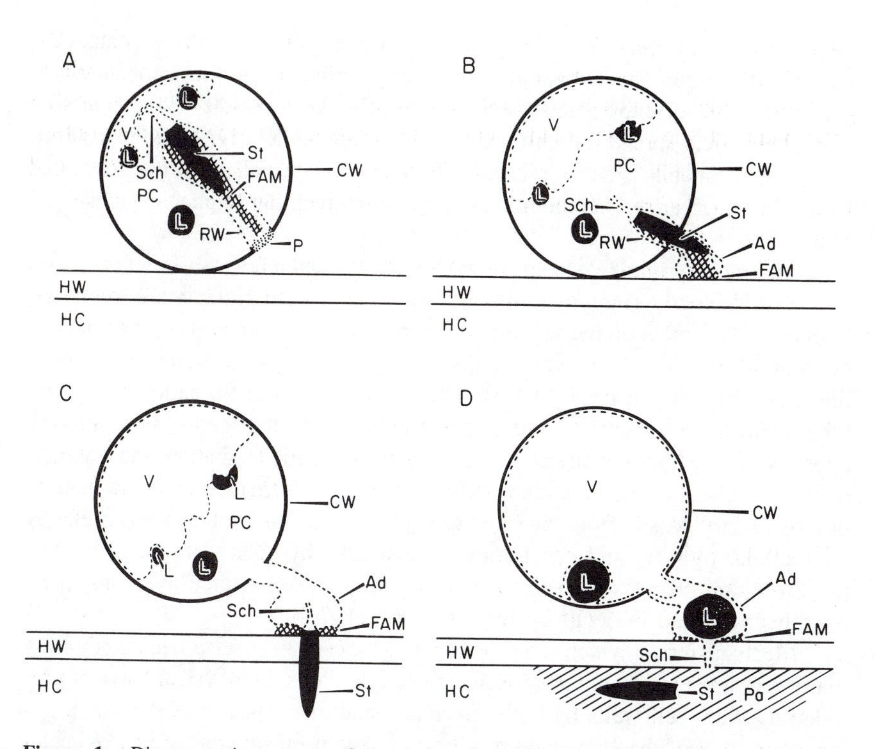


**Figure 1.** Diagrammatic summary of penetration of a cabbage root hair cell by an encysted zoospore of *Plasmodiophora brassicae*. (A) Cyst vacule (V) has not yet enlarged. (B) Vacule has enlarged and a small adehesorium (Ad) has formed. (C) Stachel (St) has punctured host wall (HW). (D) Penetration has occurred and a papilla (Pa) has been formed at the penetration site. Cyst wall (CW), fibrillar adhesive material (FAM), host cytoplasm (HC), lipid body (L), Rohr plug (P), parasite cytoplasm (PC), rohr wall (RW), Schlauch (Sch). [Reproduced with permission from Williams, P.H., *et al.*, 1973. In *Fungal Pathogenicity and the Plant's Response* (R. J. W. Byrde and C. V. Cutting, eds.), pp. 141–158. Academic Press, New York.]

including a dense, pointed structure called a stachel within a few hours. Penetration begins with the formation of a bulb-shaped structure, the adhesorium, on the root hair cell wall. From within the adhesorium the stachel is injected through the cell wall in about one second and the parasite is then injected into the cytoplasm of the root hair cell (Fig. 1). Williams (1979) has pointed out that cell wall penetration by *P. brassicae,* which is apparently a physical mechanism, may also be found to involve chemical processes.

The penetration methods of certain biotrophs, particularly rusts and mildews, are also elaborate. Most rust fungi penetrate hosts through stomates. An appressorium is formed from the germ tube of rust spores over the stomatal cavity. Shortly after, an infection thread originating from the appressorium, penetrates into the substomatal cavity where it forms a vesicle. Hyphae arising from the vesicle spread through the leaf tissue intercellularly. Finally, penetration of the host cells is accomplished with the aid of specialized structures called haustoria which arise from both infection and intercellular hyphae (Maheshwari *et al.*, 1967).

All powdery mildews, some downy mildews, and a few rust fungi penetrate host tissue directly (Ingram *et al.*, 1976). Direct penetration is accomplished with the aid of infection pegs arising from appressoria. Structural features of penetration of the lettuce downy mildew fungus, *Bremia lactucae,* have been illustrated by Sargent *et al.* (1973). The process includes formation of a germ tube from a germinating spore, differentiation of the germ tube into an appressorium which penetrates an epidermal host cell, formation and enlargement of a primary vesicle in the penetrated host cell, formation and enlargement of a secondary vesicle from the primary vesicle, invasion of neighboring cells by intercellular hyphae, and penetration of host cells by means of haustoria. The ultrastructural changes associated with the infection of lettuce by *B. lactucae* have been studied in detail by Ingram *et al.* (1976).

Mechanisms by which fungi penetrate host cell walls have been studied by many investigators. Fungi are considered to overcome the physical barriers provided by plant cell walls by both enzymatic and mechanical means. The arguments in favor of the enzymatic mechanism have been summarized by Wheeler (1975) and Aist (1976a). They include: (1) the ability of many plant pathogens to produce cell-wall-degrading enzymes, (2) partial degradation of host cell walls around penetration sites judged from the results of cytochemical tests, and (3) the presence of nonstaining haloes around penetration pegs of *Erysiphe graminis* f. sp. *hordei*. The major lines of evidence for the mechanical nature of cell wall penetration include the ability of germinating spores of certain fungi to invaginate cured epoxy resin membranes and to penetrate gold foils. The tip of the penetration peg of *E. graminis* f. sp. *hordei* was partly or totally wall-less during its penetration into susceptible barley leaves (McKeen and Rimmer, 1973). It was therefore considered unlikely that the penetration peg, without the rigidity of the cell wall, could penetrate the leaf tissue solely by mechanical force.

The overall results of studies on cell wall penetration by fungi show that depending on the host–pathogen system both enzymatic and mechanical mechanisms contribute to the penetration. While the levels of contribution of these two mechanisms in any given host–pathogen system is not known, the evidence tends to support the view that the enzymatic mechanism is generally more important than mechanical means. This question, however, cannot be resolved presently. The answer might come from the results of studies in which each penetration feature is thoroughly characterized and its significance to the overall penetration process is assessed independently by genetic analysis and micromanipulation (Aist, 1976a). A better understanding of the nature of cell wall penetration requires cautious interpretation of the results of histochemical tests. Aist (1976a) has pointed out that the occurrence of densely staining deposits in the cell wall could be due to the presence of substances unrelated to its enzymatic breakdown. Moreover, since in most cases tissues have been examined some hours after penetration the observed histochemical changes could have occurred following penetration.

The significance of cell wall modifications (lignification and papilla formation) in resistance to penetration is discussed in Chapter 12. Penetration by different fungi has been discussed by Bracker and Littlefield (1973), Aist (1976a), Ingram *et al.* (1976), and Littlefield and Heath (1979).

### 2. *Penetration of the Cuticle*

Outer surfaces of cells in epidermis, substomatal, mesophyll, and palisade areas are covered with a cuticular layer composed of cutin, waxes, cellulose, and pectic materials. Cutin is a polymer formed chiefly from hydroxy fatty acids through cross-linked peroxide and ether bonds. The structure and thickness of the cuticle is dependent upon the degree of hydroxylation and the number of hydroxylated fatty acids (Van den Ende and Linskens, 1974).

Penetration of the cuticle by pathogenic fungi is thought to be accomplished by mechanical or enzymatic means. Evidence in favor of mechanical penetration is based on the observation that epidermal membranes beneath penetration pegs of certain fungi seem to be depressed inward during penetration (Wheeler, 1975). The involvement of cutin degrading enzymes in cuticular penetration is supported by the following lines of evidence: First, many fungi are known to produce cutinases (Purdy and Kolattukudy, 1975a,b; Baker and Bateman, 1978; Lin and Kolattukudy, 1980). Second, round penetration holes with smooth edges have been observed in cuticles in areas beneath germ tubes (Staub *et al.*, 1974; Wheeler, 1975). Third, using ferritin-conjugated antibody and electron microscopy, Shaykh *et al.* (1977) showed secretion of cutinase by *Fusarium solani* f. sp. *pisi* during penetration into *Pisum sativum*. Moreover, a direct correlation was found between virulence levels of *F. solani* f. sp. *pisi* isolates expressed on intact pea surfaces and the amount of cutinase released by their germinating spores (Köller *et al.*, 1982). Finally, Maiti and Kolattukudy (1979) prevented infection of pea epicotyl with *F. solani* f. sp. *pisi* by adding to the spore inoculum either specific antibodies against fungus cutinase or diisopropyl fluorophosphate, an inhibitor of cutinases. The overall results of studies on the nature of cuticular penetration show that enzymatic activity plays a major role.

## IV. FACTORS INFLUENCING ATTRACTION, PENETRATION, AND INFECTION PROCESSES

Attraction to and penetration of plants by pathogens as well as the infection process are influenced by a number of environmental factors, including temperature, moisture, light, pH, and nutrition. Environmental parameters may affect the pathogen, the host, or host–pathogen interactions. Some of these factors are discussed below.

### A. Temperature

Temperature probably plays the most critical role in early events leading to the establishment of infection. This is because certain activities of the pathogen, such as spore germination, zoospore motility, and penetration, are sensitive to temperature. For certain pathogenic fungi temperature not only influences the rate of spore germination and the growth of the pathogen but also the type of germination.

### B. Moisture

With the exception of powdery mildews, most fungi require relatively high levels of moisture for germination and development. Free moisture on the plant surfaces helps dissolve nutrients and other metabolites and facilitates their movements in and out of pathogen and plant cells. Free water may also aid penetration of bacterial cells into plants. Penetration of both motile and nonmotile strains of *Pseudomonas phaseolicola* into water-congested bean leaves was 90- to 400-fold greater than that into noncongested leaves (Panopoulos and Schroth, 1974). Soilborne fungal pathogens, like leaf pathogens, also have specific moisture requirements. This is particularly true with zoospore-forming soil fungi where their growth, zoospore release, motility, and persistence depend on the maintenance of a proper soil moisture (see a review by Duniway, 1979).

### C. Gas Composition

The pattern of vertical distribution of different pathogenic fungi in the soil varies. Certain fungi are confined to the upper surfaces of the soil while others are more prevalent at lower depths. In many cases, the factor(s) responsible for the specific vertical distribution pattern is not known. Since temperature, moisture, and gas composition vary with soil depth, a pathogen may reside at a specific depth where the environmental factors are most favorable for its survival and activity. While the influence of soil temperature and moisture on the activity of soil-borne fungi has been studied frequently, little attention has been given to the role of soil gas composition (composition of $CO_2$,$O_2$, and $N_2$). Soil gas composition is a function of certain physical, chemical, microbiological, and environmental parameters of the soil. Therefore, some of the host and/or pathogen responses which have been attributed to temperature, moisture, or texture of the soil could have been due to the influence of the gas composition.

The effect of gas composition on the behavior of *Verticillium dahliae* was studied by Ioannou *et al.* (1977a,b). Microsclerotial formation decreased progressively *in vitro* in the presence of increasing levels of $CO_2$ and decreasing levels of $O_2$ and was totally inhibited in the presence of 10% $O_2$ and 11% $CO_2$. There was a 90% reduction in the number of microsclerotia formed on *Verticillium*-infected stems kept at 10% $O_2$ and 11% $CO_2$ compared to those subjected to normal air. The inhibition of microsclerotial production in the flooded soil was attributed largely to decreased $O_2$ and increased $CO_2$ concentration brought about by flooding. Effects of gas composition on growth, sclerotial production, germination, and infection of lettuce by *Sclerotinia minor* were studied by Imolehin and Grogan (1980). Mycelial growth and sclerotial production of the fungus were greatly reduced at $O_2$ concentrations below 4% with $CO_2$ kept at about 0.03%, and at $CO_2$ concentrations at or above 8% with $O_2$ at about 21%. Judging from these and other results it was concluded that sclerotial production and germination probably are not limited by the levels of $O_2$ and $CO_2$ commonly present in the soil and that the lack of infection of lettuce plants by the fungus at soil depths greater than 10 cm is probably not due to the presence of unfavorable gas composition. The growth and development of fungi and bacteria may also be influenced by the levels of ethylene (Smith, 1973; Smith and Cook, 1974).

*Chapter 3*

# The Role of Pathogen-Produced Cell-Wall-Degrading Enzymes in Pathogenesis



## I. INTRODUCTION

Alterations of plant cell walls and degradation of their components are common features of infection by pathogenic agents. These changes are brought about by cell-wall-degrading enzymes which are produced by almost all pathogenic and nonpathogenic fungi and bacteria. In diseases induced by certain nonbiotrophs, alterations of the cell wall are more dramatic. In such cases dissolution of cell wall components and middle lamella by pathogen-produced enzymes causes separation of cells and waterlogging of infected tissues, a process known as "maceration." Activity of pectic enzymes also causes death of plant cells by a mechanism which is not yet completely understood. This chapter discusses structure and chemistry of the plant cell walls, nature and mechanism of action of pathogen-produced cell-wall-degrading enzymes, and the role of these enzymes in plant cell wall degradation, tissue maceration, and cell death.

## II. STRUCTURE AND CHEMISTRY OF THE CELL WALL

### A. Structure of the Cell Wall

The structure and chemistry of the cell wall have been discussed by Albersheim (1969, 1976, 1978), Bateman (1976), and Bateman and Basham

(1976). The wall is an orderly, complex structure composed in young tissue mainly of polysaccharide and structural glycoprotein (Lamport, 1970), and in some older tissue, lignin (Albersheim, 1965).

The polysaccharide portion of the cell wall has traditionally been separated into pectic substances, hemicelluloses, and cellulose (Northcote, 1963), merely on the basis of their solubility in various extracting solvents. The pectic portion is primarily made of rhamnogalacturonans, galactans, and arabans (Aspinall, 1970). Cellulose is made up of long linear chains of β-1,4-linked glucose (Northcote, 1972). Hemicelluloses generally are polymers such as xylans, mannans, and a number of other heteropolymers. The cell wall is generally divided into three regions based on composition and particular architecture: middle

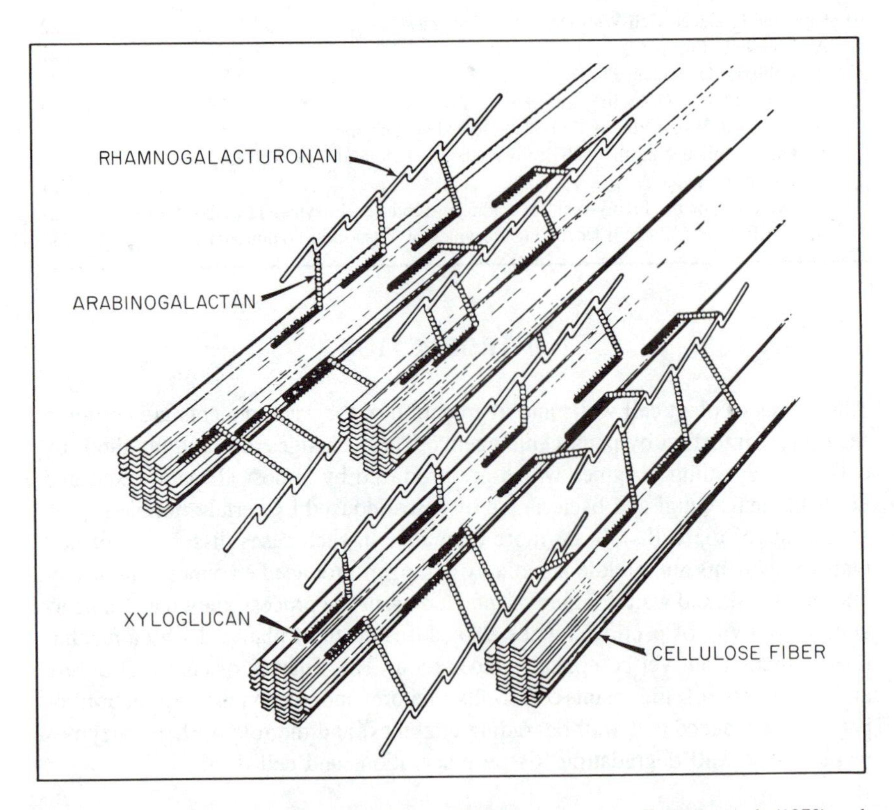


**Figure 2.** A model of the primary cell walls of dicots proposed by Keegstra *et al.* (1973) and redrawn by Albersheim (1975). Protein molecules which are not shown in this model are linked to pectic polymers (rhamnogalacturonans) through 3,6-linked arabinogalactans. [Reproduced with permission from Albersheim, P. 1975. The walls of growing plant cells. *Sci. Am.* **232:**81–95. Copyright © 1975 by Scientific American, Inc., all rights reserved.]

lamella, primary cell wall, and secondary cell wall. There is no clear demarcation between these three regions of the cell wall. Instead, the constituents of each portion may extend from one region to the next. A particular type of polysaccharide is more abundant in one region compared to other regions (Bateman and Basham, 1976).

The primary cell walls in different plants probably have similar structures (Albersheim, 1976). However, there are marked variations in the constituents of the secondary cell walls from plant to plant and in different tissues of a single plant (Talmadge *et al.*, 1973).

A model for the primary cell walls of dicots has been proposed by Keegstra *et al.* (1973). An artist's conception of Keegstra's model presented by Albersheim (1978) is shown in Fig. 2. According to this model the cellulose which occurs as microfibrils is the main component of the wall and provides the wall with structural strength. Cellulose microfibrils are extensively cross linked by noncellulosic polysaccharides such as xyloglucans, arabinogalactans, and rhamnogalacturonans. Xyloglucans, which completely cover the surfaces of cellulose microfibrils, are linked to rhamogalacturonans (pectic polymers) through arabinogalactan chains. The pectic polymers may be further linked to protein through 3,6-linked arabinogalactan chains. In this model hydrogen bonds interconnect cellulose and xyloglucan while other polymers are probably interconnected by covalent bonds. The validity of this model for the primary cell walls of dicotyledons has not been questioned in subsequent studies (Albersheim, 1978).

## B. Chemistry of Cell Wall Components

### *1. Cellulose*

Cellulose, the major structural component of the cell wall, is made up of long linear chains of β-1,4-linked glucose (Northcote, 1972). A cellulose microfibril is considered to be made of glucan chains held to each other by a great number of hydrogen bonds.

### *2. Hemicelluloses*

The most important hemicellulosic polymer is xyloglucan which is made of β-1,4-linked D-glucosidic residues with terminal branches of α-1,6-linked xylose. The polymer also contains a few molecules of galactose, arabinose, and fucose. Xyloglucans link the pectic and cellulosic fractions of the primary cell wall. Other hemicellulosic polymers include glucomannans, galactomannans, and arabinogalactans (Bauer *et al.*, 1973; Northcote, 1972).

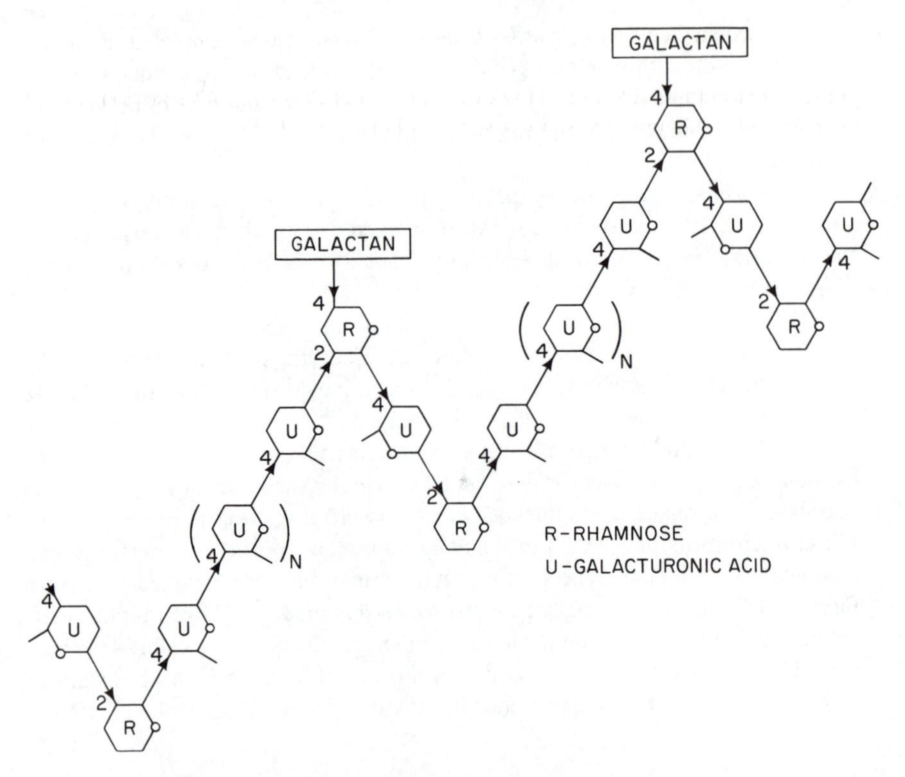


**Figure 3.** Structure of rhamnogalacturonan. The linear sequences of galacturonic acid are interrupted by rhamnose residues, giving the molecule a zig-zag shape. [Proposed by Talmadge, K. W., *et al.*, 1973. Reproduced with permission from *Plant Physiol.* **51:**158–173. The American Society of Plant Physiologists, Rockville, Maryland.]

### *3. Pectic Polymer*

The pectic polymer (rhamnogalacturonan) consists of a linear sequence of α-1,4-D-galacturonic acid interspersed with rhamnosyl residues. Unlike glucan chains of cellulose and xyloglucan, rhamnogalacturonan has a zig-zag structure with rhamnosyl residues occurring randomly at branch points probably as galacturonosyl-1,2-rhamnosyl-1,4-galacturonosyl-1,2-rhamnosyl units (Talmadge *et al.*, 1973; Albersheim, 1975) (Fig. 3).

### *4. Protein*

Cell wall protein is composed of hydroxyproline-rich glycoprotein with galactose and arabinose as the chief components of the carbohydrate portion

(Lamport *et al.*, 1973; Keegstra *et al.*, 1973). Since structural proteins in animal cells are also rich in hydroxyproline, cell wall protein has been considered to possess a structural function. However, the evidence for such a function for the primary cell wall protein is not strong (Albersheim, 1978).

### 5. *Lignin*

Lignin is a branched, three-dimensional polymer formed through oxidative condensation of three substituted cinnamyl alcohols, sinapyl, coniferyl, and *p*-hydroxy cinnamyl alcohols (Dehority, *et al.*, 1962; Freudenberg, 1968; Kirk, 1971; Gross, 1979; Hahlbrock and Grisebach, 1979) (Fig. 4). Lignin deposition in the secondary cell wall increases its structural integrity and may make other

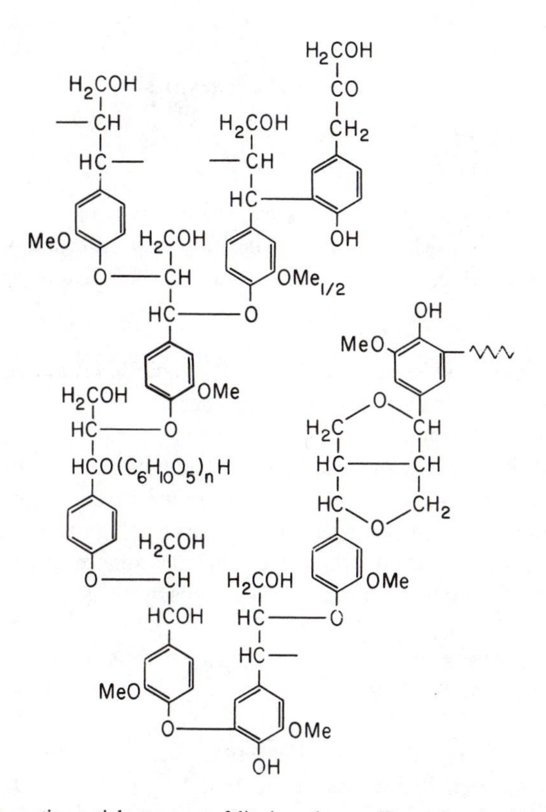


**Figure 4.** A schematic partial structure of lignin polymer. [Reproduced with permission from Freudenberg, K. 1968. In *Molecular Biology, Biochemistry, and Biophysics* (A. Kleinzeller, G. F. Springer, and H. C. Witmann, eds.), pp. 45–125. Springer-Verlag, New York.]

cell wall constituents less accessible to the cell-wall-degrading enzymes (Cowling and Kirk, 1976).

## III. PATHOGEN-PRODUCED CELL-WALL-DEGRADING ENZYMES

Many plant pathogenic and nonpathogenic organisms are capable of producing an array of cell-wall-degrading enzymes. With a few exceptions, most of these enzymes are produced extracellularly and inducibly. Some of the important members of this group of enzymes will be discussed briefly.

The literature on the nature and the function of pathogen-produced cell-wall-degrading enzymes has been reviewed by Bateman (1976), Bateman and Basham (1976), Wood (1972, 1976a), Mussell and Strand (1977), and Mount (1978). These sources should be consulted for additional information.

### A. Pectolytic Enzymes

*Nomenclature of Pectolytic Enzymes*

The terminology for pectic enzymes used in this book is based on that proposed by Bateman and Millar (1966). The following criteria have been used to classify pectic enzymes (Bateman and Millar, 1966; Rombouts and Pilnik, 1972):

1. Mechanisms by which the α-1,4-glycosidic bond of pectic polymer is split (i.e., hydrolytic or lytic cleavage).
2. Substrate specificity (i.e., pectin or pectic acid).
3. Site of the cleavage in the polymer. The prefix "exo" is used when the enzyme attacks the substrate in a terminal manner generally releasing monomeric or dimeric products (i.e., exo-polygalacturonase). The prefix "endo" is used when the enzyme attacks randomly releasing oligomeric products (i.e., endo-polygalacturonase).

Using the above criteria pectic enzymes are divided into the following three groups (Bateman and Millar, 1966):

a. *Chain-Splitting Hydrolytic Enzymes*. These enzymes break glycosidic bonds through hydrolysis. Two important members of this group are pectin methylgalacturonases, which exhibit a preference for pectin as a substrate, and polygalacturonases, which prefer pectic acid (Bateman and Millar, 1966) (Fig. 5b).

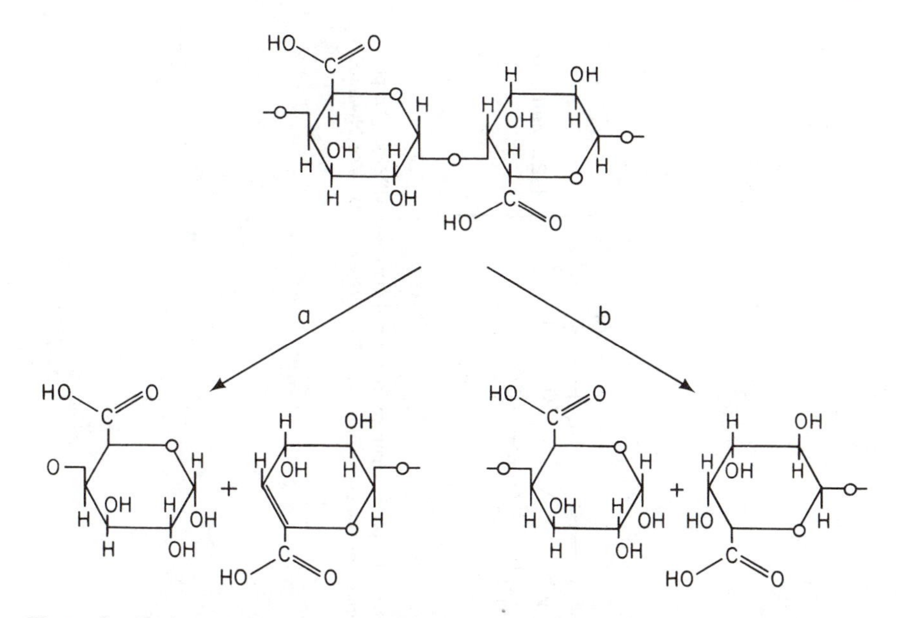

**Figure 5.** Products of lytic (transeliminative) (a) and hydrolytic (b) breakdown of pectic acid. [Adapted with modifications from Goodman, R.N., *et al.* 1967. *The Biochemistry and Physiology of Infectious Plant Disease*. D. Van Nostrand, Princeton, New Jersey. Used with permission.]

b. *Chain-Splitting Transeliminative (Lytic) Enzymes*. These enzymes break glycosidic bonds through transelimination. The reaction products of these enzymes possess an unsaturated bond between carbon 4 and 5 of the uronide residue adjacent to the α-1,4-bond cleaved (Albersheim *et al.*, 1960) (Fig. 5a).

c. *Pectinmethylesterase*. Pectinmethylesterase is produced by most plant pathogens and is also present in healthy plants bound to cell walls (Drysdale and Langcake, 1973; Bateman and Basham, 1976). The enzyme hydrolyzes the methyl ester groups of the uronic acid residues in the pectic chain forming uronic acid groups and methanol (Fig. 6). The enzyme does not break pectic chains. However, free carboxyl groups formed by the action of the enzyme might react with calcium or other multivalent cations to form rigid cross connections between pectic chains (Bateman and Millar, 1966). Incorporation of calcium into the pectic fraction appears to make the cell walls increasingly resistant to hydrolysis by cell-wall-degrading enzymes (Bateman, 1964a,b) and to physicochemical changes (Beckman, 1971).

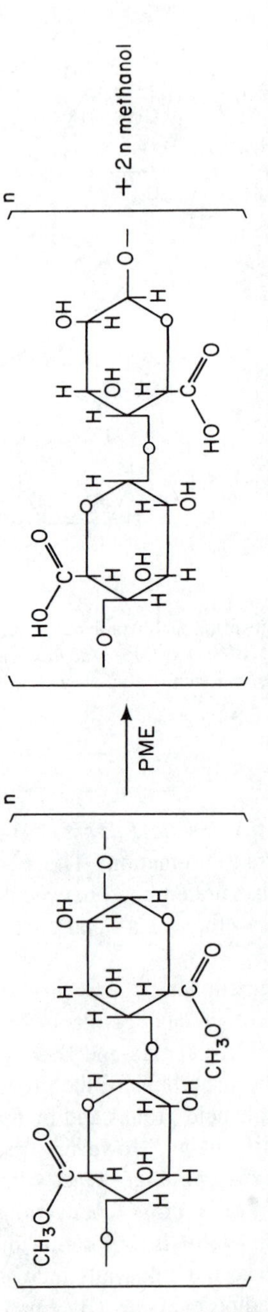


**Figure 6.** Hydrolysis of methyl ester groups of the uronic acid residues in the pectic chain by pectinmethylesterase (PME). [Adapted with modifications from Goodman, R.N., *et al*. 1967. *The Biochemistry and Physiology of Infectious Plant Disease*. D. Van Nostrand, Princeton, New Jersey. Used with permission.]

Many pathogenic and nonpathogenic fungi and bacteria are capable of producing pectolytic enzymes (Bateman and Basham, 1976; Wood, 1976a; Mussell and Strand, 1977; Mount, 1978).

## B. Cellulose-Degrading Enzymes

Enzymatic degradation of cellulose is presumably accomplished by four different enzymes referred to as $C_1$, $C_2$, $C_x$, and β-glucosidase (cellobiase) (Reese, 1963; King and Vessal, 1969; Halliwell and Mohammed, 1971). $C_1$ attacks native cellulose; $C_2$ also attacks native cellulose forming short chain β-1,4-glucans and probably exposing individual glucan strands to the action of $C_x$ enzymes which degrade them to cellobiose. Finally, cellobiose is degraded by β-glucosidase to glucose.

Higher levels of cellulase ($C_x$) have been found in carnations infected with *Erwinia chrysanthemi* (Garibaldi and Bateman, 1970) and in the wood of cricket bat willows infected with *Erwinia salicis* (Wong and Preece, 1978). The following pathogens are also known to produce cellulases: *Rhizoctonia solani* (Bateman, 1964a; Bateman, *et al.*, 1969), *Sclerotium rolfsii* (Bateman, 1969), *Botrytis* spp. (Hancock *et al.*, 1964), *Pseudomonas solanacearum* (Kelman and Cowling, 1965), *Pyrenochaeta lycopersici* (Goodenough *et al.*, 1976), *Colletotrichum orbiculare* (Porter, 1969), *Verticillium albo-atrum* (Cooper and Wood, 1980), and *Fusarium oxysporum* f. sp. *lycopersici* (Cooper and Wood, 1975).

## C. Hemicellulose-Degrading Enzymes

The nature, distribution, and roles in the pathogenesis of hemicellulose-degrading enzymes have been discussed by Bateman (1976), Bateman and Basham (1976), and more briefly by Mount (1978). Enzymes capable of degrading hemicellulosic components of the cell wall (e.g., galactans and arabans) to their monomeric constituents are produced *in vitro* by a variety of phytopathogenic microorganisms (Fuchs *et al.*, 1965; Knee and Friend, 1968; Cole and Bateman, 1969; Van Etten and Bateman, 1969; Mullen, 1974; Bateman, 1969; Hancock, 1967; Strobel, 1963). Like cellulose, the breakdown of many hemicelluloses requires the activity of a number of enzymes (Mount, 1978). The contribution of hemicellulosic enzymes to cell wall breakdown is not clear. Moreover, the role, if any, of these enzymes in the pathogenicity of microorganisms capable of producing them is not known (Bateman, 1976). A few of the hemicellulose-degrading enzymes are discussed below.

*1. Xylanase (β-Xylosidase)*

Xylose polymers (xylans) are hydrolyzed by both ''exo'' and ''endo'' forms of xylanase. The products of ''endo'' forms are xylobiose, xylotriose, and higher oligomers; those of the ''exo'' forms are xylobiose and xylose (Strobel, 1963; King and Fuller, 1968; Van Etten and Bateman, 1969). A number of plant pathogens are known to produce enzymes that split the β-1,4-xylosyl bond in the cell wall polymers. These include *Helminthosporium maydis* (Bateman *et al.*, 1973), *Rhizoctonia solani* (Bateman *et al.*, 1969), *Sclerotinia sclerotiorum* (Hancock, 1967), *Fusarium roseum* (Mullen, 1974), *Diplodia viticola* (Strobel, 1963), *Sclerotium rolfsii* (Van Etten and Bateman, 1969), and *Trichoderma pseudokoningii* (Baker *et al.*, 1977). However, only in some instances has the xylose content of the cell wall in lesions caused by these pathogens been shown to be greatly reduced (Mullen, 1974; Van Etten and Bateman, 1969; Baker *et al.*, 1977). The contribution of xylanases to cell wall breakdown is not known.

*2. Galactanases (α-Galactosidases, β-Galactosidases)*

A number of plant pathogens are known to produce galactanases in substantial quantities (Bateman and Basham, 1976; Cole and Sturdy, 1973; Knee and Friend, 1968; Mullen and Bateman, 1975; Van Etten and Bateman, 1969). Bauer *et al.* (1977) have isolated an endo-β-1,4-galactanase from *Sclerotinia sclerotiorum* capable of solubilizing a substantial amount of the carbohydrate portion (including a major galactan component) of isolated sycamore and potato cell walls. The purified enzyme, however, failed to macerate potato tuber tissue.

*3. Glucanase (β-1,3-Glucanohydrolase)*

The enzyme is produced by plants (Abeles *et al.*, 1970) and by *Colletotrichum lagenarium* (Rabenantoandro *et al.*, 1976). The possible role of this enzyme in resistance is discussed in Chapter 12.

*4. Arabinases (α-Arabinosidases)*

These enzymes are produced by a number of plant pathogens including *Pyrenochaeta lycopersici* (Goodenough *et al.*, 1976), *Sclerotinia sclerotiorum* (Hancock, 1967; Baker *et al.*, 1979), *Verticillium albo-atrum* (Cooper and Wood, 1975), and *Fusarium oxysporum* f. sp. *lycopersici* (Cooper and Wood, 1975). The activity of a purified arabinase from *S. sclerotiorum* was tested by Baker *et al.* (1979). The enzyme released arabinose from arabinan and from isolated cell walls of bean and rice plants but failed to macerate tissues from potato tubers or cucumber endocarp.

*5. Mannase (β-Mannosidase)*

Complete hydrolysis of β-1,4-mannans requires two enzymes, an endomannase and a β-mannosidase (Reese and Shibata, 1965). Both of these enzymes are produced by *Sclerotium rolfsii in vitro* (Bateman, 1976).

## D. Pattern of Production of Cell-Wall-Degrading Enzymes

Cell-wall-degrading enzymes are produced in a sequence with pectic enzymes forming first, and hemicellulose- and cellulose-degrading enzymes appearing last (Bateman and Basham, 1976; English *et al.*, 1971; Jones and Albersheim, 1972; Mullen, 1974). Pectic enzymes, which are formed first, apparently loosen the cell wall structures and thereby render other wall polymers susceptible to enzymatic attack (Bateman and Basham, 1976). Mankarios and Friend (1980) determined production patterns of cell-wall-degrading enzymes of *Botrytis allii* and *Sclerotium cepivorum* in the presence of isolated onion cell walls by measuring the pattern of release of individual sugars. The released sugars could be separated into two groups. The first group was suggested to be released by polygalacturonase, a major enzyme produced by both pathogens, and the second group by cellulase and possibly xylanase. Judging from the results of their study and those of others, Stack *et al.* (1980) concluded that the *Erwinia caratovora* has the necessary enzyme systems for depolymerizing the pectic acid fraction of plant cell walls, for utilizing breakdown products of maceration as sources of energy, and for generating essential molecules for the synthesis and activity of these enzymes. They presented a hypothetical model for the activity of the bacterial enzymes and their interactions with plant cell walls.

## E. Factors Influencing the Activity of Pathogen-Produced Cell-Wall-Degrading Enzymes

The regulation of synthesis of cell-wall-degrading enzymes has been reviewed by Bateman and Basham (1976) and Mount (1978). Some of the factors known to influence the activity of cell-wall-degrading enzymes produced by pathogens are now discussed.

*1. Effect of Inducers and Repressors*

There is evidence that the monosaccharides or oligosaccharides formed by enzymatic hydrolysis of a particular carbohydrate polymer can serve as effective inducers of specific carbohydrate-degrading enzymes (Cooper and Wood, 1975).

Cooper and Wood (1975) showed that the addition of specific sugars to carbon deficient media in which *Fusarium oxysporum* and *Verticillium albo-atrum* were grown resulted in the induction of specific cell-wall-degrading enzymes, i.e., D-galacturonic acid, D-xylose, and cellobiose induced synthesis of endopolygalacturonase, xylanase, and cellulase, respectively. However, the syntheses of the enzymes were repressed when the concentration of the inducers increased above the optimum levels. Of the enzymes studied by Cooper and Wood (1975), catabolite repression and specific induction were particularly pronounced with pectic enzymes.

Synthesis of cell-wall-degrading enzymes by some phytopathogenic microorganisms is subject to catabolite repression by sugars (Keen and Horton, 1966; Patil and Dimond, 1968; Cooper and Wood, 1975; and a few others). The mechanism of catabolite repression of endopolygalacturonate transeliminase (PGTE) synthesis by *Erwinia carotovora* has been a subject of intensive study (Hubbard *et al.*, 1978; Moran and Starr, 1969; Zucker and Hankin, 1970; Mount *et al.*, 1979). Hubbard *et al.* (1978) showed that catabolite repression of PGTE by glucose can be prevented by adenosine 3′,5′-cyclic monophosphate (cAMP) and that the cellular levels of cAMP are directly correlated with the amount of PGTE activity. This observation was later confirmed by Mount *et al.* (1979) who showed that a cAMP-deficient mutant of *E. carotovora* produced low levels of PGTE *in vitro,* even in the presence of the inducer, sodium polypectate. However, in the presence of both sodium polypectate and cAMP, enzyme production was increased to the levels produced by the parent strain. These results show that cAMP may have a regulatory role in the synthesis of PGTE.

### 2. *Effect of pH*

Production of pectate lyases and polygalacturonases is favored by an alkaline and acid environment, respectively (Bateman, 1966). Bateman (1966) showed that the patterns of production of pectic enzymes by *Fusarium solani* f. sp. *phaseoli* in two different media were different and speculated that the difference was related to pH of the cultures. He suggested that the type of enzyme obtained in culture filtrates might be a reflection of its pH stability. Hancock (1966) showed that in the alfalfa–*Colletotrichum trifolii* system the increase in pH of the diseased tissue could favor the activity of pathogen-produced polygalacturonate transeliminase (pectate lyase), which has a pH optimum between 8.5 and 9.0

### 3. *Effect of Phenolic Compounds*

Although direct evidence for the inactivation by phenolic compounds of cell-wall-degrading enzymes *in vivo* is lacking, some of these enzymes apparently can be inactivated *in vitro* by oxidized phenolics (Deverall and Wood, 1961; Patil and Dimond, 1967). Byrde *et al.* (1973) were able to isolate polygalacturonase isozymes from *Sclerotinia fructigena*-infected lesions on apple fruits containing phenolic compounds and from surrounding tissues devoid of phenolics.

#### 4. *Effect of Protein Inhibitors*

Some proteins have been reported to inhibit polygalacturonases by binding with enzymes (Uritani and Stahmann, 1961; Albersheim and Anderson, 1971).

### F. Evidence for the Involvement of Pathogen-Produced Enzymes in Pathogenesis

The following criteria have been used to establish the role of enzymes in pathogenesis:

#### 1. *Correlation between in Vitro and in Vivo Production of Enzymes*

Fairly good evidence for the involvement of cell-wall-degrading enzymes in pathogenesis may be obtained by establishing a correlation between *in vitro* and *in vivo* production of enzymes. However, for the reasons stated below, the absence of such a correlation does not constitute proof that the enzymes in question are not involved in pathogenesis.

1. Many enzymes are subject to catabolite repression and/or induction (previous section) and their production and activities might, therefore, vary *in vitro* and *in vivo* depending upon the presence or absence of inducers and repressors (Bateman and Basham, 1976).
2. In some cases, differences have been found in chemical and physical properties of a single enzyme produced by pathogens in culture and in diseased tissue (Garibaldi and Bateman, 1971; Byrde and Fielding, 1968; Hancock, 1976). A molecular form of an acid endopectate lyase was produced by several strains of *Erwinia chrysanthemi* in *Dieffenbachia* leaves but not in culture (Pupillo *et al.*, 1976). Hancock (1976) found differences in properties of endopectate lyase produced by *Hypomyces solani* f. sp. *cucurbitae* in culture and in infected squash plants. However, he did not rule out the possibility that the observed differences might have resulted from structural modification of the enzyme during purification. Differences were also found in ionic properties of a prominent endopolygalacturonase produced by *Rhizoctonia solani* in

culture and in infected cotton hypocotyls (Brookhouser *et al.*, 1980). Here again, the possibility of structural modifications during purification could not be ruled out entirely. However, such modifications were considered unlikely because enzymes from culture and from infected tissue were subjected to identical purification procedures.

### 2. *Use of Enzyme-Deficient Mutants*

In order to assess the involvement of pectolytic enzymes in cotton plants infected with *Verticillium dahliae,* Howell (1976) used mutants of the fungus deficient in pectolytic enzymes. Results of inoculations of plants with different mutants showed that mutants deficient in endopolygalacturonase, pectin methylesterase, pectin transeliminase, and pectic acid transeliminase, or all of these enzymes except pectin methylesterase, induced apparently normal symptoms of *Verticillium* wilt in plants. These results were regarded as evidence against the involvement of pectolytic enzymes in *Verticillium* wilt of cotton. On the other hand, McDonnell (1958), who investigated the pathogenicity of UV-induced mutants of *Fusarium oxysporum* f. sp. *lycopersici,* found that polygalacturonase-deficient isolates were less pathogenic in susceptible tomato plants than isolates capable of producing the enzyme. Chatterjee and Starr (1977) developed recombinant clones from a cross between a donor and a recipient strain of *Erwinia chrysanthemi* and found that $Pat^{+}$ recombinants (capable of producing polygalacturonic acid *trans*-eliminase) and not $Pat^{-}$ recombinants could macerate celery petioles, potato tubers, and carrot root slices.

Genetic manipulation of pathogens is a powerful tool available for elucidating the involvement of cell-wall and membrane-degrading enzymes in pathogenesis.

### 3. *Depletion of Cell Wall Fractions during Pathogenesis*

This is the most direct way to demonstrate the involvement of cell wall-degrading enzymes in pathogenesis (Hancock, 1967; Lumsden, 1969; Bateman and Jones, 1976; Baker *et al.*, 1977). For example, an analysis of the sugar content of the hypocotyls of healthy and *Sclerotium rolfsii*-infected bean cell walls showed that about 60% of the noncellulosic cell wall polysaccharides were removed within three days after inoculation (Bateman and Jones, 1976).

## G. The Role of Cell-Wall-Degrading Enzymes in Tissue Maceration and Cell Death

Plant cells are cemented together with intercellular materials composed mainly of pectic polysaccharides. Breakdown of these pectic polymers by pathogen-produced pectolytic enzymes leads to the loss of tissue coherence and separation of individual cells, a phenomenon known as maceration (Wood, 1967). The role of enzymes in tissue maceration was established early by DeBary (1886), Jones (1909), and Brown (1915). It is now universally accepted that tissue maceration is the result of degradation of the pectic fraction of cell wall polymers by endopectic enzymes. However, nonpectolytic enzymes such as endo-β-1,4-galactanase from *Fusarium* and *Phytophthora* (Cole and Sturdy, 1973), a factor known as "phytolysin" produced by *Dothidea ribesia* (Naef-Roth *et al.*, 1961; Kern and Naef-Roth, 1971), and a nonpectolytic protein from *P. capsici* (Yoshikawa *et al.*, 1977b) also have been implicated in tissue maceration.

In addition to tissue maceration, pectolytic enzymes cause protoplast death (Basham, 1974; Mount *et al.*, 1970). Cell death is closely associated with cell separation and is preceded by large increases in membrane permeability (Hall and Wood, 1973; Basham and Bateman, 1975a) (Fig. 7). A number of

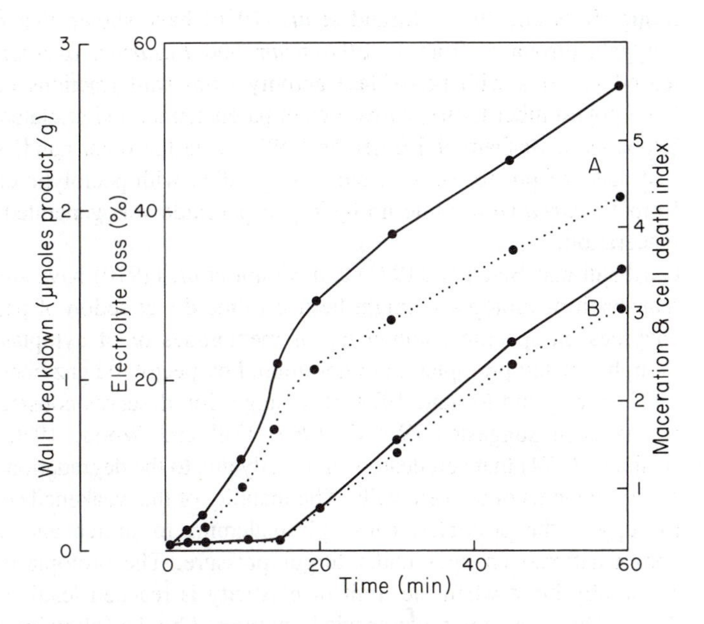


**Figure 7.** Electrolyte loss, wall breakdown, cell death, and maceration in potato disks treated with endopectate lyase from *Erwinia chrysanthemi*. (A) Wall breakdown (·····) as μmoles of unsaturated uronides released from enzyme-treated tissue, and electrolyte loss (———) as percent of electrolytes released from enzyme-treated tissue. (B) Maceration index (·····) and cell death index (———) in enzyme-treated potato disks (0 = no effect, 5 = complete maceration or cell death). In potato disks treated with heat-inactivated enzyme there was no wall breakdown, electrolyte loss was less than 2.5%/hr, and maceration and cell death indices were both 0 at the end of the experiment. [Reproduced with permission from Basham, H. G., and Bateman, D. F. 1975. *Phytopathology* **65:**141–153. The American Phytopathological Society, St. Paul, Minnesota.]

hypotheses have been advanced to explain the death of plant cells caused by pectolytic enzymes. These hypotheses, which have been discussed by Bateman (1976), Bateman and Basham (1976), Wood (1972, 1976a), and Mount (1978), are discussed here briefly:

1. Disruption of cellular organelles by pectolytic enzymes could lead to the autolysis of cytoplasm by the action of released hydrolytic enzymes (Wilson, 1973). However, phosphotidases and proteinases do not cause cell separation or death when applied to potato tubers (Mount *et al.*, 1970; Stephens and Wood, 1975).
2. Cell death may be due to a soluble reaction product of tissue maceration. However, solutions of pectolytic digests do not cause tissue maceration if the pectolytic enzymes are removed (Basham and Bateman, 1975b).
3. Protoplast death may be the result of the activity of hydrogen peroxide, a by-product of pectolytic enzyme activity in *Erwinia carotovora*-infected cauliflower floret tissue (Lund, 1973) and in *Verticillium*-infected cotton tissue (Mussell, 1973). Strand *et al.* (1976) have shown that purified polygalacturonases from *V. albo-atrum* and *Fusarium oxysporum* released proteins with peroxidase activity from wall fractions prepared from potato tuber tissue, carrot xylem parenchema, and etiolated cotton hypocotyls. Basham and Bateman (1975b), on the other hand, showed that disks of potato tissues were readily killed with pectolytic enzymes from *E. carotovora* while no hydrogen peroxide was generated during maceration.
4. Garibaldi and Bateman (1971) and Mount *et al.* (1970) have suggested that death of protoplasts might be due to the direct action of pectolytic enzymes on specific components of membranes or of cytoplasm. Although isolated protoplasts are not injured by pectolytic enzymes (Pilet, 1973; Tseng and Mount, 1974) this suggestion deserves consideration.
5. It has been suggested (Wood, 1967; Hall and Wood, 1970, 1973; Basham, 1974) that cell death is indirectly due to the degradation of pectic components of the cell walls. The inability of the weakened cell walls to support the protoplast cause plasmalemma to stretch and undergo conformational changes under turgor pressure. The protoplasts might eventually burst when the limit of elasticity is reached leading to cell death. This interesting hypothesis is supported by the following lines of evidence: First, neutral red was lost from stained protoplasts following exposure to pectolytic enzymes, possibly due to changes in plasmalemma (Hall and Wood, 1973). Second, results of electron microscopic studies of Fox *et al.* (1972) and Stephens and Wood (1975) have shown that in rotted tissues cell wall degradation is associated with bursting of cell membranes. Third, Tribe (1955) and Basham and

Bateman (1975a) have shown that plasmolyzed cells are not damaged by pectolytic enzymes. Fourth, Basham and Bateman (1975a) have found a direct correlation between the rate of release of unsaturated uronides from cell walls of tissue treated with a purified pectate lyase from *Erwinia chrysanthemi* and changes in the rate of electrolyte loss from cells. Moreover, in plasmolyzed cells subjected to purified lyase, cell wall degradation was observed in the absence of any damage to the membrane (Basham and Bateman, 1975a). Membrane damage was observed only after cells were deplasmolyzed (Fig. 8). Similar results have been obtained by Stephens and Wood (1975).

The above lines of evidence tend to support the view that cell death in plant tissues subjected to pectolytic enzymes is due to conformational changes in the

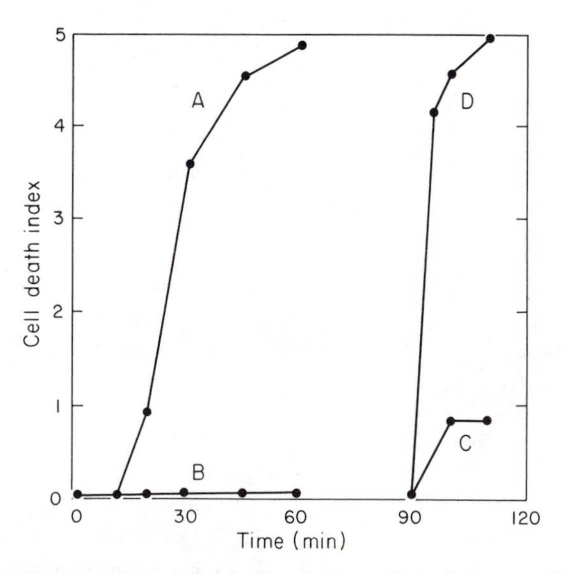


**Figure 8.** The effect of plasmolysis and deplasmolysis on cell death in potato disks treated with endopectate lyase. Potato disks were incubated with lyase in buffer or osmoticum. Disks were plasmolyzed by incubation in osmoticum for 15 min prior to the addition of enzyme. Cell death index was determined by using the Neutral Red method (0 = no cells dead, 5 = all cells dead). (A) Disks treated with lyase in buffer. (B) Disks treated with heat-inactivated lyase in buffer, lyase in osmoticum, or heat-inactivated lyase in osmoticum. After 60 min, disks in osmoticum were rinsed for 30 min in osmoticum to remove enzyme. (C) Rinsed disks, pretreated with heat-inactivated lyase, deplasmolyzed in phosphate buffer (5 mM, pH 7.5). (D) Rinsed disks, pretreated with active lyase, deplasmolyzed in phosphate buffer. [Reproduced with permission from Basham, H. G., and Bateman, D. F. 1975. *Physiol. Plant Pathol.* **5**:249–261. Copyright © 1975 Academic Press Inc. (London) Ltd.]

plasmalemma. However, these lines of evidence should be balanced against reports that the ultrastructure of the plasmalemma was only slightly altered in severely injured plant cells following treatment with pectolytic enzymes. Moreover, Kenning and Hanchey (1980) showed that ultrastructural changes observed in cytoplasm of bean hypocotyl cells infected with *Rhizoctonia solani* were not always associated with swelling of the cell walls. It was therefore concluded that ultrastructural alterations associated with this disease are not solely the consequence of weakening of cell walls and bursting of protoplasts.

*Chapter 4*

# The Role of Pathogen-Produced Toxins in Pathogenesis

## I. INTRODUCTION

Chemicals with phytotoxic activities are produced by many plant pathogenic fungi and bacteria. A few toxins are capable of producing all or most of the characteristic symptoms of diseases. Some toxins exhibit high toxicity only to the hosts of the toxin-producing pathogens and in a few cases the ability of a pathogen to produce a toxin is correlated with its potential to cause disease.

The literature on pathogen-produced toxins has been reviewed by Luke and Gracen (1972a), Templeton (1972), Durbin (1972), Graniti (1972), Wheeler (1975, 1981), Rudolph (1976), Scheffer (1976), Strobel (1977), Patil (1974, 1980), Yoder (1976, 1980), and Mitchell (1981); see also the book edited by Durbin (1981).

The nature and significance of a number of plant pathogen-produced toxins and their speculative mode of action will be discussed.

## II. DEFINITION AND CLASSIFICATION OF PATHOGEN-PRODUCED TOXINS

There is little agreement on how pathogen-produced toxins should be defined. But many seem to agree that phytotoxins are diffusable and translocatable substances produced by living organisms which are injurious in small quantities to certain plants. There is also no consensus on how pathogen-produced toxins should be classified. However, phytotoxins are generally classified on the basis of their host selectivity (specificity) and their roles in plant pathogenesis.

## III. TOXIN CATEGORIES

Three toxin categories—vivotoxin, pathotoxin, and host-selective toxin—will be discussed briefly. Of these, the term *vivotoxin* is no longer in use. However, this term has historical significance and the criteria used to define it have served as a basis for formulating modern concepts in toxin research. Toxin categories have been discussed in detail by Wood (1967), Graniti (1972), and Luke and Gracen (1972a).

## A. Vivotoxin

The term coined by Dimond and Waggoner (1953a) was defined as a substance produced by a pathogen, a host plant, or both that contributes to disease development. They proposed the following three criteria for verifying vivotoxicity: (1) reproducible isolation of the putative toxin from diseased but not healthy plants, (2) purification as well as characterization of the toxin, and (3) reproduction of at least part of the symptoms by application of the toxin to plants.

Some of the criteria used by Dimond and Waggoner (1953a) for vivotoxin were not well received. For example, Ludwig (1960) thought that the criteria were too restrictive and could not be applied to nonspecific toxins that are produced outside the host and which may predispose it to attack by pathogens. Wheeler and Luke (1963) did not consider it necessary to isolate the substance from the diseased tissue (Wood, 1967).

## B. Pathotoxin

Wheeler and Luke (1963) introduced the term *pathotoxin* for a toxic product of a pathogen, a plant, or a plant–pathogen interaction that has an important role in pathogenesis. According to Wheeler and Luke (1963), pathotoxicity is established when a pathogen-produced substance meets the following minimum criteria: (1) The toxin, when applied at concentrations found in diseased tissues, produces all the characteristic symptoms of the disease in susceptible hosts. (2) Similar host selectivity (specificity) is exhibited by the pathogen and the toxin. (3) The toxin-producing ability of the pathogen correlates with its ability to cause disease. (4) Only one toxin is involved.

Wheeler and Luke (1963) also used the term *phytotoxin* for ". . . all products of living organisms toxic to plants,"* with no reference to their involvement in plant diseases.

## C. Host-Selective (-Specific) Toxin

To emphasize the significance of host specificity of toxins, Pringle and Scheffer (1964) introduced the term *host-specific toxin,* defining it as "a metabolic product of a pathogenic microorganism which is toxic only to the host of that pathogen."† This term now is in general use. Wheeler (1975) has pointed

*Reproduced, with permission, from *Annu. Rev. Microbiol.* **17**:223–242. ©1963 by Annual Reviews, Inc.

†Reproduced, with permission, from *Annu. Rev. Phytopathol.* **2**:133–156. ©1964 by Annual Reviews, Inc.

out that so-called host-specific toxins exhibit selectivity rather than specificity because they are also active against resistant plants at high concentrations. He therefore considers the term *host-selective toxin* to be more appropriate than *host-specific toxin*. In this book the term *host-selective toxin* has been used in place of *host-specific toxin*.

## IV. EVALUATION OF THE CRITERIA USED TO ESTABLISH THE ROLE OF TOXINS IN PATHOGENESIS

As pointed out earlier, a number of criteria have been proposed to establish the role of pathogen-produced toxins in pathogenesis. These criteria and their limitations, which have been discussed by Wood (1967), Rudolph (1976), and Yoder (1980) in detail, are reviewed here briefly. The format of this section is patterned after Yoder (1980).

### A. Isolation of the Toxin from Diseased Plants

Toxins cannot always be isolated from diseased tissue because they may be inactivated rapidly by the host tissue prior to or during extraction (Wood, 1967). Moreover, the concentration of the toxin in diseased tissue may be below detectable limits (Wood, 1967). Because of these limitations this criterion is considered to be of little value for assessing the role of toxins in pathogenesis.

### B. Reproduction of Disease Symptoms by the Toxin

This criterion also is considered to have certain limitations. For example, under natural conditions toxins may not act alone. A toxin might induce disease by predisposing the plant to the detrimental effects of other products produced in the diseased tissue (Rudolph, 1976). Moreover, wilting and necrosis can be induced by a large number of natural and artificial products (Rudolph, 1976). Yoder (1980) has pointed out that a comparison of the physiological changes induced by the pathogen and by the toxin might be more meaningful than that of disease symptoms.

### C. Correlation between Pathogenicity and the Level of Toxin Produced *in Vitro*

A correlation has been found between pathogenicity and the level of toxin produced *in vitro* in a number of phytotoxic compounds. However, failure to demonstrate a direct correlation might be due to differences between *in vitro* and *in vivo* conditions (Rudolph, 1976).

### D. Correlation between Disease Susceptibility and Sensitivity to Toxins

Host-selective toxins exhibit high toxicity only to the host of the toxin-producing pathogens. However, some of the factors discussed under Section B might obscure this correlation. Yoder (1973, 1980) and Wood (1976b) have pointed out that the host selectivity of a toxin should not be regarded as a highly significant criterion for establishing the role of a toxin in pathogenesis. This is because (1) selective toxins are also active against resistant plants at relatively low concentrations, (2) certain substances with no known roles in disease exhibit apparent host selectivity, and (3) substances with limited host selectivity may be present in the culture filtrates of known saprophytes.

### E. Genetic Analysis of the Host, the Pathogen, or Both

Genetic analysis of the host, the toxin-producing pathogen, or both, may provide the most reliable and convincing evidence for the involvement of a toxin in pathogenesis (Yoder, 1980). However, some of the toxin-producing pathogens cannot be manipulated genetically. The following plant–pathogen systems have been subjected to genetic analysis: oat–*Helminthosporium victoriae;* corn–*H. maydis,* race T; corn–*Phyllosticta maydis;* sugarcane–*H. sacchari;* corn–*H. carbonum,* race 1; sorghum–*Periconia circinata;* tobacco–*Pseudomonas tabaci;* and tomato–*Alternaria alternata* f. sp. *lycopersici*. The literature on the subject has been reviewed by Yoder (1980) and Panapoulos and Staskawicz (1981).

The following criteria have also been used to establish the important role of toxins in pathogenesis (Rudolph, 1976): (1) the breaking of disease resistance by toxins (Rudolph, 1972), (2) the reduction in disease incidence by toxin inactivation (LeBeau and Atkinson, 1967), and (3) the similar effects on disease development, *in vivo* toxin production, and sensitivity to the applied toxin, by changes in environmental conditions (Durbin and Sinden, 1967).

## V. HOST-SELECTIVE TOXINS

A striking feature of host-selective toxins is their host selectivity. That is, susceptible plants are highly sensitive to the toxins while resistant plants are rather insensitive. Some of the host-selective toxins are discussed in this section. See also reviews by Scheffer (1976), Wheeler (1975), and Strobel (1976).

## A. Victorin

This toxin is produced by *Helminthosporium victoriae*, the causal organism of Victoria or *Helminthosporium* blight of oats (Meehan and Murphy, 1947; Luke and Wheeler, 1955).

### *1. Evidence for the Causal Role in Pathogenesis*

The following evidence has been cited for the causal role of victorin in pathogenesis: (1) The toxin produces all the known characteristic symptoms of the disease (Scheffer and Pringle, 1967). (2) The toxin exhibits extreme selectivity. Oat cultivars that are susceptible, moderately susceptible, and resistant to *H. victoriae* are sensitive, moderately sensitive, and insensitive to the toxin (Luke and Wheeler, 1964; Scheffer and Pringle, 1967; Kuo and Scheffer, 1970; Scheffer and Yoder, 1972). (3) All isolates of *H. victoriae* that produce the toxin in culture are pathogenic to oats while all those that fail to produce the toxin are nonpathogenic to oats (Luke and Wheeler, 1955; Scheffer and Pringle, 1967). (4) Initial colonization of the host by the pathogen is aided by the toxin (Nishimura and Scheffer 1965; Yoder and Scheffer, 1969). (5) Similar biochemical changes are induced by the pathogen and the toxin (Scheffer and Pringle, 1967) (Table 1).

### *2. Physiological Changes Caused by Victorin*

Toxin-treated tissues have been reported to show loss of electrolytes and other materials from cells, increased respiration, increased $CO_2$ fixation in the dark, and an increase in free amino acids. Other changes include decreases in growth, in protein synthesis, in the incorporation of inorganic phosphates into organic compounds, and in the closing of stomates (Scheffer, 1976).

**Table 1. Comparison of the Physiological Changes Induced by Fungal and Bacterial Infection (in General) and by Host-Selective Toxins**[a]

| Response | Infection | Victorin | PC toxin | HC toxin |
|---|---|---|---|---|
| Loss of electrolytes | + | + | + | + |
| Increase in respiration | + | + | + | + |
| Increase in dark $CO_2$ fixation | + | + | ? | + |
| Increase in free amino acid content | + | + | ? | + |

[a]A modified version of a table presented by Scheffer, R. P., and Yoder, O. C. 1972. In *Phytotoxins in Plant Diseases* (R. K. S. Wood, A. Ballio, and A. Graniti, eds.), pp. 251–272. With permission from the authors and Academic Press Inc. (London) Ltd.

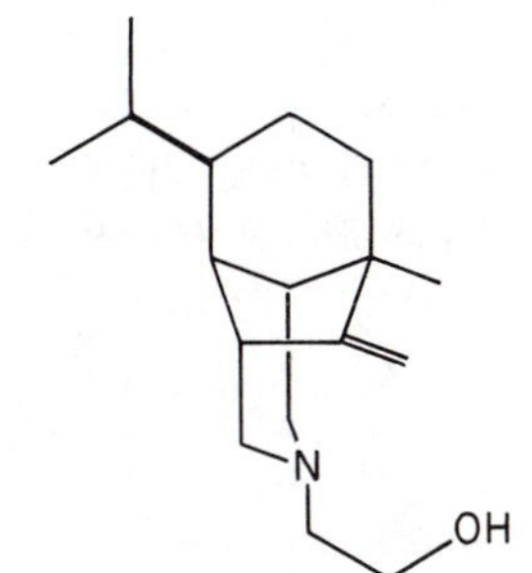


**Figure 9.** Proposed structure for victoxinine. [Dorn, F., and Arigoni, D. 1972. *J. Chem. Soc. Chem. Commun.* **1972:**1342–1343.]

### *3. Victorin-Induced Changes in Cell Walls*

Drastic changes in the cell wall structure have been reported to precede detectable changes in membrane structure in victorin-treated oat roots (Luke *et al.*, 1966; Hanchey *et al.*, 1968; Hanchey, 1980). Victorin-induced cell wall lesions in oat roots may result from activation of host cell wall enzymes (Hanchey, 1980).

### *4. Chemical Nature of Victorin*

The toxin is considered to be made of two components: a peptide and a heterocyclic moiety (Pringle and Braun, 1957). The heterocyclic moiety, known as victoxinine, is considerably less toxic than victorin and is not host selective. The empirical formula of victoxinine, according to Dorn and Arigoni (1972), is $C_{17}H_{29}NO$ (Fig. 9). Purification and characterization attempts have been hampered by lability of the toxin (Wheeler, 1975).

## B. T Toxin

This toxin is produced by *Helminthosporium maydis* race T, the causal organism of southern corn leaf blight (Hooker *et al.*, 1970). Much of the early literature on T toxin is controversial, possibly because many of the studies have been done using high concentrations of impure toxin preparations.

### *1. Evidence for the Causal Role in Pathogenesis*

Leaves from Texas male-sterile cytoplasm (T cytoplasm) corn plants which are susceptible to race T of the fungus are at least 4000–8000 times more sensitive to a highly purified T toxin than resistant normal cytoplasm (N cytoplasm) leaves (Kono and Daly, 1979). Evidence for the involvement of T toxin in disease severity has also been provided by the results of mutational analyses of corn plants (Gengenbach *et al.*, 1977) and recombinational analyses of *H. maydis* isolates (Yoder and Gracen, 1975).

### 2. *Physiological Changes Caused by T Toxin*

Physiological changes induced by T toxin include selective ion leakage in sensitive plants (Gracen *et al.*, 1972; Arntzen *et al.*, 1973b; Halloin *et al.*, 1973; Keck and Hodges, 1973), drastic changes in mitochondria from susceptible T cytoplasm corn (Miller and Koeppe, 1971; Bednarski *et al.*, 1977; Gregory *et al.*, 1978; Matthews *et al.*, 1979; Payne *et al.*, 1980b), root growth inhibition (Hooker *et al.*, 1970; Arntzen *et al.*, 1973b; Bhullar *et al.*, 1975), changes in photosynthesis (Arntzen *et al.*, 1973b; Bhullar *et al.*, 1975), stomatal closure as well as changes in transpiration (Arntzen *et al.*, 1973a), and alterations in dark $CO_2$ fixation (Bhullar *et al.*, 1975). Payne *et al.* (1980a) have also reported that a highly purified toxin preparation from the fungus was active in all the above processes in susceptible T cytoplasm corn, while the toxin, at concentrations 1000 times greater, did not have any effect on resistant N cytoplasm corn.

### 3. *Chemical Nature of T Toxin*

Karr *et al.* (1974) suggested that the toxin consisted of a sterol and related glycosides, although the structure of the toxin was not determined. On the other hand, Aranda *et al.* (1978) described the toxin as an acetate ester of mannitol covalently linked to two N-formyl-L-valine residues. Kono and Daly (1979) have recently isolated a highly active and specific toxin from *H. maydis* race T. The concentration of the toxin required to kill an entire first true leaf of susceptible T cytoplasm corn was 4–8 ng while 30 μg of the toxin was ineffective against resistant N cytoplasm corn leaves. According to Kono and Daly (1979) the highly purified toxin consists of a mixture of linear $C_{35}$ to $C_{45}$ polyketols, each having apparent host selectivity. The possible structure of the principal component presented by Kono and Daly (1979) is shown in Fig. 10.

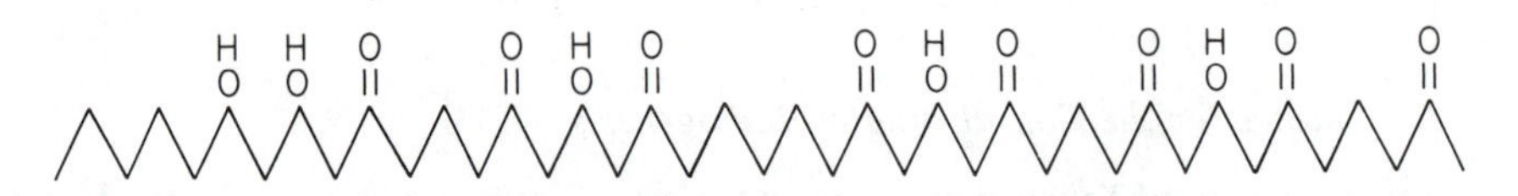

**Figure 10.** Proposed structure for band 1 toxin from *Helminthosporium maydis*, race T. [Redrawn with permission from Kono, Y., and Daly, J. M. 1979. *Bioorg. Chem.* **8**:391–397. Academic Press, New York.]

The Daly group (Payne *et al.*, 1980a,b) has also reported that the purified toxin affected ion balance, dark $CO_2$ fixation, coleoptile elongation, and mitochondrial oxidation of susceptible T cytoplasm corn at concentrations much lower than that reported for toxins isolated by Karr *et al.* (1974) and by Aranda *et al.* (1978). The greater toxicity of the toxin prepared by Kono and Daly (1979), compared to toxins isolated by Karr *et al.* (1974) and Aranda *et al.* (1978) was cited by Kono and Daly (1979) as evidence for the purity of their preparation upon which the evidence for the structure shown in Fig. 10 is based. The higher toxicity of the toxin purified by the Daly group compared to that of other preparations also has been confirmed by Gregory *et al.* (1980).

## C. HC Toxin

*Helminthosporium carbonum* race 1, the causal organism of corn leaf spot, has been reported to produce a toxin known as HC toxin.

### *1. Evidence for the Causal Role in Pathogenesis*

The following evidence has been cited for the causal role of HC toxin in pathogenesis: (1) Sensitivity to the toxin is correlated with susceptibility to the pathogen (Scheffer and Ullstrup, 1965; Scheffer and Pringle, 1967; Kuo and Scheffer, 1970). (2) Physiological changes characteristic of the disease can be reproduced by toxin treatment (Kuo and Scheffer, 1970). (3) The toxin is required for successful colonization of susceptible tissue (Comstock and Scheffer, 1973). (4) The progeny of crosses between *H. victoriae* and *H. carbonum* segregated on a 1 : 1 : 1 : 1 ratio for production of victorin, HC toxin, both toxins, or neither toxin and for pathogenicity to oats, corn, both, or neither host, respectively (Scheffer *et al.*, 1967). It was therefore concluded that for each fungus pathogenicity is correlated with the ability to produce its toxin.

### *2. Physiological Changes Caused by HC Toxin*

Sensitive plant tissue shows an increase in electrical potential across plasmalemma within 5 min of exposure to the toxin (Gardner *et al.*, 1974), followed by an increase in uptake of certain solutes (Yoder and Scheffer, 1973). The toxin at low concentrations stimulates growth of susceptible roots. Root growth, however, is inhibited when high concentrations of the toxin are used (Kuo *et al.*, 1970).

### *3. Chemical Nature of HC Toxin*

According to Pringle (1970, 1971, 1972) HC toxin is a cyclic substituted polyamide.

## D. HS Toxin

This toxin is produced by *Helminthosporium sacchari,* the causal organism of eyespot disease of sugarcane (Steiner and Strobel, 1971).

### *1. Evidence for the Causal Role in Pathogenesis*

Susceptibility of sugarcane clones to the pathogen is fairly correlated with their sensitivity to the toxin (Steiner and Byther, 1971; Steiner and Strobel, 1971; Strobel and Steiner, 1972). Based on the results of an assay using drops of toxin solution on leaves of sugarcane clones, Strobel (1974b) reported a statistically significant correlation between susceptibility to the disease and sensitivity to the toxin in 149 of 182 sugarcane clones. Scheffer and Livingston (1980) tested sensitivity of 17 sugarcane clones to a pure toxin preparation using a bioassay based on toxin-induced loss of electrolytes from leaf tissue. In general, sensitivity or insensitivity to the toxin correlated with susceptibility or resistance to the fungus. However, the correlation did not hold for 3 clones of the 17 tested. They suggested that the toxin may determine pathogenicity of the fungus to some clones but not to others and that the toxin should be used with caution in screening for disease resistance. The evidence for the toxin's causal role in pathogenesis does not seem to be as strong as that for other selective toxins discussed earlier. The mode of action of the toxin is discussed later in this chapter.

### *2. Chemical Nature of HS Toxin*

Steiner and Strobel (1971) characterized the toxin as 2-hydroxy cyclopropyl-α-D-galactopyranoside and called it helminthosporoside. However, Livingston and Scheffer (1981), working with a highly purified preparation of the toxin, have recently reported that it contains an oligosaccharide made of β,1→5 galactofuranose units (probably 5 units) and a $C_{15}H_{21}$ moiety (probably a sesquiterpene) connected to the reducing end of the oligosaccharide. Livingston and Scheffer (1981) were able to separate the toxin from three closely related nontoxic fractions which contained galactose and a $C_{15}H_{21}$ moiety. Judging from the results of comparative analyses they concluded that the characterized toxin was not the same as the one described by Steiner and Strobel (1971).

## E. PC Toxin

*Periconia circinata,* the causal agent of milo disease or *Periconia* blight of grain sorghum, produces a host-selective toxin designated as PC toxin (Scheffer and Pringle, 1961). The toxin was reported to exist as three or more closely related compounds; each proved to be a selective toxin on susceptible sorghum (Scheffer, 1976).

### *1. Evidence for the Causal Role in Pathogenesis*

The following evidence has been cited for the causal role of PC toxin in pathogenesis: (1) Only sorghum genotypes that are susceptible to the pathogen show sensitivity to the toxin (Scheffer and Pringle, 1961; Schertz and Tai, 1969). (2) Toxin-producing isolates of the fungus cause extensive damage to susceptible sorghum lines, while non-toxin-producing isolates are weak parasites (Odvody *et al.*, 1977). (3) The purified toxin from culture filtrates of the fungus inhibits root growth of susceptible sorghum seedlings by 50% at a concentration of 1 ng/ml but has no effect on root growth of resistant seedlings at concentrations up to 2 μg/ml (Wolpert and Dunkle, 1980).

Judging from the results of their study discussed below, Dunkle and Wolpert (1981) have suggested that the development of milo disease symptoms is not necessarily the result of electrolyte loss. Five- to 7-day-old sorghum seedlings were exposed to low concentrations of the toxin, insufficient to produce symptoms and causing only small electrolyte leakage. Ten hours later the seedlings were treated with concentrations of PC toxin sufficient to induce severe milo disease symptoms, yet a marked electrolyte loss was not detected. Seedlings pretreated and not pretreated with the lower concentrations of the toxin developed symptoms at an equal rate. Treatment of both susceptible and resistant seedlings with fusaric acid and citrinin resulted in a greater loss of electrolytes than treatment of susceptible seedlings with PC toxin. However, distinct milo disease symptoms developed in susceptible seedlings treated with PC toxin 2–3 days earlier than the mild chlorosis which developed in seedlings exposed to fusaric acid or citrinin.

### *2. Physiological Changes Caused by the Toxin*

The physiological changes induced by PC toxin are similar to those caused by victorin (Scheffer, 1976). However, sulfhydryl- and carbonyl-binding compounds which protect oat tissues against victorin (Gardner and Scheffer, 1973) do not protect sorghum tissues against PC toxin (Gardner *et al.*, 1972).

### *3. Chemical Nature of the Toxin*

Working with a partially purified preparation of the toxin, Pringle and Scheffer (1963, 1966) concluded that the toxin was a peptide of four common amino acids. Another toxin with nearly equal activity was later detected in a culture filtrate of the fungus (Pringle and Scheffer, 1967). More recently Wolpert and Dunkle (1980) have achieved greater purification of the toxin fractions by using more powerful analytical tools. They concluded that the toxins which were resolved into two fractions were low-molecular-weight acidic compounds containing multiple residues of aspartic acid and one or more polyamine residues.

## F. PM Toxin

*Phyllosticta maydis,* which causes yellow leaf blight of corn, produces a host-selective toxin known as PM toxin (Scheffer, 1976).

### *Evidence for the Causal Role in Pathogenesis*

As with T toxin, corn cultivars carrying Texas male-sterile cytoplasm are more susceptible to *P. maydis* and more sensitive to PM toxin produced by the fungus than cultivars carrying non-male-sterile cytoplasm. Moreover, certain physiological responses induced by T toxin and PM toxin are very similar (Comstock *et al.,* 1973; Yoder, 1973). Despite these similarities it is not known if T toxin and PM toxin are similar chemically (Wheeler, 1975).

## G. AK Toxin

A host-selective toxin found in culture filtrates of *Alternaria kikuchiana,* the causal organism of black spot disease of Japanese pear, was first reported by Tanaka (1933). Of the three toxins isolated from the fungus, two have been purified (Nishimura *et al.,* 1979). The toxins were found to be highly selective causing disease symptoms only on leaves of susceptible pears at concentrations as low as 0.01–0.1 μg/ml. No symptoms were produced on leaves of resistant pears or on those of nonhost plants at concentrations as high as 100 μg/ml (Nishimura *et al.,* 1974). Loss of electrolytes and damage to plasmalemma have been detected within 5 and 60 min, respectively, after exposure of the sensitive tissues to the toxins (Park *et al.,* 1976; Nishimura *et al.,* 1979).

## H. AM Toxin

*Alternaria mali,* the causal agent of *Alternaria* blotch of apple, produces several toxins known as AM toxins that are selective for certain apple varieties. Three of these toxins have been characterized as depsipeptides (Okuno *et al.,* 1974; Ueno *et al.,* 1975a,b) (Fig. 11). The resistant apple cultivars tolerate the

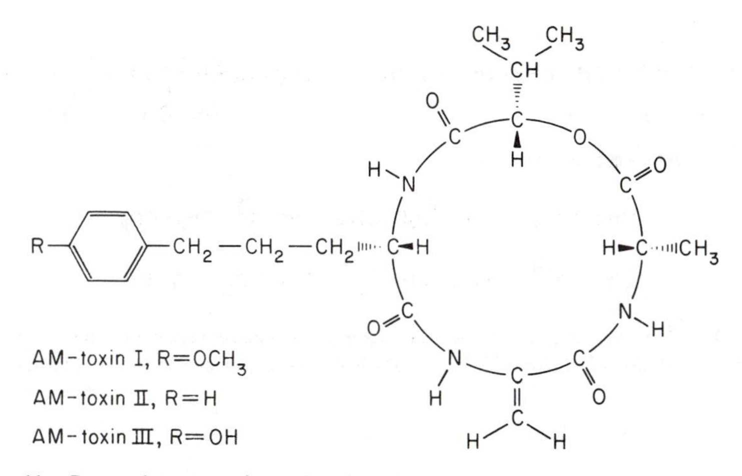


**Figure 11.** Proposed structure for AM toxins. [Ueno, T., *et al.* 1975. *Agric. Biol. Chem.* **39:**1115–1122.]

toxins at concentrations about 10,000 times higher than those that cause damage to susceptible cultivars (Nishimura *et al.*, 1979).

Judging from the results of ultrastructural studies Park *et al.* (1981) concluded that the target sites for the toxin in susceptible tissue may be located on the plasmalemma–cell wall interface and on the chloroplasts.

## I. AA Toxin

The stem canker disease of tomato, caused by *Alternaria alternata* f. sp. *lycopersici,* is controlled in the host by a single gene, and symptom expression is due to host-selective toxins (AA toxins) (Gilchrist and Grogan, 1976).

### *1. Evidence for the Causal Role in Pathogenesis*

Fractionation of culture filtrates of the pathogen yielded two ninhydrin-positive fractions, TA and TB, capable of producing disease symptoms on excised tomato leaves at concentrations of 10 ng/ml or less (Bottini and Gilchrist, 1981; Bottini *et al.*, 1981). Reproduction of the disease symptoms is under quantitative control by the same genetic locus that regulates the host–pathogen interaction. However, the two alleles at this locus segregate as a complete dominant for resistance to the pathogen but express incomplete dominance for tolerance to the toxins. The sensitivity of $F_1$ (*Rr*) genotype to the toxin is intermediate com-

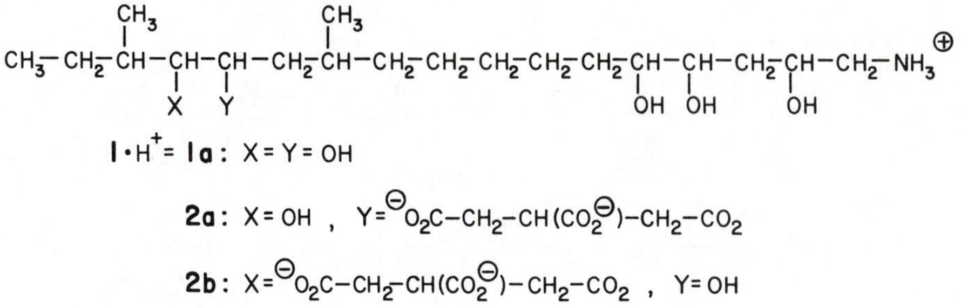

$$CH_3-CH_2-CH(CH_3)-CH(X)-CH(Y)-CH_2-CH(CH_3)-CH_2-CH_2-CH_2-CH_2-CH_2-CH(OH)-CH(OH)-CH_2-CH(OH)-CH_2-NH_3^{\oplus}$$

$\mathbf{1}\cdot H^+ = \mathbf{1a}$: $X = Y = OH$

$\mathbf{2a}$: $X = OH$ , $Y = {}^{\ominus}O_2C-CH_2-CH(CO_2^{\ominus})-CH_2-CO_2$

$\mathbf{2b}$: $X = {}^{\ominus}O_2C-CH_2-CH(CO_2^{\ominus})-CH_2-CO_2$ , $Y = OH$

**Figure 12.** Proposed structure for AA toxin. [Reproduced with permission from Bottini, A. T., *et al.* 1981. *Tetrahedron. Lett.* **22:**2723–2726. Copyright © 1981 Pergamon Press Inc., Elmsford, New York.]

pared to either parent, being 50 times more sensitive than the resistant parent (RR) and 20 times less sensitive than the susceptible (rr) parent (Gilchrist and Grogan, 1976). Plants representing eight solanaceous species and at least one representative of each of eight other plant families were resistant to the disease and insensitive to the purified toxin preparations at concentrations that were toxic to susceptible (rr) tomato plants. The toxin was produced *in vitro* only by isolates pathogenic on tomato (Gilchrist and Grogan, 1976). Moreover, toxin production paralleled growth rates in a number of *A. alternaria* isolates obtained from various hosts.

### 2. *Chemical Nature of the Toxins*

The toxin was resolved into two major fractions, TA and TB, by a combination of gel permeation chromatography, isoelectric focusing, and thin-layer chromatography. The results of NMR and high-resolution mass spectroscopic analysis of TA (Fig. 12, 1a) revealed two esters of 1,2,3-propanetricarboxylic ( = tricarballyic) acid and the novel 1-amino-11,15-dimethylheptadeca-2,4,5,13,14-pentol with equivalent toxicity (Bottini *et al.*, 1981). Analysis of CMR spectra led to the conclusion that the two forms (2a and 2b) represent single-site esterification at $C_{13}$ and $C_{14}$ and appear to occur in a ratio of 4 : 3 in the TA mixture. Preliminary evidence indicates that TB consists of two components with the same carbon skeletons as 2a and 2b but which lack the $C_5$ hydroxyl and differ in stereo-chemistry at one or more of the chiral centers from $C_{11}$ to $C_{15}$.

Less information is available on other host-selective toxins including those produced by *Alternaria citri* (Kohmoto *et al.*, 1979) and *A. fragariae* (Nishimura *et al.*, 1978).

## VI. NONSELECTIVE TOXINS

Nonselective toxins do not exhibit selectivity toward the hosts of the pathogens that produce them. Some of the more important members of this group will be discussed. The literature on some of these toxins has been reviewed by Rudolf (1976) and Stoessl (1981).

### A. Amylovorin

A wilt-inducing polysaccharide, amylovorin, was described originally as a product of an interaction between susceptible host and *Erwinia amylovora* because it could only be isolated from bacterial ooze exuded from inoculated slices of immature apple fruits but not from cultures of the bacterium (Goodman *et al.*, 1974). However, Beer *et al.* (1977) and Beer and Woods (1978) were subsequently able to isolate the polysaccharide from axenic culture filtrates of the bacterium. The polysaccharide isolated from culture filtrates was identical to that present in bacterial ooze on the basis of chromatographic and serological tests (Beer *et al.*, 1977).

Amylovorin has not been considered here as a selective toxin because the evidence for its selectivity is not very strong (Beer and Aldwinckle, 1976; Ayers *et al.*, 1977; Goodman *et al.*, 1978). However, amylovorin plays a role in pathogenicity (Ayers *et al.*, 1979) and the ultrastructural changes induced by it are the same as those induced by the bacterium (Huang and Goodman, 1976). Ayers *et al.* (1979) showed that a high-molecular-weight polysaccharide from culture filtrates of *E. amylovora* and a polysaccharide isolated from ooze on infected apple fruits were serologically identical. The polysaccharides from the culture filtrate and from ooze caused wilting in shoots of susceptible apple cultivars. Similar results have been obtained by Suhayda and Goodman (1981).

Huang and Goodman (1975) suggested that wilting induced by amylovorin resulted from increased membrane permeability to solutes. Sjulin and Beer (1978), on the other hand, reported that amylovorin-induced wilt in cotoneaster shoots is the result of increased resistance to water flow, while *E. amylovora*-induced wilt results from a disruption of membrane integrity.

### B. Tentoxin

A toxin known as tentoxin produced by *Alternaria tenuis* causes variegated chlorosis in seedlings of cucumber, cotton, and a few other plants (Templeton,

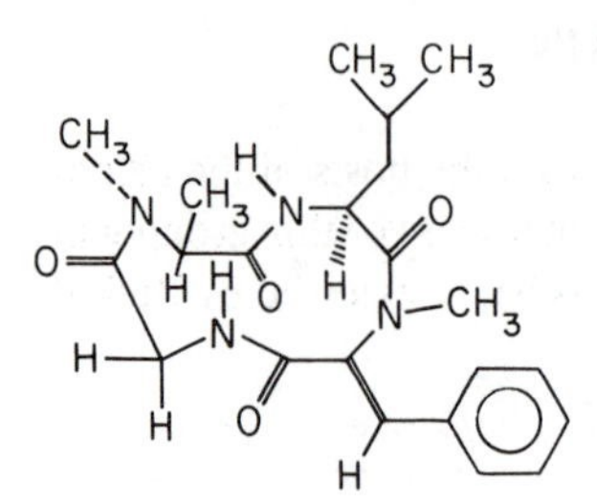

**Figure 13.** Proposed structure for tentoxin. [Proposed by Rich, D. H., and Bhatnagar, P. K. 1978. *J. Am. Chem. Soc.* **100:**2212–2218.]

1972). The toxin is a cyclic tetrapeptide (Templeton *et al.*, 1967). The latest structure for tentoxin proposed by Rich and Bhatnagar (1978) is shown in Fig. 13. The mode of action of tentoxin will be discussed later in this chapter.

## C. Phaseolotoxin

*Pseudomonas phaseolicola,* the causal agent of halo blight of bean, produces a number of toxins in culture. These toxins have been studied by Patil (1974), Patil *et al.* (1972, 1974, 1976), Mitchell and Parsons (1977), Mitchell and Bieleski (1977), and Mitchell (1967a,b, 1978).

Patil *et al.* (1976) isolated four components from the culture filtrates of *P. phaseolicola,* one of which was identified as *N*-phosphoglutamic acid and was named phaseotoxin. The toxin extracted by Patil and co-workers is capable of causing chlorosis in susceptible bean leaves, inhibiting ornithine carbamoyltransferase, accumulating ornithine in plant tissue (Patil *et al.*, 1970), preventing hypersensitive response (Patil and Gnanamanickam, 1976; Rudolph, 1972), and inhibiting or suppressing phytoalexin formation in bean leaves inoculated with certain incompatible strains of *P. phaseolicola* (Gnanamanickam and Patil, 1977a,b; Patil and Gnanamanickam, 1976).

Mitchell (1976a) isolated a toxin from culture filtrates of *P. phaseolicola* and characterized it as ($N^{\delta}$-phosphosulphamyl)ornithylalanylhomoarginine and named it phaseolotoxin (Fig. 14). Recent NMR spectral data are fully consistent

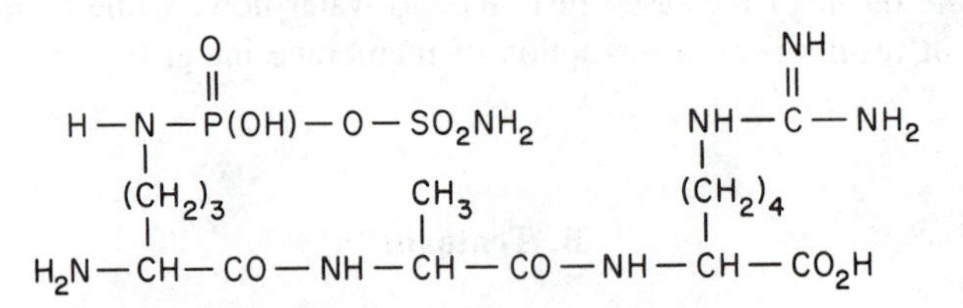

**Figure 14.** Proposed structure for phaseolotoxin. [From Mitchell, R. E. 1976. *Phytochemistry* **15:**1941–1947.]

with the proposed structure (R.E. Mitchell, personal communication). A minor component found to be an analogue, (2-serine)-phaseolotoxin [($N^{\delta}$-phosphosulphamyl)ornithylserylho moarginine], was also isolated from culture filtrates of the bacterium (Mitchell and Parsons, 1977). The evidence for the involvement of phaseolotoxin and (2-serine)phaseolotoxin in the development of the disease symptoms is strong. Both toxins induced chlorosis in bean leaves at low concentrations (1 μg/ml), ornithine accumulation in the chlorotic tissue (presumably due to the inhibition of ornithine carbamoyltransferase) and systemic chlorosis in young trifoliate leaves (Mitchell and Bieleski, 1977; Mitchell, 1978). Furthermore, phaseolotoxin produced in infected bean tissues was shown to be metabolized by peptidases present in plant tissue to $N^{\delta}$-phosphosulphamyl)ornithine (Psorn). Psorn was identified as the main functional toxin responsible for accumulation of ornithine and chlorophyll breakdown (Mitchell and Bieleski, 1977).

Differences among bacterial strains do not seem to account for the discrepancy regarding the chemical identity of the toxins, because the toxin investigated by Mitchell is produced by several bacterial isolates collected from different geographic areas (Mitchell, 1978). The discrepancy may be due to the differences in methods employed by Patil and Mitchell to isolate and purify the toxin from diseased tissues. See a review by Mitchell (1981).

Phaseolotoxin is not a selective toxin. Evidence for the nonselectivity of phaseolotoxin has been provided by the results of studies by Ferguson and Johnston (1980), who showed that application of the toxin to 10 plant species, including bean, resulted in an accumulation of ornithine and inhibition of ornithine carbamoyltransferase in all. The treatment also induced chlorosis in all but one species. Moreover, the toxin is produced by both race 1 and race 2 of *P. phaseolicola* which have different host ranges.

### D. Tabtoxin and Tabtoxinine-β-lactam

Tabtoxin and tabtoxinine-β-lactam are produced by *Pseudomonas tabaci,* the causal agent of wildfire of tobacco and closely related species, and cause chlorosis when applied to leaves (Sinden and Durbin, 1970; Durbin, 1971). Tabtoxin is a dipeptide containing tabtoxinine-β-lactam [2-amino-4-(3-hydroxy-2-oxo-azocyclobutan-3-y1)-butanoic acid] linked to either threonine or serine (Stewart, 1971; Taylor *et al.*, 1972) (Fig. 15). Free tabtoxinine-β-lactam that occurs in culture filtrates of *P. tabaci* exhibits activities similar to those of tabtoxin, including chlorosis and accumulation of ammonia in chlorotic tissue in treated tobacco leaves (Durbin *et al.*, 1978). Chlorosis was evident earlier in tabtoxinine-β-lactam-treated tissues compared to those treated with tabtoxin (Uchytil and Durbin, 1980).

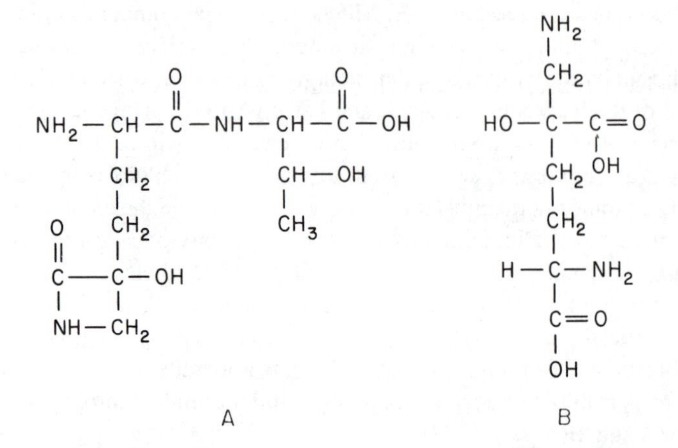


**Figure 15.** Proposed structures for tabtoxin (A) and tabtoxinine (B). [From Stewart, W. W. 1971. *Nature* **229:**172–178; and Taylor, P. A., *et al.* 1972. *Biochim. Biophys. Acta* **286:**107–117.]

Results of a recent study by Uchytil and Durbin (1980) have shown that tabtoxinine-β-lactam, but not purified tabtoxin, inhibited glutamine synthetase *in vitro* and that both plants and *Pseudomonas* spp. possess peptidases capable of hydrolyzing tabtoxin to tabtoxinine-β-lactam. Judging from the results of the above studies Uchytil and Durbin (1980) concluded that tabtoxinine-β-lactam might be more important than tabtoxin in induction of chlorosis and that hydrolysis of tabtoxin in diseased tissues by peptidases might be a prerequisite for its biological activity.

## E. Syringomycin and Syringotoxin

Most pathogenic strains of *Pseudomonas syringae,* a pathogen of stone fruits and plants in 33 other families, produce in culture a nonselective necrogenic toxin known as syringomycin which is toxic to plants and to a wide spectrum of microorganisms (DeVay *et al.,* 1968; Sinden *et al.,* 1971). The toxin causes symptoms very similar to those caused by the bacterium (DeVay *et al.,* 1968; Gross and DeVay, 1977a).

Gross and DeVay (1977b) studied the correlation between syringomycin production by 75 field isolates of *P. syringae* and their pathogenicity on various hosts. The toxin was produced *in vitro* by all isolates pathogenic to either corn, cowpea, or both, but not by isolates nonpathogenic to these hosts. These results show that syringomycin may play a role in pathogenicity of corn and cowpea, two of the many hosts of the bacterium. Some nonpathogens of other host plants produced the toxin, a characteristic reflecting the occurrence of ecotypes within the *P. syringae* species (Gross and DeVay, 1977b). The primary effect of syringomycin appears to be on cell membranes (Backman and DeVay, 1971).

*P. syringae* produces another toxin, named syringotoxin (Gross and DeVay, 1977b; Gonzalez *et al.*, 1981). The toxin, which is a wide-spectrum biocide and phytotoxin was found to be unique to the citrus isolate of *P. syringae* that causes citrus blast (Gonzalez *et al.*,1981). Syringomycin and syringotoxin are small peptides containing a few amino acids (Sinden *et al.*, 1971; Gross *et al.*, 1977; Gonzalez *et al.*, 1981).

## F. Coronatine

A non-host-selective toxin named coronatine has been isolated from culture filtrate of a virulent isolate of *Pseudomonas coronafaciens* pv. *atropurpurea*, the causal agent of chocolate spot disease of Italian ryegrass (Ichihara *et al.*, 1977a,

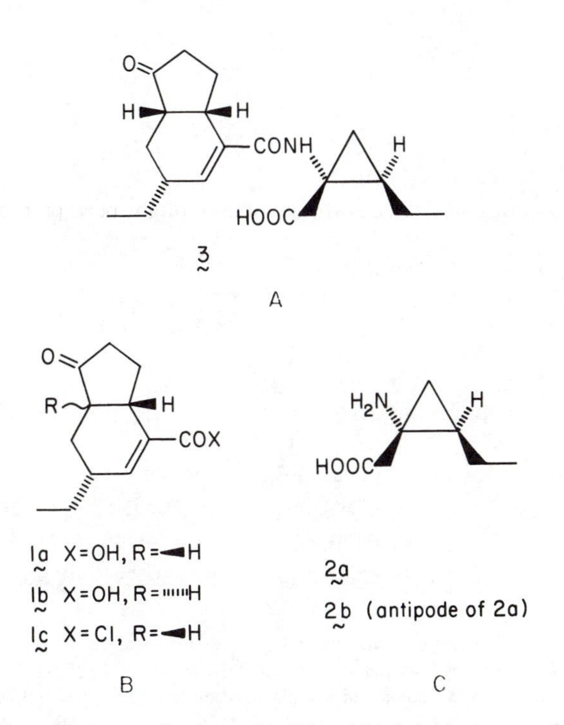


**Figure 16.** Proposed structure for coronatine (A) and its derivatives, coronafacic acid (B) and coronamic acid (C). [From Ichihara, A., *et al.* 1977. *Tetrahedron Lett.* **3**:269–272.]

b), and from extracts of diseased tissue (Nishiyama *et al.*, 1977). Coronatine is also produced by *P. glycinea* and *P. syringae* pv. *atropurpurea in vitro* (Mitchell and Young, 1978; Mitchell, 1982). The toxin, which is thought to have auxinlike properties, was reported to cause chlorotic and necrotic halos similar to those induced by *P. coronafaciens* pv. *atropurpurea* in both host and nonhost plants when injected into the leaves. The purified toxin was reported to consist of two major fragments, $C_{12}H_{15}O_2$ and $C_6H_{10}O_2N$, joined by an amide linkage (Ichihara *et al.*, 1977a,b) (Fig. 16).

## G. Fusicoccin

This toxin and its derivatives are produced by *Fusicoccum amygdali,* which causes wilting and desiccation of twigs on almond and peach trees (Graniti, 1962; Ballio *et al.*, 1968). Symptoms closely resembling those caused by fungus infection have been produced by introduction of fusicoccin into the xylem of almond, peach, and a wide range of other plants.

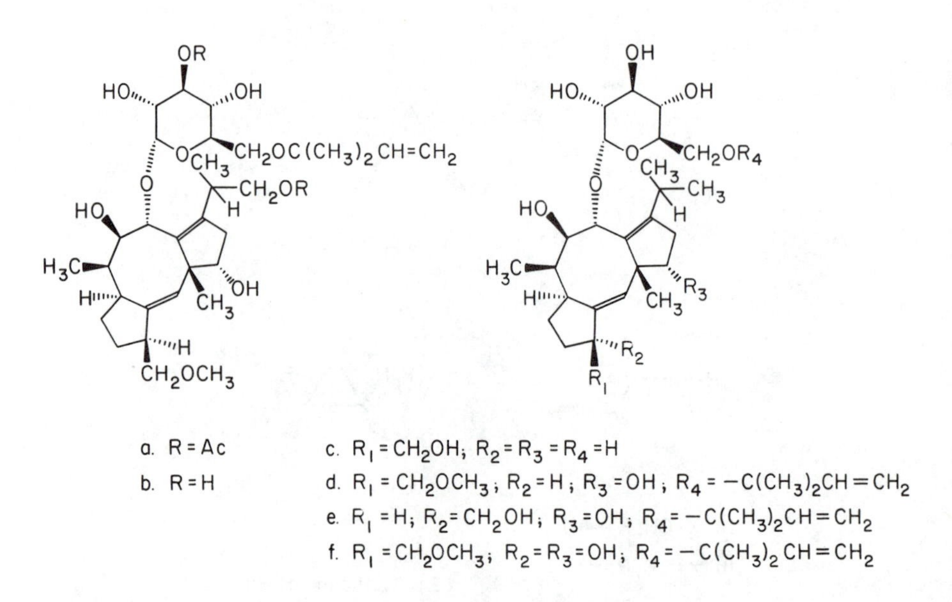


**Figure 17.** Proposed structures for fusicoccin A (a), dideacetlyfusicoccin (b), and deoxyfusicoccins (c–f). [Proposed by Ballio *et al.* (1972. *Experientia* **28:**1150–1151), Ballio (1978. *Ann. Phytopathol.* **10:**145–156), confirmed and extended by Randazzo *et al.* (1978. *Gazz. Chim. Ital.* **108:**139–142).]

The toxin causes a variety of physiological changes in plants including increases in $H^+$ and $Na^+$ efflux, and $K^+$ influx; changes in electrical potential, stimulation of the uptake of several anions, glucose, sucrose, and amino acids; and stimulation of respiration, dark $CO_2$ fixation, plant cell enlargement, and stomatal opening (Turner and Graniti, 1976; Marré, 1979). In some cases, fusicoccin mimics the action of natural plant hormones (Marré, 1979). The mode of action of the toxin has been studied by a number of investigators (Turner and Graniti, 1969, 1976; Squire and Mansfield, 1972; Marré *et al.*, 1973a,b, 1974a,b; Marré, 1977) and reviewed by Marré (1979).

The toxin was found by Ballio *et al.* (1968, 1972) and Ballio (1978) to be a glycoside of carbotricyclic terpene (Fig. 17).

## H. Fumaric Acid

Fumaric acid is produced by species of *Rhizopus* which cause hull rot disease of almonds characterized by leaf blighting and killing of twigs (Mirocha, 1972). Disease symptoms can be reproduced by treating fruit mesocarps with the toxin. Mirocha (1972) has provided evidence that fumaric acid formed in diseased tissue is metabolized via the epoxysuccinate cycle (Foster, 1958) to nonphytotoxic oxalacetate.

## I. Oxalic Acid

Oxalic acid, which is produced by *Sclerotinia sclerotiorum* both *in vitro* and *in vivo* (Maxwell and Lumsden, 1970), has been implicated in pathogenesis (Maxwell and Lumsden, 1970; Lumsden and Dow, 1973; Rai and Dhawan, 1976; Hancock, 1977; Noyes and Hancock, 1981). Results of a study of Noyes and Hancock (1981) have provided strong evidence for the involvement of oxalic acid in wilting in sunflower infected with *S. sclerotiorum*. First, the development of unilateral wilting in split-stem sunflower plants could be stopped by removal of stem lesions (Fig. 18). Second, the level of oxalic acid in wilted leaves of infected plants was about 15 times higher than that in healthy plants. Oxalic acid was also present in tracheal fluid of *Sclerotinia*-infected plants. Third, symptoms identical to those observed in diseased plants were induced in healthy plants by application of oxalic acid to hypocotyls. Finally, the results of preliminary tests using leaf cell suspension bioassays showed that sunflower cultivars resistant to *Sclerotinia* wilt were more tolerant to oxalic acid than susceptible cultivars at the flowering stage.

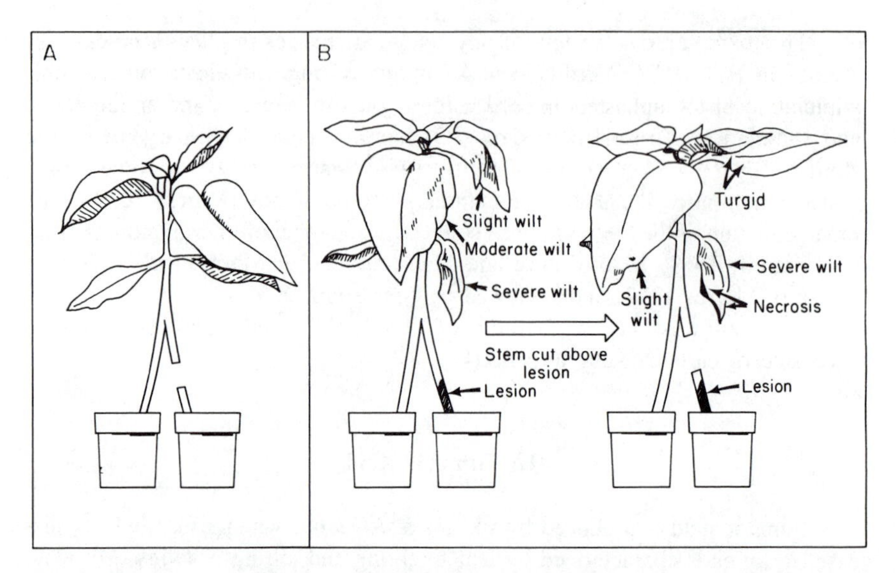


**Figure 18.** Diagrammatic representation of results of split-stem experiments with *Sclerotinia sclerotiorum*. (A) Noninoculated control, shows no wilting after half-stems were severed. (B) Infected, shows unilateral wilting after lesion development in inoculated half-stems and occurrence of partial recovery when half-stems are severed above lesions. [Reproduced with permission from Noyes, R. D., and Hancock, J. G. 1981. *Physiol. Plant Pathol.* **18**:123–132. Copyright © 1981 Academic Press Inc. (London) Ltd.]

## J. Fusaric Acid

The exact role of fusaric acid, which is reported to be produced by several species of *Fusarium* in culture and in diseased tissue, is not known. Kuo and Scheffer (1964) evaluated fusaric acid as a factor in the development of *Fusarium* wilt and concluded that the toxin was not responsible for the wilt symptoms and that it could play no more than a secondary role in disease development. The toxin has been characterized as 5-*N*-butylpicolinic acid (Kern, 1972) (Fig. 19).

## K. Ethyl and Methyl Acetate

Ethyl acetate is produced by *Ceratocystis paradoxa* and *C. adiposum* at levels that are toxic to sugarcane tissue and causes inhibition of sugarcane buds and shoots (Kuo *et al.*, 1969). In their search for the factor(s) responsible for brown-

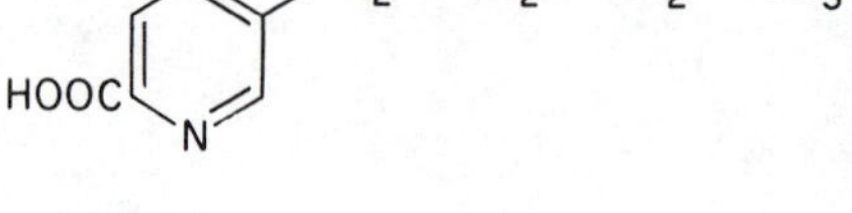

$CH_2-CH_2-CH_2-CH_3$
HOOC N

**Figure 19.** Fusaric acid. [Yabuta, T., and Hayashi, T. 1939. *J. Agric. Chem. Soc. Jpn.* **15**:257–266.]

ing and other symptoms of black rot of cotton caused by *Thielaviopsis basicola,* Tabachnik and DeVay (1980) detected methyl acetate and several other simple aliphatic alcohols and esters in culture filtrate of the fungus. Methyl acetate also was detected in diseased cotton roots at concentrations 10 times greater than those in healthy roots and caused browning of the tissue and other typical symptoms of the disease in healthy roots at concentrations as low as 0.126 mM. Despite this, no correlation was found between methyl acetate levels produced by the fungus isolates *in vitro* and their virulence.

### L. Ammonia

Ammonia which is produced in diseased tissue and tissues undergoing hypersensitive reactions, has been implicated as having a causal role in pathogenesis. In general, however, ammonia accumulation is the result rather than the cause of injury to the tissue. Recently Bashan *et al.* (1980) have reported that ammonia produced by *Pseudomonas tomato,* a pathogen of tomato, both *in vitro* and *in vivo,* caused necrosis in healthy tomato leaves. Moreover, inoculation of tomato leaves with *P. tomato,* and not *P. fluorescens,* a saprophyte, resulted in increases in the levels of ammonia and pH in the leaves. Production of toxic quantities of ammonia in infected leaves was reported to precede electrolyte leakage and symptom development. While the search for chlorosis-inducing factors other than ammonia in culture filtrates of two *P. tomato* isolates used in the study was not successful, the possibility that other isolates of *P. tomato* might produce chlorosis-inducing factors was not ruled out.

High-molecular-weight substances are produced by a number of wilt-inducing pathogens. A number of these substances are known to cause wilting when introduced into stems and petioles of excised plants, presumably by increasing resistance to water flow in treated plants. These substances are not considered toxins by many because they generally do not exhibit toxicity to plant cells and their activity is purely physical due to their large size (Van Alfen and Allard-Turner, 1979). The roles of high-molecular-weight substances in wilting are discussed in Chapter 6.

All of the toxins discussed above are produced by nonbiotrophs. However, toxins also have been implicated as having a role in interactions between wheat and two biotrophs, *Puccinia graminis* f. sp. *tritici* (Olien, 1957; Silverman, 1960) and *P. recondita* (Jones and Deverall, 1978).

## VII. MODE OF ACTION OF TOXINS

In many cases the mechanism responsible for toxin-induced changes in plants is not known. The mode of action of toxins is considered as one of the most challenging and perhaps most fruitful aspects of toxin research. Understanding the mechanism of the action of toxins, particularly selective toxins, would not only provide a better appreciation of host–pathogen interactions and resistance mechanisms, but may also help elucidate some obscure aspects of membrane transport and the action of hormones in plants. Following is a brief discussion of the results of some studies related to the mode of action of toxins. The literature on the subject has been reviewed recently by Daly (1981).

### A. Effect on Mitochondria

Mitochondria from corn plants susceptible to *Helminthosporium maydis,* race T, but not those from resistant plants, exhibited drastic changes following treatment with partially purified *H. maydis* T toxin (Miller and Koeppe, 1971; Gengenbach *et al.,* 1973; Bednarski *et al.,* 1977; Gregory *et al.,* 1978, 1980; York, 1978). Mitochondria from susceptible plants exposed to the T toxin became swollen, inner membranes became disrupted, and phosphorylation was uncoupled from oxidation (Miller and Koeppe, 1971; Gengenbach *et al.,* 1973). Bednarski *et al.* (1977) showed reductions in total as well as labile phosphates in susceptible Texas male-sterile cytoplasm (T cytoplasm) roots of corn but not in resistant normal cytoplasm (N cytoplasm) roots treated with T toxin. ATP level was reduced in dark-grown susceptible corn mesophyll protoplasts treated with T toxin or oligomycin, an inhibitor of mitochondrial ATP synthesis, while no change in ATP levels was detected in resistant corn mesophyll protoplasts (Walton *et al.,* 1979). Payne *et al.* (1980b) found that a highly purified toxin prepared by Kono and Daly (1979) uncoupled and stimulated NADH and succinate respiration but did not significantly uncouple or inhibit malate respiration. They suggested that the toxin may act at two separate sites. However, judging from the result of their study using the same toxin preparation, Gregory *et al.* (1980) concluded that all of the observed effects of the toxin are related to a single-action site. The purified toxin was highly active against mitochondria and protoplasts from susceptible T cytoplasm corn but had no effect on resistant N cytoplasm corn (Payne *et al.,* 1980a; Gregory *et al.,* 1980).

### B. Effect on Membrane Permeability

Changes in membrane permeability have been reported in plant tissues exposed to the following toxins: victorin (Wheeler and Black, 1963; Samaddar and Scheffer, 1968, 1971), PC toxin (Gardner *et al.,* 1972, 1974), *Alternaria kikuchiana* toxin (Otani *et al.,* 1973), T toxin (Miller and Koeppe, 1971), syringomycin (Backman and DeVay, 1971), a lipopolysaccharide produced by

*Pseudomonas lachrymans* (Keen and Williams, 1971), and a number of other toxins. Fusicoccin-induced stimulation of net $H^+$ efflux and $K^+$ influx and effects of other toxins on permeability are discussed in Chapter 5.

### C. Effect on Photosynthesis

Application of fusicoccin, which is produced by *Fusicoccum amygdali,* to leaves of beans caused inhibition of photosynthesis and photorespiration while dark respiration doubled (Heichel and Turner, 1972). High concentrations of tentoxin produced by *Alternaria tenuis* caused stimulation of the coupling factor 1 ATPase activity, while low concentrations of the toxin caused inhibition (Steele *et al.,* 1978). Tentoxin at low concentrations inhibited ATPase and photophosphorylation in toxin-sensitive species through binding to chloroplast coupling factor 1 ATPase (Steele *et al.,* 1976, 1978). *Pseudomonas tabaci* toxin (tabtoxin) was reported to inhibit the activity of ribulose 1,5-bisphosphate carboxylase from tobacco plants (Crosthwaite and Sheen, 1979).

### D. Effect on Enzymes

The inhibition of ornithine carbamoyltransferase, ribulose 1,5-bisphosphate carboxylase, and glutamine synthetase by phaseolotoxin, tabtoxin, and tabtoxinine-β-lactam, respectively, was referred to earlier.

### E. Growth Regulator Effect

Few toxins exhibit growth regulator properties. This subject is discussed in Chapter 10.

## VIII. TOXIN BINDING SITES

Except for tentoxin, which binds chloroplast coupling factor 1, and phaseolotoxin, tabtoxin, and tabtoxinine-β-lactam, which are known to inhibit certain specific enzymes, probably by binding to the enzymes, little information is available on the identity and location of toxin binding sites in plant cells. Toxin-induced metabolic changes occur with a lag of about 30 min (Wheeler, 1976a), and this led to the suggestion that the toxin binding site is located within the cytoplasm (e.g., mitochondria, chloroplasts) and not on the cell wall or plasmalemma. The evidence in favor of the view that the toxin binding sites are located on the cell walls is discussed in Chapter 5. As was pointed out earlier, there is strong evidence that T toxin produced by *Helminthosporium maydis* interferes specifically with mitochondria of susceptible plants.

The receptor sites for certain toxins have been proposed to be located on the plasmalemma of susceptible cells. This is because changes in the permeability of plasmalemma occur in cells soon after toxin treatment (Samaddar and Scheffer, 1968, 1971; Gardner *et al.*, 1974; Novacky and Hanchey, 1974) and wall-less protoplasts lyse after exposure to victorin (Samaddar and Scheffer, 1968; Novacky and Hanchey, 1974).

Strobel (1974a,b) and Strobel and Hapner (1975) consider a protein associated with the plasmalemma of susceptible but not resistant sugarcane clones to be the primary binding site for helminthosporoside, a toxin isolated from *H. sacchari*. Strobel (1974a,b) has shown that helminthosporoside is preferentially bound *in vivo* to tissues of susceptible and not resistant sugarcane cultivars. A protein isolated from membranes of susceptible tissue was capable of binding the toxin. A protein isolated from resistant clones that was serologically indistinguishable from the protein of susceptible clones did not bind the toxin (Strobel, 1974a,b). The protein is considered to bind to galactoside and to function as a galactoside transport. Upon complexing with the toxin, conformational changes presumably occur in the binding protein in the membrane. This in turn triggers activation of other membrane-bound proteins, including the K and Mg ATPases, culminating in an imbalance of electrolyte exchange, resulting in cell death (Strobel, 1974a).

The data presented above were thought to support the hypothesis that resistance to *H. sacchari* is due to the inability of a protein in the cell membrane of resistant sugarcane clones to bind with the toxin. However, the results of subsequent studies have shown that the ability to bind the toxin is not specific to susceptible cultivars. Kenfield and Strobel (1977) have shown that tobacco and mint, which are nonhosts of *H. sacchari,* bind to the toxin as much as or even more than susceptible sugarcane. More recently, Kenfield and Strobel (1981) compared the binding proteins from mint, tobacco (two nonhosts of *H. sacchari*), and two clones of sugarcane with different sensitivity to helminthosporoside and found that these proteins are composed of a mixture of size species consisting of a high-molecular-weight multimer and low-molecular-weight oligomers. The number of oligomers in the proteins depends upon the concentration of the proteins. The lower the concentration of the protein the greater the number of oligomers and the greater the binding activity of the protein. The galactoside-binding proteins from the above four sources were suggested to be similar but not identical with respect to composition and function. It was concluded that the proteins from different plants susceptible or resistant to *H. sacchari* are potentially capable of binding with galactoside. However, the binding *in vitro* and probably *in vivo* is regulated (modified) by a number of factors such as protein concentration, glycolipids, and heterologous subunits of each binding protein.

Larkin and Scrowcroft (1981), who studied the interaction of the toxin with sugarcane leaf cells by determining conductivity changes, concluded that the kinetics of the interaction agrees with the plasmalemma-bound toxin-binding protein hypothesis proposed by Strobel and associates (Strobel, 1973; Strobel and Hess, 1974). The toxin interaction followed Michaelis–Menton hyperbolic saturation kinetics and did not show positive or negative cooperative effects. The last finding is not consistent with earlier reports that the plasmalemma-bound binding protein possesses two or more binding sites and its binding exhibited negative cooperative effect (Strobel, 1973; Strobel and Hess, 1974).

Larkin and Scrowcroft (1981) found that significantly different concentrations of the toxin were required to saturate binding sites on suspension cultures of two highly susceptible sugarcane cultivars. According to these workers the observed differences between the two sugarcane cultivars cannot easily be explained on the basis of a simplistic binding protein hypothesis. The difference may reflect differential affinity of the binding protein from the two above cultivars for the toxin. Alternatively, leaf cell membranes of the two cultivars may contain different concentrations of the binding protein.

The hypothesis advanced by Strobel is very interesting. However, in view of the above exceptions and the controversy regarding some aspects of the study (Daly, 1976d; Wheeler, 1976b), results should be extended and confirmed in additional studies in order to establish its validity. Moreover, it would be worthwhile to determine if a fairly purified toxin recently isolated from *H. sacchari* (Livingston and Scheffer, 1981) and different from helminthosporoside studied by Strobel (1974a) binds with the same membrane-bound protein which presumably binds with helminthosporoside.

## IX. MECHANISM OF PLANT RESISTANCE TO TOXIN-PRODUCING PATHOGENS

The mechanisms responsible for the resistance of plants to toxin-producing pathogens are not known. The controversial nature of the binding site hypothesis advanced by Strobel (1974a) to explain the resistance of sugarcane clones to *H. sacchari* was described above. Attempts to identify specific binding sites for other toxins so far have not been successful (Scheffer, 1976).

There is presently no evidence in favor of the view that toxin inactivation contributes to plant resistance to toxin producing pathogens (Scheffer, 1976). Resistant and susceptible plants inactivated victorin (Wheeler, 1969) as well as PC and HC toxins (Scheffer and Yoder, 1972) at equal rates during the first few hours after exposure to the toxins. The literature on detoxification has been reviewed by Patil (1980) and more briefly by Scheffer (1976).

*Chapter 5*

# Alterations in Permeability Caused by Disease



## I. INTRODUCTION

Changes in permeability are thought to be the earliest host response to pathogenic agents (Thatcher, 1939, 1942, 1943; Gaümann, 1958; Wheeler and Black, 1963). The consequence of permeability changes is the release of large amounts of electrolytes and other substances from cells and the inability of cells to accumulate mineral salts.

The objective of this chapter is to outline in general terms the information available on the structure of the cell membranes, the nature of pathogen-induced alterations in cell permeability, and speculative mechanisms proposed to explain permeability changes in diseased tissue. The available evidence is also examined to ascertain whether changes in permeability in diseased tissue are the initial events in pathogenesis and whether permeability alterations are the cause or the result of infection. The literature on the subject has been reviewed by Wheeler (1975, 1976a, 1978), Rudolf (1976), and Hanchey and Wheeler (1979). This chapter is patterned after reviews by Wheeler (1975, 1976a, 1978).

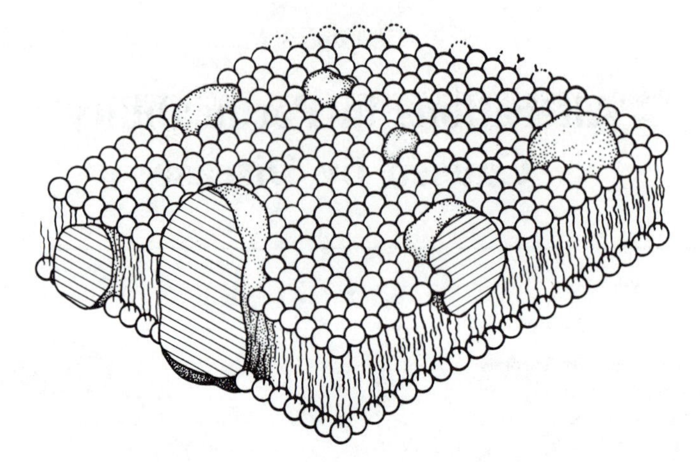

**Figure 20.** Schematic three-dimensional and cross-sectional proposed view of "fluid mosaic" model of membrane structure. Integral proteins (solid bodies) are partially or totally embedded in the lipid bilayer. [Reproduced with permission from Singer, S. J., and Nicolson, G. L. 1972. *Science* **175**:720–731. The American Association for the Advancement of Science, Washington, D.C.]

## II. COMPOSITION AND STRUCTURE OF MEMBRANES

Lipids and proteins are major components of cell membranes (Korn, 1969; Singer, 1974). Several structural models of bacterial and animal membranes have been proposed (Branton, 1969; Singer, 1974). According to the popular "fluid mosaic" model proposed by Singer and Nicolson (1972) (Fig. 20) the membrane is made of a lipid bilayer which is interrupted by globular proteins in a mosaic pattern. The ionic groups of phospholipid molecules in the lipid bilayer are oriented toward the exterior with their nonpolar ends toward the interior. In this model the integral proteins are assumed to move laterally without affecting membrane integrity.

Current views on membrane structure are based on studies of animal and microbial systems. However, judging from some chemical and structural data (Korn, 1969; Branton, 1969), plant and animal membranes seem to be fairly similar. The structure, chemistry, and biogenesis of plant cell membranes have been studied by Korn (1969) and Branton (1969) and reviewed by Anderson (1973), Morré (1975), and Wickner (1980).

## III. CELL PERMEABILITY THEORIES

Three theories have been advanced to explain the movement of ions and substances in and out of cells.

### A. Membrane Theory

This theory, which is widely accepted, particularly by scientists in western countries, holds that the influx or the efflux of substances through the cells is mainly controlled by the plasmalemma (plasma membrane).

### B. Sorption Theory

According to this theory, permeability of any substance is a function of its solubility in protoplasmic fluids and its binding by cell colloids and is not the function of membranes (Troshin, 1966; Ling, 1969; Ling and Ochsenfeld, 1973).

### C. Cytotic Theory

This theory assumes that the movement of substances into plant cells is through cytosis or vesicular transport (Holter, 1960; Cocking, 1970).

For a more detailed discussion of cell permeability theories see Wheeler (1975, 1976a).

## IV. CHANGES IN CELL PERMEABILITY CAUSED BY DISEASE

Alterations in membrane permeability are the first detectable changes in response to infection by pathogens (Thatcher, 1942; Wheeler and Black, 1963; Lai *et al.*, 1968; Hancock, 1968, 1969, 1972; Ohashi and Shimomura, 1976) and by host-selective pathogen-produced toxins produced by *Helminthosporium victoriae* (Wheeler and Black, 1963; Luke *et al.*, 1969), *Periconia circinata* (Scheffer, 1976), *Helminthosporium carbonum* (Comstock and Scheffer, 1973), *Alternaria kikuchiana* (Otani *et al.*, 1973), *A. mali* (Kohmoto *et al.*, 1976), and *H. maydis* (Gracen *et al.*, 1972). Several nonselective phytotoxins also cause increases in membrane permeability (see Rudolph, 1976, for references). Higher rates of electrolyte loss also have been detected in tissues undergoing hypersensitive response (Goodman, 1968; Pellizzari *et al.*, 1970; Turner, 1976; Weststeijn, 1978). Moveover, it is now known that chain-splitting pectolytic enzymes cause drastic changes in membrane permeability before bringing about maceration or cell death (Mount *et al.*, 1970). These observations are the basis for the unabated interest of plant pathologists in infection-induced changes in permeability (Wheeler, 1978).

Since changes in permeability induced by *H. victoriae* toxin (victorin) are entirely identical to those occurring in diseased tissue, this toxin has been used by many workers to study the nature of pathogen-induced alterations in membrane permeability (Wheeler, 1975). Changes in oat tissues treated with victorin

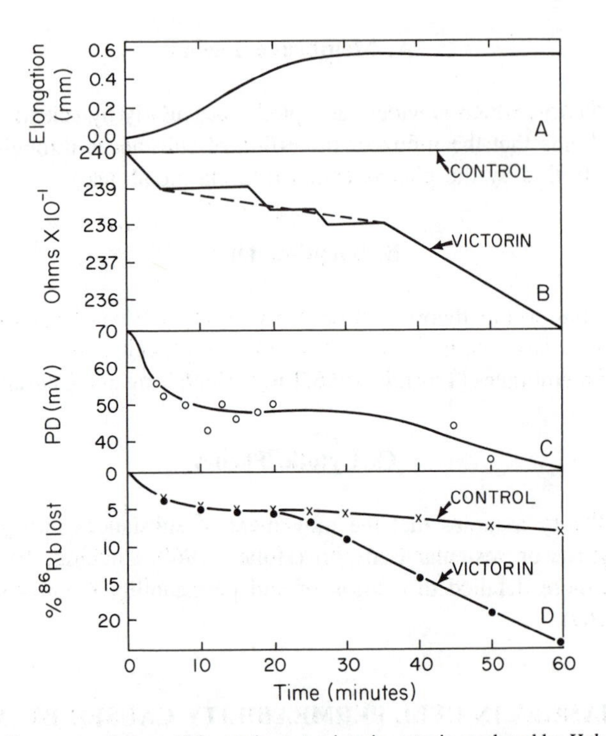


**Figure 21.** Responses of susceptible oat tissue to victorin, a toxin produced by *Helminthosporium victoriae*. (A) Increase in elongation of coleoptiles. (B) Loss of electrolytes from coleoptiles measured as a decrease in resistance of a bathing solution. The dashed line represents the average rate during a period of fluctuation. (C) Alterations in electrochemical potential (PD) induced by victorin. (D) Release of $^{86}$Rb from roots labeled with the isotope. [Reproduced with permission from Wheeler, H. 1976. In *Encyclopedia of Plant Physiology, New Series, Physiological Plant Pathology* (R. Heitefuss and P. H. Williams, eds.) Vol. 4, pp. 413–429. Springer-Verlag, New York.]

include: (1) a temporary increase in coleoptile elongation (Fig. 21A), (2) loss of electrolytes within a short time after toxin treatment (Fig. 21B,D), (3) increase in water permeability judged from the time required for plasmolysis and deplasmolysis, (4) inhibition of uptake and accumulation of mineral salts and other materials, (5) a drop in electrochemical potentials (Fig. 21c), (6) false plasmolysis, and (7) cell death (Wheeler, 1975; Scheffer, 1976). These results constituted the foundation for the hypothesis that toxin-induced permeability changes are the earliest response to the toxin and that the other physiological changes observed in diseased and toxin-treated tissues are the consequence of altered permeability (Wheeler and Black, 1963). The hypothesis that the plasmalemma is the initial target of victorin is supported further by the following observations: (1) The bursting and lysis of membrane-bound protoplasts after exposure to victorin (Samaddar and Scheffer, 1968, 1971) and (2) the ability of certain compounds that bind with the components of the protoplasm or cell membrane to counteract partially the toxic effects of victorin in susceptible tissue (Samaddar and Scheffer, 1971). However, as pointed out by Wheeler (1975, 1978), there has been no unequivocal evidence that the disease-inducing ability of pathogens or their products is due to their effects on cell permeability or that the membrane is the initial target of pathogen-produced toxins. Patterns of loss of electrolytes loss (Saftner and Evans, 1974) (Fig. 21B) and $^{86}$Rb (Keck and Hodges, 1973) (Fig. 21D) in victorin-treated tissue in short-term experiments show that the initial brief period of electrolyte loss during the first 5 min of exposure to victorin is followed by a leveling off for 20–30 min prior to a second increase in the rate of loss. Wheeler (1978) has observed that this pattern of change in electrolyte loss and similar changes in electrochemical potential (Novacky and Hanchey, 1974) (Fig. 21C) does not reflect the type of changes expected if the plasmalemma was the initial target of the toxin. The initial brief period of electrolyte loss is too short for complete washout of the cytoplasm. The transient leveling off following an initial decline in the potential (Fig. 21C) might be due to activation of a temporary repair mechanism capable of restoring the initial disruption of the plasmalemma (Wheeler, 1978). Novacky and Karr (1977) have suggested that changes in electrochemical potential in response to toxins other than victorin might also be due to interference with processes involved in maintenance of cell permeability and not to toxin–plasmalemma interactions.

According to Wheeler (1975) some of the characteristic features of electrolyte losses in tissues treated with victorin might be more easily explained on the basis of sorptional rather than membrane theory. These include the observed changes in permeability after a lag of about 20–30 min in short-term experiments (Fig. 21D) and losses of carbohydrates and other substances in victorin-treated tissue at an essentially constant rate in long-term experiments (Fig. 22).

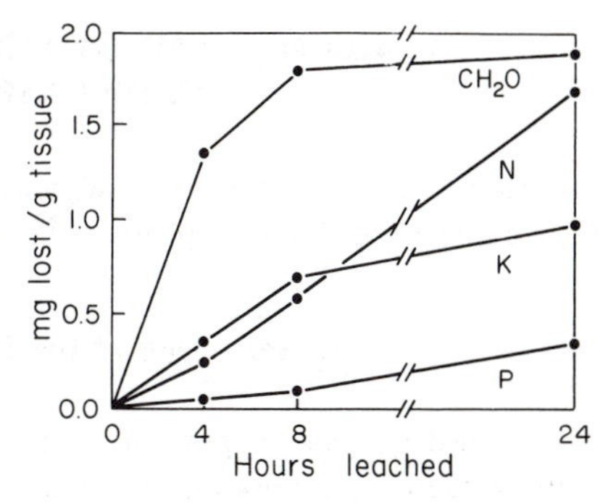


**Figure 22.** Loss of carbohydrates ($CH_2O$), nitrogenous materials (N), potassium (K), and phosphorus (P) from oat leaves pretreated with victorin, a toxin produced by *Helminthosporium victoriae*, and then shaken in water. [Reproduced with permission from Wheeler, H. 1975. *Plant Pathogenesis*. Springer-Verlag, Berlin, New York.]

It has been suggested that permeability changes observed in response to infection might be the consequence of minor or major disruption in the membrane structure and not the result of changes in permeation across the membrane. However, as Wheeler (1978) has pointed out, differences in the rates of release of electrolytes and other solutes observed in toxin-treated tissues (Black and Wheeler, 1966) can best be explained on the basis of changes in permeation rather than disruption of the membrane structure. Moreover, gross physical disruption of the plasmalemma reported in tissues invaded by pathogenic fungi (Ehrlick and Ehrlick, 1971), bacteria (Goodman, 1972, 1976), viruses (Esau, 1968), and nematodes (Paulson and Webster, 1970), do not appear to occur at early stages of pathogenesis (Wheeler, 1976a). In victorin-treated oat leaves (Hanchey *et al.,* 1968) and in *Helminthosporium maydis* T-toxin-treated corn leaves (White *et al.,* 1973) the tonoplast was the first membrane to be disrupted.

Another possibility suggested by Wheeler (1975, 1976a, 1978) is that the initial effect of victorin is on the cell wall and not on the plasmalemma and that plasmalemma permeability is affected about 30 min after exposure to victorin. This suggestion seems to be supported by the reports that isolated oat protoplasts treated with victorin do not respond as quickly as intact tissues (Samaddar and Scheffer, 1968; Easton and Hanchey, 1972). Moreover, the rate of uptake of labeled compounds was identical in leaf protoplasts from resistant and susceptible oats during the first 30 min after exposure to victorin (Rancillac *et al.,* 1976).

Judging from the results of the above studies it is evident that the initial target of pathogen-produced toxins has not yet been identified and it is not clear if permeability changes have a primary role in pathogenesis. Moreover, the question of how changes in membrane structure and function may influence disease development has not been adequately answered. Whether or not altered permeability in diseased tissue is an initial event in pathogenesis is a question of paramount importance, mainly because the answer to the question would provide us with an indirect clue as to whether permeability change is the cause or the consequence of the disease.

## V. MECHANISMS OF PERMEABILITY CHANGES IN DISEASED PLANTS

Four hypotheses have been advanced to explain the mechanism of permeability changes in diseased plants.

### A. Activation of the Plasmalemma ATPase

According to this hypothesis, toxin-induced changes in permeability are due to the effects of toxins on membrane-bound enzymes, particularly ATPase, which is considered to be involved in the activity of ion pumps. Fusicoccin, a phytotoxin produced by *Fusicoccum amygdali*, stimulates energy-linked net $H^+$ efflux and $K^+$ influx across the plasmalemma in cells of practically all higher plants so far tested (Marré, 1979). Marré (1977, 1979) has postulated that this response to the toxin results from activation of a plasmalemma ATPase which mediates electrogenic $H^+$ out $K^+$ in exchange. The hypothesis that fusicoccin activates plasmalemma ATPase is supported by the observation that fusicoccin stimulates ATPase activity *in vitro* (Beffagna *et al.*, 1977; Lurie and Hendrix, 1979). Moreover, fusicoccin-stimulated $H^+/K^+$ exchange is inhibited by lowering of ATP level in the tissue (Marré *et al.*, 1973a,b) and by inhibitors of ATPase activity *in vitro* (Marré, 1979). Although activation of an ATPase-coupled $H^+/K^+$ exchange mechanism located on the plasmalemma might be responsible for stimulation of $H^+/K^+$ exchange by fusicoccin, it appears that the process does not involve binding of the toxin to an ATPase. Stout and Cleland (1980) have shown that in oat roots a plasmalemma-bound protein other than ATPase may be a primary receptor of fusicoccin. However, they did not exclude the possibility that the fusicoccin-binding protein might be associated in some way with an ATPase ion-transport complex. Resolution of this question, according to Stout and Cleland (1980), requires reconstitution of a model membrane system containing both ATPases and the fusicoccin-binding protein.

The reported influence of *Helminthosporium maydis* T toxin on membrane-bound ATPase (Tipton *et al.*, 1973) has not been confirmed (Scheffer, 1976). Evidence for the presence on the plasma membrane of sugarcane cells of a protein capable of binding specifically to a toxin produced by *Helminthosporium sacchari* was discussed in Chapter 4. Binding of the toxin to the protein has been postulated to cause changes in membrane proteins including ATPase, culminating in membrane permeability changes (Strobel, 1974a). However, despite the initial report, the binding does not seem to be very specific (see Chapter 4).

According to Hancock (1969) increased rates of uptake and accumulation of certain metabolites in squash hypocotyls infected with *Hypomyces solani* f. sp. *cucurbitae* may be the result of changes in carrier systems involved in the transport of metabolites.

## B. Effects of Pathogen-Produced Toxins on Specific Components of the Transport System

Arntzen *et al.* (1973a,b) have suggested that *Helminthosporium maydis* T-toxin-induced permeability alteration is due to the effect of the toxin on specific components of the transport system rather than a general disruption of membranes. This view has also been supported by Wheeler and Ammon (1977).

## C. Disruption of Energy Supply for the Maintenance and Repair of Membranes

According to this hypothesis, toxin-induced membrane dysfunction is a result of disruption of processes involved in energy generation required for maintenance and repair of the membrane (Wheeler, 1978). For example, tentoxin produced by *Alternaria tenuis* has been reported to bind with chloroplast coupling factor I and to inhibit ATPase activity and photophosphorylation in toxin sensitive plant species (Steele *et al.*, 1976, 1978).

## D. Degradation of Lipid and Protein Components of Membranes by Pathogen-Produced Enzymes

The following plant pathogens are known to produce phospholipase B or C which catalyzes the hydrolysis from diacylglycerophosphoryl compounds of fatty acids and phosphoryl moiety, respectively: *Thielaviopsis basicola* (Lumsden and Bateman, 1968), *Sclerotium rolfsii* (Tseng and Bateman, 1969), *Botrytis cinerea* (Shepard and Pitt, 1976), *Erwinia carotovora* (Tseng and Mount, 1974), and *Phoma medicagenis* var. *Pinodella* (Plumbley and Pitt, 1979). A lipoxygenase capable of catalyzing oxidation of certain fatty acids has been reported to increase in homogenates of tobacco leaves infected with *Erysiphe cichoracearum* (Lupu *et al.*, 1980).

Proteases are also produced by *Rhizoctonia solani* (Van Etten and Bateman, 1965) and three *Fusarium* species (Urbanek and Yirdaw, 1978). Moreover, Stephens and Wood (1975) and Tseng and Mount (1974) have reported that both phospholipases and proteases can cause bursting of the plasmalemma of isolated plant cell protoplasts. These enzymes, however, are incapable of killing intact plant cells.

There is presently no good evidence that lipid or protein components of membranes are degraded by pathogen-produced enzymes (Bateman and Basham, 1976). Moreover, observed changes in the levels of phospholipids in *Uromyces phaseoli* bean leaves (Hoppe and Heitefuss, 1975a, b) were thought to occur too late to account for permeability alterations (Wheeler, 1978).

*Chapter 6*

# Alteration in Water Relations Caused by Disease



## I. INTRODUCTION

Changes in water relations occur in plants infected by many pathogens, particularly those invading vascular tissues. The major impact of water deficits is reduced productivity. As Kozlowski (1978) has pointed out, the difference in productivity of crops grown with and without irrigation in arid regions is indicative of the great impact of water stress on plants.

The basic objectives of water relations research, as it relates to plant diseases are (1) to determine the mechanisms by which pathogens affect the water status of the host, (2) to assess the effect of pathogen-induced water stress on plant productivity, and (3) to determine the effect of water stress on pathogens themselves, including their disease-inducing ability, survival, and competitive potentials. The subject has recently been reviewed by Duniway (1976), Talboys (1978), Zimmerman and McDonough (1978), Ayres (1978, 1981), and

Kozlowski (1978). The format of this chapter, which deals with the physical aspects of water relations in diseased plants, is patterned somewhat after a review by Duniway (1976).

## II. DEFINITION OF A FEW TERMS

### A. Water Potential

Water potential ($\Psi$) is a quantitative term describing the status and dynamics of water in a system such as soil or plant. It is the difference between the free energy of water in a system and that of pure water per unit volume or mass of water at the same temperature and atmospheric pressure. $\Psi$ can be expressed as potential energy per unit volume (ergs/$cm^3$) or per unit mass (joules/kg). Since energy per unit volume is dimensionally equivalent to pressure, $\Psi$ can be measured by the unit of ''bars,'' which is an international unit of pressure and is equal to each of the following units: 0.987 atm, 75.0062 cm Hg, 1019.716 g $cm^{-2}$; or 14.5038 lb $in^{-2}$. Water is at a negative water potential (−bars) when its work capacity is less than that of pure water (e.g., when salt or sugars are dissolved in it or when it is absorbed to a matrix such as cell walls or soil particles). Water is at a positive water potential (+bars) when its work capacity is more than that of pure water (e.g., water which is kept under pressure). $\Psi$ is also directly related to the driving force for water movement because the driving force is determined by gradients in the free energy of water.

### B. Osmotic, Pressure, and Matric Potentials

The water potential of pure water at atmospheric pressure is arbitrarily considered to be zero and may be changed by one or more of the following factors: (1) the presence of solute in water [solute potential or osmotic potential ($\Psi_s$)], (2) the negative or positive hydraulic pressures such as the negative pressures generated in the xylem or the positive pressure identified with cell turgor [pressure potential ($\Psi_p$)], and (3) adsorption or capillary forces such as those generated among soil particles or in the cell wall matrix [matric potential ($\Psi_m$)]. The magnitude of water potential is the sum of components of water potential, that is, $\Psi = \Psi_s + \Psi_p + \Psi_m$.

### C. Relative Water Content

Relative water content (RWC), which is a convenient reference point of water status of plant tissue (Barrs, 1968), is determined by the following formula:

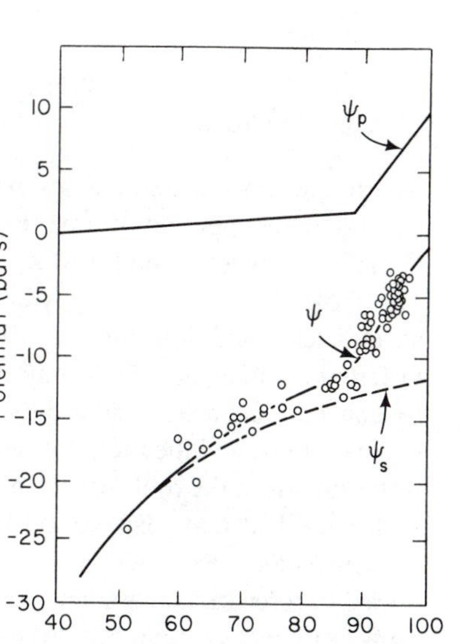


**Figure 23.** Changes in solute potential ($\Psi_s$), pressure potential ($\Psi_p$), and water potential ($\Psi$) of a pepper leaf as functions of relative water content. [Reproduced with permission from Gardner, W. R., and Ehlig, C. F. 1965. *Plant Physiol.* **40:**706–710. The American Society of Plant Physiologists, Rockville, Maryland.]

$$\text{RWC} = \frac{(\text{Fresh weight of the tissue}) - (\text{oven-dry weight})}{(\text{Fully turgid weight}) - (\text{oven-dry weight})} \times 100$$

Fully turgid weight is determined by floating leaf disks on water until the tissue is fully turgid and maximum weight has been achieved. The relations between RWC and other components of water potential are of great value for assessing the water relations of plant tissues (Fig. 23).

Water potential and some of its components can be determined by pressure chamber (Boyer, 1967; Wenkert *et al.*, 1978), thermocouple psychrometer (Barrs, 1968; Brown, 1970; Papendick and Campbell, 1975; Young, 1978), and beta gauge (Jones, 1973).

## III. WATER MOVEMENT IN PLANTS

### A. Water Movement in Healthy Plants

Movement of liquid water in soil–plant systems is along gradients of decreasing $\Psi$ and is controlled, for the most part, by a negative $\Psi$ generated by the evaporation of water from the leaf tissues. Root presssure that arises from a solute potential might contribute occasionally to the upward flow of water.

Water movement in the plant can best be described by an equation analogous to Ohm's Law as follows:

$$F = \frac{\psi_{\text{root}} - \psi_{\text{leaf}}}{R_{\text{root}} + R_{\text{stem}} + R_{\text{leaf}}}$$

In the above equation, $F$ is the rate of water flow in plants or water loss from a unit leaf area per unit of time (e.g., cm $\text{day}^{-1}$); $\Psi$ and $R$ represent water potential and resistance to liquid water flow for the indicated plant part, respectively. The above description can only be applied to a steady-state or at least to a nearly steady-state condition (Slatyer, 1967; Wardlaw and Passioura, 1976). However, different components of resistance to water movement in plants are not necessarily constant and are influenced by factors such as root temperature and aeration, soil variables, and the rate of water uptake (Slatyer, 1967). A major resistance to water flow is in the root, particularly across the endodermis. Small xylem vessels within the leaf may also exhibit large resistance to water flow (Slatyer, 1967).

Resistance to water flow can be estimated by measuring the pressure required to force certain amounts of water to flow across excised root or shoot sections or by measuring rate of water flow at a constant pressure (Duniway, 1977). Movement of radioisotopes (Dimond and Waggoner, 1953b) and dye solutions (Scheffer and Walker, 1953) in roots and stems has also been used for this purpose. It is useful to compare values for resistance to water flow obtained by these methods with those obtained by Poiseuille's Law (Duniway, 1971b; Dimond, 1966).

## B. Water Movement in Diseased Plants

Infection of plants by certain root pathogens results in changes in water movement. Changes in water transport in diseased plants can be due to interference with water absorption (Ghabrial and Pirone, 1967) as a result of root decay (Sterne *et al.*, 1978) or to an abnormally high resistance to water flow in xylem elements. For example, in the following plant–pathogen systems, resistance to water flow in roots or stems and twigs of infected plants has been found to exceed that in healthy plants: root resistance in black shank of tobacco (Schramm and Wolf, 1954), stem resistance in *Verticillium* wilt of tomato (Threlfall, 1959), twig resistance in oak wilt (Gregory, 1971), stem and root resistance in *Phytophthora* root rot of safflower (Duniway, 1977), and in *Phymatotrichum*-infected cotton (Olsen *et al.*, 1982).

It is important to judge increased resistance to water flow against the magnitude of the total resistance to water flow through the plant. In *Fusarium*-infected tomato plants, stem resistance was approximately 500 times that of healthy plants. However, despite its magnitude stem resistance represented only a fraction of the total resistance between ground level and leaves because petioles and leaves were the major sites of resistance (Duniway, 1971a). Since the resistance to water movement in leaves of healthy plants is relatively large (Begg and Turner, 1970; Boyer, 1971; Duniway, 1971a), pathogen-induced changes in leaf resistance may contribute more to total resistance than changes in root and stem resistance. The following factors may contribute to the increased resistance to water flow:

### *1. Pathogen-Produced High-Molecular-Weight Substances*

Husain and Kelman (1958) were the first to propose that wilting caused by *Pseudomonas solanacearum* in tomato plants was due to reduction in water flow in xylem elements resulting from accumulation of pathogen-produced high-molecular-weight substances in water conducting elements. High-molecular-weight glycopeptides and polysaccharides are now known to be produced by a number of wilt-inducing pathogens. Wilting may be induced in excised plants a few hours after the cut ends of their stems are placed in solutions containing these materials. These substances generally are not considered toxins because they do not exhibit toxicity to plant cells and their activity is purely physical due to their large size (Van Alfen and Allard–Turner, 1979). The following pathogens are known to produce high-molecular-weight susbstances:

*a. Cerotocystis ulmi.* A glycopeptide is produced *in vitro* by *C. ulmi,* the causal agent of Dutch elm disease (Salemink *et al.*, 1965). Stem water conductance and leaf water potential are reduced in elm seedlings treated with crude preparations of the glycopeptide (Van Alfen and Turner, 1975b) and more recently with the purified material (Strobel *et al.*, 1978).

*b. Corynebacterium insidiosum.* *C. insidiosum,* the causal agent of alfalfa wilt, produces a glycopeptide in culture (Ries and Strobel, 1972a,b) capable of causing wilt in leaves and stems of plants. The mechanism of wilting in alfalfa plants infected with the bacterium has been studied extensively by Van Alfen and Turner (1975a) who concluded that the material interferes with water movement through pit membranes. However, in a recent study Van Alfen and McMillan (1982) found no correlation between resistance of alfalfa to the bacterium and wilt caused by a large wilt-inducing glycopeptide produced by the bacterium *in vitro*.

*c. Corynebacterium sepedonicum.* A glycopeptide with a molecular weight of about 22,000 has been isolated from the culture of *C. sepedonicum,* the causal agent of ring rot of potato (Strobel, 1967, Strobel *et al.*, 1972). The glycopeptide causes rapid wilt in young potato and tomato cuttings similar to that caused by bacterial infection.

*d. Corynebacterium michiganense.* Less information is available on the wilt-inducing glycopeptide which is produced by *C. michiganense,* the causal agent of bacterial canker of tomato (Rai and Strobel, 1969).

*e. Phytophthora* spp. Water-soluble β-glucans produced by *Phytophthora* spp. cause wilting in a number of higher plants (Keen *et al.*, Woodward *et al.*, 1980). Keen *et al.* (1975) have suggested that the responses to the β-glucans appear to be due to cellular toxicity rather than plugging of vascular systems.

The wilt-inducing potential of some of the high-molecular-weight materials described above has generally been attributed to their ability to cause an increase in resistance to water flow in excised plants when the cut ends of their stems were placed in toxin solutions. According to Zimmerman and McDonough (1978) such experiments are not informative. Since vessels have been opened at the cut ends, any suspension is expected to increase resistance to water flow in such a situation. Van Alfen and Allard-Turner (1979) have demonstrated that the xylem-elements of alfalfa, particularly those in leaflet veins, were obstructed by picomole quantities of dextrans. The ability of dextrans to interfere with vascular conductance was correlated directly to their molecular weight. The artifactual nature of tests involving the use of plant cuttings to bioassay macromolecular plant-wilting products has been discussed further by Van Alfen and McMillan (1982).

*2. Vascular Gel Produced by the Action of Cell-Wall-Degrading Enzymes*

Xylem vessels of roots, stems, and leaves of plants infected with wilt pathogens are often occluded by gel-like substances. These materials, which fill vessels prior to wilt development (Beckman and Halmos, 1962; Heale and Gupta, 1972; Robb *et al.*, 1975; Misaghi *et al.*, 1978), are thought to originate from exposed primary walls and middle lamellae at pit membrane or perforation plates (Beckman, 1969; Vandermolen *et al.*, 1977; Wallis, 1977). The occlusion of the xylem elements by these gel-like materials is thought to be responsible for the observed increases in resistance to water flow in the infected roots and stems (Dimond, 1970) and might contribute to wilting.

There is uncertainty regarding the involvement of pathogen-produced cell-wall-degrading enzymes in wall softening and production of gel-like materials in vessels of plants infected with wilt pathogens. Howell (1976) found that mutants of *Verticillium dahliae* deficient for endo-poly-galacturonase (endo-PG) and endo-pectin lyase (endo-PL) were capable of inducing wilt symptoms in cotton plants, suggesting that these enzymes may not be involved in pathogenesis. McDonnell (1958), on the other hand, found that polygalacturonase-deficient isolates of *Fusarium oxysporum* f. sp. *lycopersici* were less pathogenic in tomato plants than isolates capable of producing the enzyme. Cooper and Wood (1980) presented evidence for the possible involvement of pectolytic enzymes in *Verticillium* wilt of tomato. They isolated endo-PL from vascular tissues of susceptible tomato cuttings three days after infection by *V. albo-atrum* and before symptom development. The enzyme was detected at much lower levels in a near-isogenic resistant cultivar. The level of endo-PG in infected tissue was only slightly above that in healthy tissue. Cooper and Wood (1980) were able to induce disease symptoms, including vascular gel formation, by supplying stems and leaves with partially purified endo-PL and endo-PG at concentrations equal to those found in infected plants. The controversy surrounding the role of pathogen-produced cell-wall-degrading enzymes in induction of wilt could be due to the use of unsuitable media and the diverse genetic background of test isolates (Cooper and Wood, 1980).

#### 3. *Degradation of Components of Conducting Tissue*

Based on his ultrastructural studies of tomato plants infected with *Corynebacterium michiganense,* Wallis (1977) concluded that wilting of infected plants is due to degradation of components of conducting tissue by bacterial enzymes rather than to the occlusion of xylem elements. However, in vascular diseases caused by fungal pathogens the vessels generally are not visibly degraded at the onset of wilting.

#### 4. *The Pathogen Itself*

In tomato plants infected with *Pseudomonas solanacearum,* bacterial cells, together with their extracellular products released into vessels following tylosis collapse, are probably the major cause of vessel occlusion and wilting (Wallis and Truter, 1978). However, the contribution of spores of fungal wilt pathogens to the increased resistance to water flow is probably very negligible.

#### 5. *Embolism*

Embolism, which can occur as a result of breaking of the water column, may develop in vessels of plants undergoing excessive water stress and may contribute to a reduction in water flow (Zimmerman and McDonough, 1978). The occurrence of embolism in plants infected with wilt pathogens and its impact on water relations of diseased plants is unknown.

## IV. TRANSPIRATION

### A. Transpiration in Healthy Plants

Transpiration is controlled by resistance to diffusion of water vapor and concentration of water vapor inside and outside the leaf. The transpiration rate is calculated according to the following equation:

$$\underset{[\text{g cm}^{-2}\ \text{sec}^{-1}]}{\text{Transpiration rate}} = \frac{C_l - C_a}{R_l - R_a}$$

where $C_l$ and $C_a$ (g cm$^{-3}$) are respective concentrations of water vapor at the evaporating surfaces within the leaf and in the air. $R_l$ and $R_a$ (sec cm$^{-1}$) represent diffusive resistances inside and outside the leaf, respectively. The major resistance to water-vapor diffusion is considered to be provided by the cuticle and stomatal pores. The degree of stomatal opening is the most important factor contributing to transpiration rate. The literature on transpiration and water movement in plants recently has been reviewed by Davies (1981).

## B. Transpiration in Diseased Plants

Transpiration rates of plants infected with vascular pathogens are similar to those of healthy plants prior to development of visible symptoms, but drop substantially as wilt symptoms develop (Dimond, 1955; Duniway and Slatyer, 1971; Duniway, 1971b; Harrison, 1971; Talboys, 1968; Threlfall, 1959). However, in-

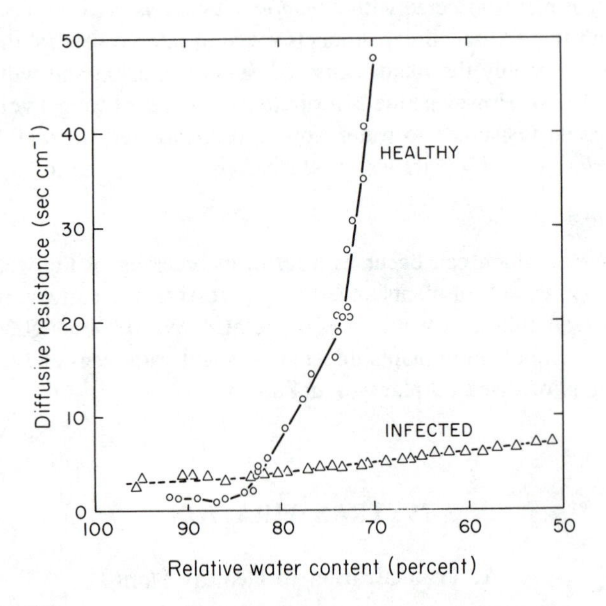


**Figure 24.** Diffusive resistance of healthy and rust-infected bean leaves determined in the light as relative water content of the leaves dropped. [Reproduced with permission from Duniway, J. M., and Durbin, R. D. 1971. *Plant Physiol.* **48:**69–72. The American Society of Plant Physiologists, Rockville, Maryland.]

fection of red oak seedlings by *Ceratocystis fagacearum* caused an abrupt increase in stomatal resistance (decrease in transpiration) three days before development of the visible wilt (TeBeest *et al.*, 1976).

In rust-infected plants, transpiration is reduced prior to sporulation and is increased with the onset of sporulation (Duniway and Durbin, 1971a). Duniway and Durbin (1971a,b) found that the diffusive resistance in a healthy bean leaf increased steadily with a drop in relative water content while that of an infected leaf with 30 uredosori per $cm^2$ remained rather unchanged (Fig. 24). The increased transpiration in diseased leaves with the onset of sporulation was attributed to damage to the cuticle.

Transpirational patterns in plants infected with powdery mildews are similar to those described for rusts (Majernik, 1971).

Transpiration rate is influenced by leaf water potential, mainly because low leaf water potential causes stomates to close (Duniway, 1971b). The behavior of stomates is also controlled by the concentration of $CO_2$, potassium, soluble sugars, and abscisic acid in the leaf tissue. Disease-induced alterations in any of these parameters are likely to modify the behavior of stomates and therefore cause changes in leaf water potential and transpiration rate.

## V. DISRUPTION OF MEMBRANE PERMEABILITY IN LEAF CELLS BY PATHOGEN-PRODUCED SUBSTANCES

Wilting may result from alterations in membrane permeability in both mesophyll and guard cells by pathogen-produced metabolites. The loss of water and solute from affected cells may cause changes in osmotic and water potentials of leaves culminating in stomatal dysfunction and altered rate of transpiration. Wilting in *Fusarium*-infected tomato (Collins and Scheffer, 1958) and in *Verticillium*-infected chrysanthemum (Hall and Busch, 1971) has been attributed to changes in leaf cell membrane permeability. Infection of field peas by powdery mildew (*Erysiphe pisi*) caused reduction in water and osmotic potentials, relative water content, and turgid weight at the onset of wilting (Ayres, 1977). The loss of turgor and fresh weight in infected tissue was associated with the loss of solutes from the cells.

## VI. WATER POTENTIAL AND DISEASE SEVERITY

Water potential may influence both host and pathogen as well as the outcome of their interaction. Low water availability has been reported to increase resistance of barley to powdery mildew infection (Ayres and Woolacott, 1980) and to reduce wheat resistance to *Fusarium* root rot (Papendick and Cook, 1974).

The growth and yield of wheat infected with leaf rust fungus (*Puccinia recondita* f. sp. *triticina*) were considerably reduced at soil water potentials of −2.5 bars, which favored the growth of the pathogen, and soil water potentials of −10.25 bars, due to increased water loss in already stressed plants. At −4.5 bars the growth and yield were only slightly reduced, because at this water potential value the host resistance was most fully expressed (Van der Wal *et al.*, 1975). Ayres (1977) has shown that lowering the water potential of pea leaves resulted in reduced sporulation of *Erysiphe pisi*. These results have been interpreted to mean that, under some field conditions, the combined detrimental effects of disease and water stress would be most pronounced when water stress is mild enough to allow development of the pathogen in the infected plant, but sufficient to accentuate the severity of the damage.

## VII. ASSESSMENT OF THE NATURE OF WATER RELATIONS CHANGES IN DISEASED PLANTS

A comparison of the pattern of changes in one water relations parameter as a function of changes in another parameter in plants subjected to pathogen-induced and artificially induced water stress may provide useful information as to the mechanism of water relations changes in healthy and diseased plants. Relationships among the following parameters have been used successfully by Duniway (1975, 1977) to determine the nature of water relations changes in *Phytophthora*-infected safflower:

1. *Relationship between water potential and percentage of wilted leaves in water-stressed diseased and healthy plants*. Mechanism of wilting at the leaf-tissue level may be the same in healthy and diseased plants when plants subjected to disease-induced and artifically induced water stress (by withholding water) wilt at nearly the same tissue water potential.
2. *Relationships between water potential and solute potential and between water potential and relative water content*. Comparisons of the pattern of changes of solute potential and relative water content as a function of water potential in plants subjected to disease-induced and artifically induced water stress may help determine if disease-induced wilting is the consequence of changes in the solute potential of leaf cells resulting from alterations in membrane permeability.
3. *Relationship of leaf diffusive resistance to leaf water status*. The nature of the relationship of leaf diffusive resistance to leaf water status in plants subjected to disease-induced and artificially induced water stress may provide clues as to the role of transpirational changes in disease-induced stress.

4. *Comparison of resistance to liquid water flow in healthy and diseased plants*. Comparison of the results of the steady-state measurements of resistance to liquid water flow in healthy and diseased plants may show whether or not resistance to water flow contributes to disease-induced water stress.

*Chapter 7*

# Disease-Induced Alterations in Carbohydrate Metabolism



## I. INTRODUCTION

Carbohydrate metabolism in plants is controlled largely by photosynthesis and respiration which occur in distinct subcellular sites. In chloroplasts, light energy is used to generate ATP and reducing power (NADPH) which in turn are used to assimilate $CO_2$ and produce photosynthetic intermediates. Additionally, chloroplasts have enzymes for metabolizing starch and the excess chemical energy produced during daylight hours. The exchange of intermediates between the chloroplast and the cytoplasm is regulated by specific membrane-bound translocation molecules and the dihydroxyacetone phosphate (DHAP)/inorganic phosphate translocator molecule that exchanges one DHAP produced in photosynthesis for

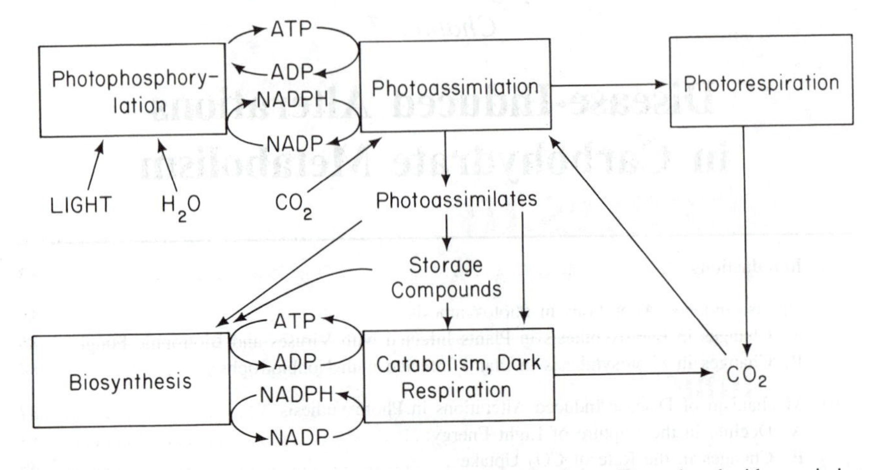


**Figure 25.** The key processes governing carbon balance in plants. [Reproduced with permission from Kosuge, T. 1978. In *Plant Disease: An Advanced Treatise* (J. G. Horsfall and E. B. Cowling, eds.), Vol. 3, pp. 85–116. Academic Press, New York.]

a cytoplasmic inorganic phosphate molecule (Heber, 1974). The cytoplasmic DHAP is then converted into various intermediates by glycolysis and is ultimately respired by mitochondria to produce ATP and reducing power (NADH). Normally, because of the absence or low activity of translocation molecules, there is little exchange of ATP, reducing power, and sugars between the chloroplast and the cytoplasm.

A more generalized view of the relationships between photosynthesis and respiration can be obtained by examination of Fig. 25, presented by Kosuge (1978). The figure also makes it easier to discern the existence of a delicate balance in the processes that control the flow of carbon in plants and maintain a steady state between the energy-generating and energy-utilizing events. It is evident that any disease-induced changes in one or more of these key processes or those controlling the sink–source relationship can disrupt the orderly flow of carbon throughout the plant.

From the economical point of view, the ultimate consequence of disease-induced deleterious changes in mechanisms controlling carbon balance is reduction in yield due to starvation or death of infected plants. Results of studies on disease-induced changes in carbohydrate metabolism have been reviewed by Wheeler (1975), Daly (1976a), Frič (1976), Kosuge (1978), and Kuć (1978). In this chapter the effect of disease on the processes governing carbohydrate metabolism will be discussed briefly.

## II. DISEASE-INDUCED ALTERATIONS IN PHOTOSYNTHESIS

Infection of plants by pathogens is known to alter one or more of the processes controlling photosynthesis, that is, photophosphorylation (formation of ATP), production of reducing power (NADPH), photoassimilation ($CO_2$ uptake), photorespiration ($CO_2$ evolution), and the movement of substances in and out of the chloroplasts.

Most studies in this area have been related to photoassimilation while few have addressed photophosphorylation and photorespiration.

### A. Changes in Photosynthesis in Plants Infected with Viruses and Biotrophic Fungi

#### *1. Changes in Photophosphorylation*

Energy-generating processes in plants, such as photophosphorylation, are not expected to undergo drastic changes in response to infection by viruses and biotrophs. This is because the synthesis of viruses and development of biotrophs in plants require energy. In a few cases studied thus far, photophosphorylation has not been altered drastically in plants infected with viruses and biotrophic fungi. In turnip yellow mosaic virus, rapid synthesis of the virus coincided with a rise in photophosphorylation (Goffeau and Bove, 1965). Wynn (1963) did not find a difference in photophosphorylation rate in chloroplasts isolated from healthy or rust-infected oats. However, Montalbini and Buchanan (1974) found a 10–30% decline in noncyclic photophosphorylation of *Vicia faba* infected with *Uromyces fabae*. Cyclic photophosphorylation, on the other hand, was not appreciably affected. In sugarbeets infected with powdery mildew, noncyclic photophosphorylation was inhibited 50% while cyclic photophosphorylation was unchanged (Magyarosy *et al.*, 1976). In the same system, the activities of cytochrome *f*, cytochrome b559 (both high and low potential forms), and cytochrome b6 were decreased by approximately one third in infected plants compared to healthy plants (Magyarosy and Malkin, 1978). Reduction, in noncyclic photophosphorylation in the *Vicia faba–Uromyces fabae* system may be the result of disease-induced alterations of components of the electron transport chain (Montalbini *et al.*, 1981).

#### *2. Changes in Apparent or Net Rate of Photosynthesis*

A drop in photosynthesis is a characteristic response to rust and mildew infection (Allen, 1942; Livne, 1964; Scott and Smillie, 1966; Black *et al.*, 1968;

Edwards, 1970b; Magyarosy *et al.*, 1976; Mignucci and Boyer, 1979; Ellis *et al.*, 1981; Gordon and Duniway, 1982). In a few cases, photosynthesis has been reported to increase during the early stages of infection by biotrophs, followed by a decline (Allen, 1942; Sempio, 1950; Livne, 1964; Yarwood, 1967; Aust *et al.*, 1977).

In certain plants infected with tobacco mosaic virus the apparent photosynthetic rate increased or remained unchanged (Owen, 1957b; Zaitlin and Hesketh, 1965; Doke and Hirai, 1969; Doke, 1972; Magyarosy *et al.*, 1973; Smith and Neales, 1977). However, the rate in other virus–plant systems was reduced (Owen, 1957a; Hampton *et al.*, 1966; Tu and Ford, 1968). The apparent inconsistency in the above studies has been attributed to differences in the protocol of experiments, in environmental conditions (leaf temperature, ambient RH, resistances to $CO_2$ diffusion, and $CO_2$ concentration), in the time when measurements were made following inoculation, and in the basis for comparison. In virus-infected variegated *Tolmiea menziesii* plants the area-based photosynthetic rate of variegated leaves was less than that of control at low light intensities but approached control rates at higher light intensities. On a chlorophyll basis, variegated leaves had higher photosynthetic rates compared to controls (Platt *et al.*, 1979). Infection of *Brassica peleinensis* by TYMV caused a change in the initial pattern of $CO_2$ fixation characterized by reduced $CO_2$ incorporation into 3-phosphoglycerate and increased incorporation into malate and aspartate (Bedbrook and Matthews, 1972). Although the observed changes may appear as a shift from the $C_3$ to the $C_4$ type of metabolism, they are probably due to an increase in PEP carboxylase and a decrease in ribulose bisphosphate carboxylase activities (Bedbrook and Matthews, 1972; Kosuge, 1978).

As Daly (1967a) has pointed out, the measurement of apparent, or net, rates of photosynthesis might be misleading. For example, Black *et al.* (1968) were able to reduce the magnitude of photosynthesis inhibition caused by *Albugo candida* infection of radish from 60% to 30% by correcting for changes in dark respiration and for dry weight increase in diseased tissue. The rate of net photosynthesis also depends on photorespiration rate. The higher the rate of photorespiration the lower the rate of net photosynthesis. It is therefore important that in studies on the effect of infection on photosynthesis, consideration be given to changes in photoassimilation, photorespiration, and dark respiration.

Ribulose 1,5-bisphosphate carboxylase/oxygenase mediates both photoassimilation and photorespiration by its dual capacity as a carboxylase and oxygenase (Jensen and Bahr, 1977). However, carboxylase and oxygenase activities of the enzyme are not regulated separately (Jensen and Bahr, 1977). The enzyme is often only partially active in intact plant tissue (Perchorowicz *et al.*, 1981). Therefore, in studies of the nature of disease-induced changes in photosynthesis, consideration should be given to the effect of infection on activation of the enzyme. Despite its important role in photosynthesis this key enzyme has rarely been studied in diseased tissue. In tobacco plants infected with TMV, the reduction in $CO_2$ assimilation was accompanied by inhibition of the synthesis of the enzyme (Hirai and Wildman, 1969).

### *3. Changes in Dark $CO_2$ Fixation*

In general, dark $CO_2$ fixation is increased in plant tissues infected with rusts and powdery mildews at the time of sporulation. However, much of the increase has been attributed to fungal activity. Increases in dark $CO_2$ fixation have also been observed in plant tissue treated with the pathotoxin victorin (Luke and Freeman, 1965) and HC toxin (Mirocha, 1972). The literature on this subject has been reviewed by Mirocha (1972).

### *4. Changes in the $CO_2$ Compensation Point*

Infection of bean leaves by *Uromyces appendiculatus* resulted in increases in the $CO_2$ compensation point and in respiration and photosynthesis rates compared to those of the control group (Raggi, 1978). The increase in the $CO_2$ compensation point in the above system was thought to be the result of increased respiration and decreased photosynthesis (Raggi, 1980). Factors which influence the $CO_2$ compensation point have been discussed by Jackson and Volk (1970). Carbon dioxide compensation point of several hundred species of plants has been measured by Krenzer *et al.* (1975).

## B. Changes in Photosynthesis in Plants Infected with Nonbiotrophs

Unlike biotrophs, nonbiotrophs are capable of developing on dead and damaged tissue and the development of these pathogens does not depend on the maintenance of the normal metabolism of the host. Infection of plants by nonbiotrophs generally is associated with extensive damage to the tissue and reduction in photosynthesis at early stages of infection. Significant inhibiton of photosynthesis has been reported in wheat inoculated with *Septoria nodorum* (Scharen and Krupinsky, 1969; Krupinsky *et al.*, 1973) and in ragi inoculated with *Helminthosporium nodulosum* (Vidhyasekaran, 1974).

# III. MECHANISM OF DISEASE-INDUCED ALTERATIONS IN PHOTOSYNTHESIS

In many cases the mechanism of disease-induced alterations in photosynthesis is not known. The following factors have been suggested to be responsible for such changes:

## A. Decline in the Capture of Light Energy

In some cases reduction in photosynthetic rate may be the consequence of a decline in the ability of infected leaves to capture light energy efficiently. This decline may be due to one or more of the factors discussed below (see also Kosuge, 1978).

### *1. Leaf Orientation and Leaf Surface Area*

The efficiency of plants in capturing light energy for photosynthesis, among other things, depends on the leaf orientation relative to the sun and the leaf surface area. An example of disease-induced alterations in leaf orientation is leaf epinasty, a characteristic symptom of certain diseases. Leaf surface area is reduced in certain diseases characterized by leaf malformation (e.g., twisting, curling) and defoliation (Kosuge, 1978).

### *2. Destruction of Photosynthetic Tissue*

Chlorosis and necrosis associated with certain diseases may cause irreversible damage to the chloroplast. Tissues thus affected are incapable of capturing enough light for energy generation and the formation of reducing power required for photosynthesis (Kosuge, 1978).

Virus infection has been reported to cause a variety of degenerative changes in chloroplasts (Gerola *et al.*, 1969; Esau, 1968; Tomlinson and Webb, 1978; Fraser and Matthews, 1979). Yellowing of leaves in virus-infected plants might not always be due to a breakdown of chlorophyll. For example, according to Montalbini *et al.* (1978), leaf yellowing in sugarbeets infected with sugarbeet yellows virus might be due to the decomposition of carotenes by particle-bound enzyme(s).

### *3. Presence of Fungal Mycelium on Leaf Surfaces*

In certain diseases such as powdery mildew the decline in the rate of photosynthesis has been attributed to a reduction of light reaching leaves due to the presence of surface mycelium of the fungus. However, this factor was not considered to be responsible for the inhibition of photosynthesis observed in powdery mildew-infected soybeans (Mignucci and Boyer, 1979) and apples (Ellis *et al.*, 1981). Inhibition of photosynthesis in powdery mildew-infected apples was not reversed by removal of fungal mycelium from leaf surfaces.

## B. Changes in the Rate of $CO_2$ Uptake

Alteration in photosynthetic rate in diseased tissue may be the consequence of a shift in the rate of $CO_2$ uptake brought about by changes in the behavior of stomates (their opening and closure) and in the mesophyll conductance to $CO_2$. Decreases in photosynthesis as a consequence of reduced stomatal aperture have been reported in leaves of tomato infected with *Fusarium oxysporum* f. sp. *lycopersici* (Duniway and Slatyer, 1971). Reduced rate of photosynthesis in sugarbeets infected with sugarbeet yellows virus was also attributed mainly to a drop in $CO_2$ uptake caused by a partial closure of stomates (Hall and Loomis, 1972). A drop in net photosynthesis in powdery mildew-infected sugarbeet leaves was attributed mainly to a decline in the mesophyll conductance to $CO_2$ rather than to a reduction in stomatal aperture (Gordon and Duniway, 1982).

Infection by certain pathogens also causes abnormal opening of stomates. Abnormal opening of stomates in both light and dark in tissues treated with fusicoccin, a toxin produced by *Fusicoccum amygdali,* was referred to earlier (see Chapter 4).

Other factors that may contribute to a decline in photosynthetic rate in diseased tissue are increases in cell mortality and in the rate of light-induced respiration and dark respiration.

## C. Effect of Pathogen-Produced Toxins

Disease-induced alterations in photosynthetic rate may be caused by toxins produced by certain pathogens. High concentrations of *Alternaria tenuis* toxin (tentoxin), which induces chlorosis in certain plant species, caused stimulation of coupling factor 1 ATPase activity while low concentrations of the toxin caused inhibition (Steele *et al.*, 1978). The toxin inhibited photophosphorylation in lettuce, a sensitive species. In radish, an insensitive species, 20 times more tentoxin was required for a 50% inhibition of photophosphorylation (Steele *et al.*, 1976). Tabtoxin produced by *Pseudomonas tabaci* was reported to inhibit ribulose 1,5-bisphosphate carboxylase activity in tobacco plants (Crosthwaite and Sheen, 1979). In soybean plants infected with the powdery mildew fungus, *Microsphaera diffusa,* decreases in the rates of net photosynthesis and transpiration were considered to be due to inhibition of chloroplast activity by a translocatable factor released by the pathogen or the host and not related to cell mortality, senescence, changes in the rates of dark respiration, or to altered rates of light-induced respiration (Mignucci and Boyer, 1979). Increased dark $CO_2$ respiration in certain toxin-treated plant tissues was referred to earlier.

## IV. DISEASE-INDUCED ALTERATIONS IN TRANSPORT OF PHOTOASSIMILATES

Accumulation of photoassimilates at infection sites is a common feature of diseases caused by biotrophic fungi (Pozsár and Király, 1966; Billett and Burnett, 1978; Doodson *et al.*, 1965; Edwards, 1971; Livne and Daly, 1966; Zaki and Durbin, 1965) and viruses (Carroll and Kosuge, 1969; Israel and Ross, 1967; Esau and Hoefert, 1971; Cohen and Loebenstein, 1975; Tu, 1977). Starch grains were seen in chloroplasts of tobacco leaves systemically infected with tobacco mosaic virus (Carroll and Kosuge, 1969). In wheat leaves infected with barley yellow dwarf virus, the concentration of soluble carbohydrates increased 400% in infected plants compared to that in control plants (Jensen, 1972).

Changes in the level of carbohydrates can also be induced in plant tissue by application of host-selective toxins. Treatment of susceptible oat leaves with victorin, a host-selective toxin produced by *Helminthosporium victoriae*, resulted in the accumulation of starch in chloroplasts and mesophyll cells.

The significance of starch accumulation in infected tissue is not clearly understood. Such accumulation may be conducive to the progress of the disease by providing the pathogen with materials needed for its development.

## V. MECHANISM OF DISEASE-INDUCED ALTERATIONS IN PHOTOASSIMILATE TRANSPORT

The following hypotheses have been advanced to explain the mechanism of changes in the transport of photoassimilates:

1. Altered photoassimilate transport may result from the presence of a concentration gradient between pathogen and host tissue resulting from removal of substrates by a growing pathogen (Livne and Daly, 1966) or from sequestration of carbon into starch or parasite products (Inman, 1962; Schipper and Mirocha, 1969).
2. Disease-induced alterations in the transport of photoassimilates may be due to long-distance transport of substrates from uninfected to infected areas (Pozśar and Király, 1966). Cotyledons of healthy flax seedlings, which are normally exporters of assimilates, became importers of assimilates upon infection with *Melampsora lini* and the "sink capacity" of the susceptible cotyledons increased at sporulation (Clancy and Coffey, 1980). In some cases accumulation of photoassimilates in the infected sites might be due to the failure of infected leaves to export carbon (Doodson *et al.*, 1965; Zaki and Durbin, 1965; Livne and Daly, 1966). Durbin (1967) found that accumulation of $^{45}Ca$ in the infected sites was

the result of decreased movement away from, rather than increased transport to, the infection sites. Smith and Neales (1980) also reported that dual infection of young peach trees with prune dwarf and *Prunus* necrotic ringspot viruses caused reduced translocation from infected leaves and reduced transport of assimilates over long distances to the stems and roots.

3. Changes in the pattern of carbohydrate accumulation in virus-infected tissue may be caused by a limited availability of orthophosphate in virus-infected tissue brought about by an accelerated demand for viral nucleic acid synthesis. Orthophosphate deficiency causes a reduction in the movement of triose phosphate from the chloroplast to the cytoplasm and an inhibition of transport of photoassimilates (T. Kosuge, personal communication).
4. Hormones, particularly cytokinins, have been implicated by Shaw and Hawkins (1958) in short-distance movement of carbohydrates among rust pustules. Cytokininlike substances have been detected in rusted leaves (Dekhuijzen and Staples, 1968). In healthy leaves, application of cytokinins causes mobilization of nutrients (Thimann *et al.,* 1974).
5. Long-distance transport of photosynthetic products from leaves to storage organs could be interrupted by the damage to phloem caused by some viruses (Jensen, 1972; Panopoulos *et al.,* 1972).
6. Alterations in starch content of diseased tissue have been attributed to changes in the activity of certain enzymes. In rice plants infected with *Lochliobolus miyabeamus,* starch accumulation has been correlated with a decrease in the activity of β-amylase (Tanaka and Akai, 1960). Fluctuations in the starch content of wheat leaves inoculated with *Puccinia striiformis* were attributed by MacDonald and Strobel (1970) to changes in the activity of ADP-glucose pyrophosphorylase resulting from variations in the levels of activators and inhibitors of the enzyme in diseased tissue during infection. Finally, changes in carbohydrate content of diseased tissue may be due to the activity of acid invertase which is increased in plants infected with certain biotrophic fungi (Long *et al.,* 1975; Callow *et al.,* 1980). The enzyme may also help provide a continuing supply of hexoses for the nutritional needs of the biotrophs.

## VI. DISEASE-INDUCED INCREASES IN RESPIRATION

Increased respiration is a characteristic early response of plants to infection by many pathogens. A brief burst of respiration during the early stages of infection might not be of any consequence. However, continued respiration at an increased rate would result in depletion of reserve carbohydrates, causing starvation and accumulation of undesirable by-products (Kosuge, 1978). Infection of bean hypocotyl by *Rhizoctonia solani* has been reported to result in an estimated increase in the rate of respiration of about 3–6 μmoles $CO_2$ $hr^{-1}g^{-1}$ fresh weight of tissue (Bateman and Daly, 1967). Kosuge (1978) has calculated that the above infected tissue would lose 2% of its dry weight (considering 80% water) assuming that respiration rate of the tissue is maintained for 24 hr and respiration utilizes storage carbohydrates and photoassimilates.

## VII. MECHANISM OF DISEASE-INDUCED INCREASES IN RESPIRATION

Despite numerous studies, the mechanism of respiratory increase in diseased tissue is not fully understood. Following is a brief account of the proposed mechanisms taken from more detailed treatments of Wheeler (1975) and Daly (1976a).

### A. Uncoupling of Phosphorylation from Electron Transport

Increased respiration in diseased tissue originally was attributed by Allen (1953) to the uncoupling of phosphorylation from electron transport brought about by the activity of pathogen-produced toxins. The uncoupling hypothesis was based on the evidence that the ratio of anaerobic to aerobic $CO_2$ production was significantly lower in diseased than in healthy tissues. However, after an extended debate, the uncoupling hypothesis was eventually dismissed on the basis of the following evidence:

1. The uncoupling agent, 2,4-dinitrophenol (DNP), does not invoke a response in rusted wheat leaves similar to that caused by infection.
2. The respiratory quotient (ratio of $CO_2$ evolution to oxygen uptake), which increases in tissues uncoupled by DNP, does not change in diseased tissue.
3. Respiratory control ratios (ratio of oxygen consumption of the tissue without added ADP to that of the tissue with added ADP) as well as P/O ratios (ratio of phosphorus atoms esterified to oxygen atoms utilized), which are expected to change in uncoupled tissue, are not changed in coupled mitochondria from victorin-sensitive oat leaves treated with the phytotoxin victorin.
4. Finally, the reported increases in growth rate and synthetic activity in plants infected with biotrophic fungi (Daly, 1967) are not consistent with

the uncoupling hypothesis. Moreover, the energy charge value (defined later in this chapter), which decreases in uncoupled tissue, remains virtually unchanged in rust-infected wheat (Hoppe, 1973) and in *Meloidogyne*-infected cotton plant (Misaghi *et al.*, 1975).

### B. Increased Activity of Pentose Phosphate Pathway

The respiratory rise in diseased tissue may be the consequence of an increase in the activity of the pentose phosphate pathway (PPP). The $C_6/C_1$ ratio (the ratio of $^{14}CO_2$ from glucose labeled at carbon 6 to glucose labeled at carbon 1) decreases markedly in plants infected with rusts and mildews (Shaw and Samborski, 1957; Daly and Sayre, 1957), viruses (Bell, 1964; Dwurazna and Weintraub, 1969), *Rhizoctonia solani* (Bateman and Daly, 1967), and in oat leaves treated with victorin (Rawn, 1974). The decrease in $C_6/C_1$ ratio is considered by some as evidence for the contribution of the pentose phosphate pathway to the rise of respiration in plants in the above host–parasite systems. This is because the $C_6/C_1$ ratio remains near 1.0 in tissue in which respiration is predominantly via glycolysis and the TCA cycle. Moreover, rusted safflower tissue is not as sensitive to fluoride (an inhibitor of glycolysis) as healthy tissue (Daly and Sayre, 1957). Rawn (1977) reported that increased respiration in victorin-treated oat leaves was accompanied by simultaneous increases in the activities of both the pentose phosphate and the glycolysis–TCA pathways. He therefore suggested that in victorin-treated oat leaves the pentose phosphate pathway is not primarily responsible for the increased respiration.

### C. Increased Activity of Noncytochrome Oxidases

A considerable fraction of the increase in oxygen uptake in diseased tissues, particularly those involving necrosis, might be due to the increased activity of peroxidases and other noncytochrome oxidases, such as polyphenol, ascorbic acid, and glycolic acid oxidases (Delon, 1974; Frič, 1976; Grzelinska, 1969; Retig, 1974). However, precise estimates of the participation of these noncytochrome oxidases in oxygen uptake have not been made. In tomato leaves infected with *Phytophthora infestans,* activities of ascorbic acid oxidase, polyphenol oxidase, and peroxidase increased; those of catalase and cytochrome oxidase remained unchanged; and glycolic acid oxidase activity decreased (Brenneman and Black, 1979).

It is rather unlikely that a single mechanism is responsible for the increased rate of respiration in diseased tissue. The respiratory rise is probably caused by a number of factors, each contributing somewhat to the overall increase.

While the identity of the factor(s) which triggers respiratory increases in diseased tissue has not been established, pathogen-produced toxins have been the prime suspect (Allen, 1953).

Limitations of some of the methods used to assess respiratory and photosynthetic activities of diseased tissues have been discussed by Daly (1976a).

## VIII. REGULATION OF CARBOHYDRATE METABOLISM

Disease-induced alterations in carbohydrate metabolism may be due to the disturbance of a number of regulatory systems which help maintain the metabolic steady state of healthy plant cells. Metabolic regulation in plant–pathogen interactions has been discussed by Kosuge and Gilchrist (1976), Kosuge (1978), and Kosuge and Kimpel (1981).

According to Kosuge (personal communication), fluctuation in the level of orthophosphate may account for many of the observed changes in carbohydrate metabolism in infected plants. This is because orthophosphate is now known to provide effective control over carbohydrate metabolism by its ability to inhibit adenosine diphosphate glucose (ADPG) synthase, a key enzyme in starch biosynthesis (Preiss and Kosuge, 1976). Moreover, starch synthesis is promoted by a high ratio of 3-phosphoglycerate (an activator of ADPG synthase) to orthophosphate, while a low ratio promotes starch turnover and ATP production. Orthophosphate also regulates export of photoassimilates from the chloroplast (Fliege *et al.*, 1978; Flugge *et al.*, 1980) and is required for a steady production of ATP via photophosphorylation and for turnover of starch in the dark (Heldt *et al.*, 1977). According to Kosuge (personal communication), shortage of orthophosphate in virus-infected plants, caused by demands for the synthesis of virus RNA, may account for many of the alterations in carbohydrate metabolism. A low level of orthophosphate would impede starch turnover and production of ATP in the chloroplast. The ultimate consequence of these changes is reduction in photosynthesis commonly associated with virus infection.

Another regulatory device is the balance among concentrations of AMP, ADP, and ATP in the cells referred to by Atkinson (1968, 1971) as energy charge (EC). The value of EC [EC = (0.5 ADP + ATP)/(AMP + ADP + ATP)] is 1 in fully charged cells (all adenylates in the form of ATP) and 0 in fully discharged cells (all adenylates in the form of AMP). Atkinson (1971) noted that in cells growing under optimal conditions concentrations of adenylates may fluctuate but the EC value remains constant at about 0.86 (Fig. 26). In cells growing under suboptimal conditions the EC value drops below 0.8. The constancy of the EC value reflects the metabolic steady state which exists in actively growing cells. Despite its usefulness as an index of metabolic status of the cell, the EC value of diseased tissue has been studied only in very few cases. While adenylate

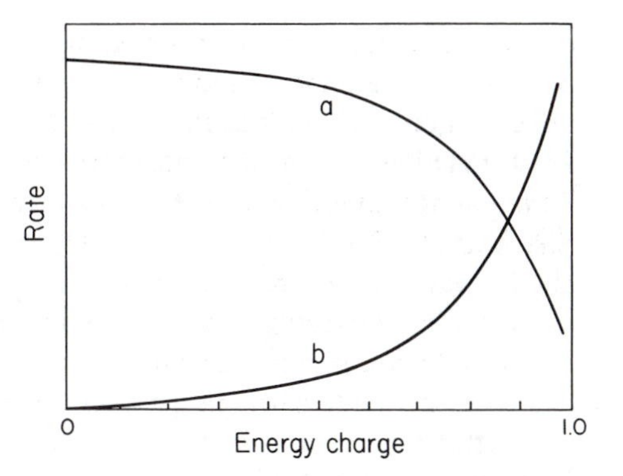


**Figure 26.** Responses of regulatory enzymes involved in ATP-generating (a) and ATP-utilizing (b) processes to the energy charge. The energy charge value of cells growing under optimal conditions and steady metabolic state remains constant at about 0.86, a point corresponding closely to the interaction of the two curves. [Reproduced with permission from Atkinson, D. E. 1968. *Biochemistry* **7**:4030–4034. Copyright © 1968, American Chemical Society, Washington, D.C.]

concentrations increased in wheat leaves infected with *Puccinia graminis* f. sp. *tritici,* compared to healthy leaves, the EC value remained unchanged (Hoppe, 1973). Marked increases in the adenylate content of cotton hypocotyls infected with *Meloidogyne incognita* also were not associated with appreciable changes in the EC values of either susceptible or resistant plants (Misaghi *et al.,* 1975). The stability of EC in rust- and nematode-infected plants is perhaps an indication of a delicate metabolic balance that must exist for the successful interaction of the pathogens and their hosts (Misaghi *et al.,* 1975).

Compartmentation is another regulatory device that helps maintain a steady metabolic state in plant cells. Compartmentation effectively separates metabolic centers in the cells, allowing simultaneous operation of energy-utilizing and energy-generating processes. Disease-induced alterations in subcellular organelles, particularly when they are accompanied by changes in their membranes, would allow unchecked movement of metabolites and cofactors in and out of organelles, resulting in loss of regulatory control imposed by compartmentation (Kosuge and Gilchrist, 1976; Kosuge, 1978). The ultimate consequence of these ultrastructural changes in chloroplasts and other cellular organelles is the reduction or even inhibition of energy-generating processes. Necrosis associated with infection and with hypersensitive responses has been attributed to the rupturing of lysosomal membranes and mixing of lysosomal acid hydrolases and their substrates (Pitt and Galpin, 1973; Wilson, 1973).

## IX. SIGNIFICANCE OF DISEASE-INDUCED ALTERATIONS IN RESPIRATION

An important, unresolved question is whether respiratory rises in diseased tissue serve a useful function in disease reaction. It has been suggested that increased respiration is essential for the development of incompatible responses in certain plant–pathogen interactions, mainly because biosynthesis of many compounds which are thought to have a role in resistance (e.g., phytoalexins, phenolics) requires carbon and energy which are furnished via respiration. However, Daly (1976a) has pointed out that reported earlier and greater increases in respiration in incompatible compared to compatible interactions (Samborski and Shaw, 1956; Tomiyama *et al.*, 1959; Uritani and Akazawa, 1959) cannot be cited as evidence that respiratory increases are necessary for the establishment of incompatible interactions, mainly because resistant and susceptible plant varieties with diverse genetic backgrounds have been used for such comparisons. For example, in a study where near-isogenic wheat lines were used, identical respiratory patterns were found in compatible and incompatible lines up to six days after inoculation with race 56 of *P. graminis tritici,* after which the rate continued to increase only in the compatible interaction (Antonelli and Daly, 1966).

It has been suggested that respiratory rises may also contribute to the establishment of compatible responses by providing the pathogens with carbon sources and ATP needed for their development. For example, Daly and Sayre (1957) found that increased respiration, host growth, and fungus development in hypocotyls of susceptible safflower infected with *Puccinia carthami* were well correlated.

## X. FUTURE STUDIES

It is generally agreed that much of the work on alterations in carbon metabolism in diseased tissue requires a reappraisal in light of new knowledge of carbon metabolism and of host–pathogen interactions which was not available when those studies were undertaken. Innovative studies are, therefore, needed to test the old hypotheses and find answers to such unresolved frequently asked questions as the following: (1) What is the degree of contribution of parasites to respiratory increases in infected tissue, particularly in compatible interactions? (2) Is the respiratory increase in response to infection associated with the expression of compatibility or incompatibility? (3) What is the function of noncytochrome oxidases in healthy and diseased tissue, and what is the basis for the observed changes in the activities of these oxidases in diseased tissue? (4) What are the bases for the stimulation or inhibition of respiration and photosynthesis in diseased tissue? (5) Are disease-induced alterations in photosynthesis essential for, or the consequence of, disease development?

*Chapter 8*

# Pathological Alterations in Transcription and Translation



## I. INTRODUCTION

Changes in transcription and translation patterns occur during the initial stages of host–parasite interactions, particularly in diseases caused by biotrophs. These changes are highly significant as they may reflect expression of genes that determine disease reaction. Disease-induced alterations in transcription and translation have not been fully explored despite their potential for elucidating the molecular aspects of compatibility and incompatibility and clarifying the genetics of host–parasite interactions. The literature on this subject has recently been reviewed by Uritani (1976), Heitefuss and Wolf (1976), Chakravorty and Shaw (1977), Samborski *et al.* (1978), and Tani and Yamamoto (1979). In the following pages some of the published works on this subject are briefly discussed.

## II. DISEASE-INDUCED ALTERATIONS IN TRANSCRIPTION

Plants infected with certain pathogens, particularly biotrophs, are known to undergo changes in the transcription process, including alterations in the activity of RNA polymerases, as well as ribonucleases, in the level of RNA, and in the structure and function of chromatin.

## A. Changes in Chromatin

Composition of chromatin is altered in cucumber as a result of infection with cucumber mosaic virus (Kato and Misawa, 1971). Changes in structure and function of chromatin have also been observed in rust-infected wheat leaves (Bhattacharya *et al.*, 1965; Shaw 1967; Flynn *et al.*, 1976).

## B. Changes in the Activity of RNA Polymerases

Plant viruses and bacteriophages alter host metabolism to their advantage by producing new virus-specific RNA polymerases or by altering the specificity of host RNA polymerases. It is therefore likely that the specificity of plant RNA polymerases is also altered in plant–fungus systems (Samborski *et al.*, 1978). Studies on RNA polymerases in plants infected with fungi have been limited to two preliminary reports showing significant changes in the template preference of RNA polymerase in wheat leaves during the first 2–4 days after inoculation with the stem rust (Flynn *et al.*, 1976; Scott *et al.*, 1976).

## C. Changes in the Activity of Ribonucleases

Alterations in RNase activity have been reported in plants infected with viruses (Reddi, 1959; Diener, 1961; Wyen *et al.*, 1971), bacteria (Reddi, 1966), and fungi (Pitt, 1970; Sachse *et al.*, 1971; Chakravorty *et al.*, 1974a,b). In plants infected with rusts (Scrubb *et al.*, 1972; Chakravorty *et al.*, 1974b; Harvey *et al.*, 1974) and powdery mildews (Chakravorty *et al.*, 1978), significant alterations have also been found in certain properties of RNase including thermal stability and substrate preference. Chakravorty and Scott (1979) found significant differences in catalytic properties of the major component of the pH 5-insoluble RNase fraction from barley leaves isolated 48 hr after inoculation with the powdery mildew fungus compared to that from healthy plants. The observed differences were thought to be induced by the fungus. In susceptible plants infected with virulent races of rust fungi, changes in RNase activity are generally attributed to the formation of new classes of RNase molecules not found in healthy plants (Scrubb *et al.*, 1972; Chakravorty *et al.*, 1974a,b).

## D. Changes in the Level of RNA

The reported increases in the RNA content of plant tissue in response to infection (Tani *et al.*, 1971, 1973; Williams *et al.*, 1973; Heitefuss and Wolf,

1976; Tani and Yamamoto, 1979) may reflect alterations in transcription processes in infected plant tissue as discussed earlier. The increases in RNA synthesis are generally more pronounced, or occur earlier, in incompatible compared to compatible interaction, and, according to some, may signal initiation of the synthesis of metabolites involved in resistance. Bhattacharya and Shaw (1968) detected a marked increase in RNA content of the host nuclei in the wheat–*Puccinia graminis* f. sp. *tritici* system. These increases were more rapid but less pronounced in the resistant cultivars. Tani and Yamamoto (1978), studying the oat–*P. coronato* f. sp. *avenae* system, reported increases in the synthesis of messenger RNA (mRNA) in resistant tissue 12–16 hr after inoculation. Changes in RNA synthesis were not detected in susceptible cultivars within the same period following inoculation. In the same system, cell necrosis and increased synthesis of cytoplasmic ribosomal RNA (rRNA) occurred at about 28 hr after inoculation (Tani *et al.*, 1973). Judging from the results of a microautoradiographical study, Tani and Yamamoto (1979) concluded that most of the observed changes in RNA synthesis occurred in the host. Results of labeling studies of Yoshikawa *et al.* (1977a) on soybean infected with *Phytophthora megasperma* f. sp. *glycinea* (Kuan and Erwin) (formerly *P. megasperma* var. *sojae*) also showed greater increases in the synthesis of mRNA in incompatible compared to compatible combinations. *De novo* synthesis of mRNA and protein in the incompatible combination in the soybean–*P. megasperma* f. sp. *glycinea* system was considered to be a prerequisite for phytoalexin formation (Yoshikawa *et al.*, 1978).

### E. Changes in Plant Nuclei

Hadwiger and Adams (1978) have reported changes in the structure of nuclei and nucleoproteins of pea cells infected with compatible and incompatible isolates of *Fusarium solani*. They suggested that the observed changes might reflect alterations in the transcription pattern in the above host–parasite system.

## III. DISEASE-INDUCED ALTERATIONS IN TRANSLATION

### A. Changes in Enzymes

Increases in the activity of certain enzymes, particularly those associated with aromatic biosynthesis and energy generation (photosynthesis, pentose phosphate pathway, glycolysis) have been reported in diseased tissues (Uritani, 1971, 1976). These changes are expected in view of the reported alterations in transcription processes of infected tissue described earlier. Increased activity of certain enzymes associated with aromatic biosynthesis, such as phenylalanine ammonia-lyase, polyphenol oxidase, and peroxidase, in incompatible interactions parallels increases in the biosynthesis of certain phenolics and phenolic derivatives, including some that have been implicated in resistance. However, increased activity of these enzymes and accumulation of their products are not specific to incompatible interactions since similar changes are also observed in certain compatible interactions. Moreover, the activity of enzymes associated with aromatic biosynthesis and with energy generation is also increased in cut-injured tissues (Uritani, 1971, 1976). Pathogen-produced alterations in phenol metabolism and the roles of phenolic derivatives in resistance are discussed in Chapters 9 and 12, respectively.

## B. Changes in Proteins

Increases in proteins have also been demonstrated in certain compatible and incompatible interactions. Pure *et al.* (1980) found that greater quantities of some discrete size classes of polypeptides were synthesized *in vitro* by polysomes isolated from the leaves of a susceptible wheat variety inoculated with *Puccinia graminis* three days after inoculation than by polysomes from healthy leaves. A fraction which might have been a new host protein was detected by Yamamoto *et al.* (1976) in the extract of oat leaves infected with a compatible race of crown rust fungus.

Some investigators have attempted to identify products of resistance genes by comparing protein synthesis patterns in compatible and incompatible interactions. Von Broembsen and Hadwiger (1972) used a double labeling technique to examine protein synthesis patterns associated with the expression of specific resistance genes in two susceptible and four resistant near-isogenic lines of flax inoculated with the flax rust fungus, *Melampsora lini*. They found significant differences in the synthesis of various size classes of soluble proteins 6–18 hr after inoculation. The rate of protein synthesis increased in incompatible interactions very soon after infection while the rate remained constant or decreased in compatible interactions. Yamamoto *et al.* (1976) and Tani and Yamamoto (1979) have also reported increases in protein synthesis during the first 14–20 hr after inoculation in incompatible interaction in the oat–*P. coronata* f. sp. *avenae* system but not in a compatible interaction. *In vitro* and *in vivo* synthesis of 21 proteins was increased in peas following exposure to compatible and incompatible isolates of *Fusarium solani f.* sp. *pisi* and to chitosan from fungal cell walls within 6 hr, the time required for the occurrence of resistance response. The pattern of changes in proteins was not significantly different in compatible and incompatible combinations within 2–8 hr after inoculation (Waggoner *et al.*, 1982). Increased protein synthesis in incompatible interactions in certain host–parasite systems has been considered to reflect accelerated biosynthesis of enzymes and other proteins involved in plant defense. This conclusion is supported by the reports that timely application of inhibitors of protein synthesis to inoculated incompatible tissues causes a reduction in resistance response (Vance and Sherwood, 1976; Yoshikawa *et al.*, 1978b; Tani and Yamamoto, 1979; Heath, 1979).

## IV. CONCLUSIONS

The significance of the observed changes in transcription and translation patterns in compatible and incompatible interactions is not well understood. Infection-induced alterations in RNA and protein syntheses in incompatible interactions are considered by some to reflect activation of resistance genes in plants. On the other hand, increased activity of certain enzymes, particularly those involved in energy generation, in compatible interaction may be due to accelerated metabolic activity of the host required to satisfy the nutritional and energy needs of the pathogen.

It has frequently been pointed out that, when interpreting the results of studies on changes in the activities of enzymes associated with aromatic biosynthesis and energy generation in diseased tissue, two important points should be taken into consideration. First, some of the changes observed may be due to nonspecific responses to infection-induced injuries unrelated to disease. Second, it is difficult to determine the degrees of contribution of the host and of the pathogen to changes observed in diseased tissue, particularly in diseases caused by biotrophs.

*Chapter 9*

# Alterations in Phenol Metabolism Caused by Disease



## I. INTRODUCTION

Phenolic compounds (Fig. 27) are common constituents of many different plants (Pridham, 1965). They include simple phenols, coumarins, most flavonoids, certain amino acids, prosthetic groups of some enzymes, plant pigments, and complex derivatives such as lignins. Phenolic substances are known to participate in a number of physiological processes which are essential to growth and development, such as oxidation–reduction reactions, lignification, and stimulation, as well as inhibition of auxin activity. Phenols and their oxidation products (quinones) are also potent uncouplers of oxidative phosphorylation, inhibitors of enzymes, and chelators of metal cofactors.

Phenolic compounds occur in a variety of simple and complex forms. Simple phenols such as cinnamic, coumaric, caffeic, ferulic, protocatechuic, chlorogenic, and quinic acids exhibit antimicrobial activities. Certain coumarins and flavonoids are also toxic to microorganisms and plants (Chapter 12). Mono- and ortho-dihydroxyphenols occur generally in the form of sugar esters and glycosides, and many form toxic substances upon hydrolysis. Phenols with more than one hydroxyl group on a benzene ring are known as polyphenols.

Phenolics are oxidized by oxidative enzymes such as peroxidases, phenol oxidases, and laccase. Peroxidases catalyze oxidation of a number of reduced compounds including phenolics with $H_2O_2$ as an oxidant while oxidative activities of phenol oxidases and laccase require oxygen.

Students of plant disease physiology are interested in phenolic compounds because a large number of those with antimicrobial properties accumulate in plant tissue as a result of infection or injury (see phytoalexins in Chapter 12).

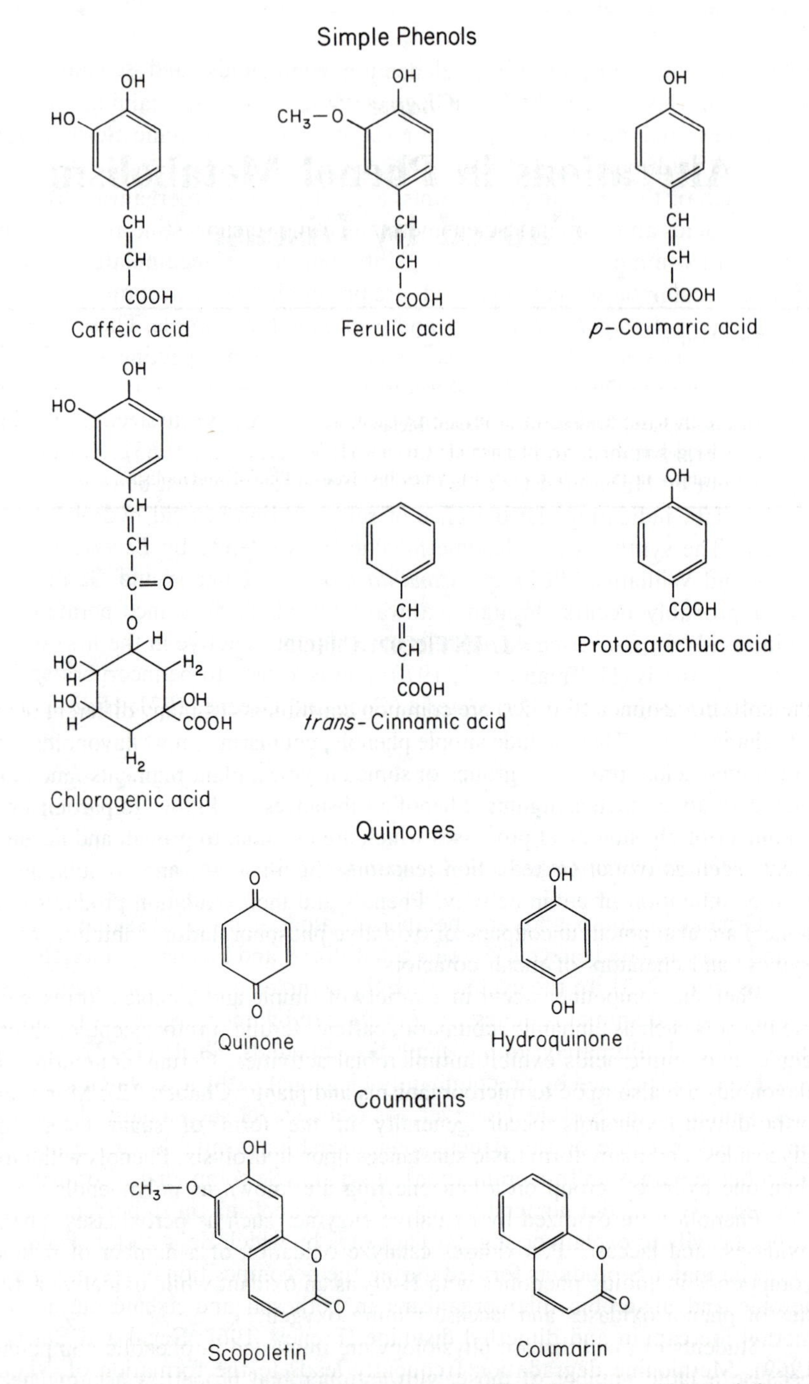


**Figure 27.** Structures of some phenolic compounds. [Partially taken with permission of Macmillan Publishing Co., Inc., from *Plant Physiology* by R. G. S. Bidwell. Copyright © 1979 R. G. S. Bidwell.]

## II. BIOSYNTHETIC PATHWAYS OF PHENOLIC COMPOUNDS

Phenolic compounds are synthesized via three different pathways: shikimic acid, acetate–malonate, and acetate–mevalonate (Neish, 1964; Kosuge, 1969). However, most phenolic substances apparently are synthesized through the shikimic acid pathway (Fig. 28). A key enzyme in the shikimic acid pathway of phenol metabolism is phenylalanine ammonia-lyase (PAL) which reductively deaminates L-phenylalanine to *trans*-cinnamic acid (Neish, 1964; Kosuge, 1969; Creasy and Zucker, 1974; Camm and Towers, 1977; Floss, 1979; Kosuge and Kimpel, 1981). Another enzyme, tyrosine ammonia-lyase, is present at high concentrations in gramineae and at low concentrations in dicotyledonous plants. The enzyme catalyzes the conversion of L-tyrosine to *p*-coumaric acid (4-hydroxy-cinnamic acid) and like PAL provides a phenylpropane carbon skeleton for the biosynthesis of certain important compounds, such as flavonoids, certain phytoalexins, and lignin (Kosuge, 1969; Kosuge and Kimpel, 1981; Hahlbrock and Griesbach, 1979). While several enzymes which participate in the bacterial shikimic acid pathway have been detected in plants (Saijo and Kosuge, 1978), certain steps in the pathway in plants are thought to differ from those in microorganisms.

Biosynthesis of phenolic compounds via the shikimic acid pathway is controlled by regulation of the activities of PAL and a few other enzymes. Kosuge (1969) has discussed the possibility that the activity of PAL may be regulated by a protein, probably a proteolytic enzyme capable of destroying PAL. Since erythrose phosphate and phosphenolpyruvate serve as precursors in the shikimic acid pathway, factors influencing the activities of the pentose phosphate and glycolytic pathways which supply these precursors indirectly influence the activity of the shikimic acid pathway (Kosuge, 1969).

Biosynthesis of phenolic compounds via the acetate–malonate pathway involves head-to-tail condensation of acetate units probably similar to that of fatty acid synthesis. This pathway of phenolic synthesis appears to participate in the synthesis of ring A of the phytoalexin pisatin, an isoflavone in peas, and 6-methyoxy mellein in carrot root tissue (Kosuge, 1969).

Metabolism and degradation of phenolic compounds in plants have been reviewed by Barz and Hoesel (1979).

## III. DISEASE-INDUCED ALTERATIONS IN PHENOL METABOLISM

### A. Changes in the Level of Phenolic Compounds

Phenolic compounds and their oxidation products (quinones) are accumulated in plants in response to infection and injury (Farkas and Király, 1962; Mace, 1964; Cruickshank and Perrin, 1964; Kuć, 1964; Rubin and

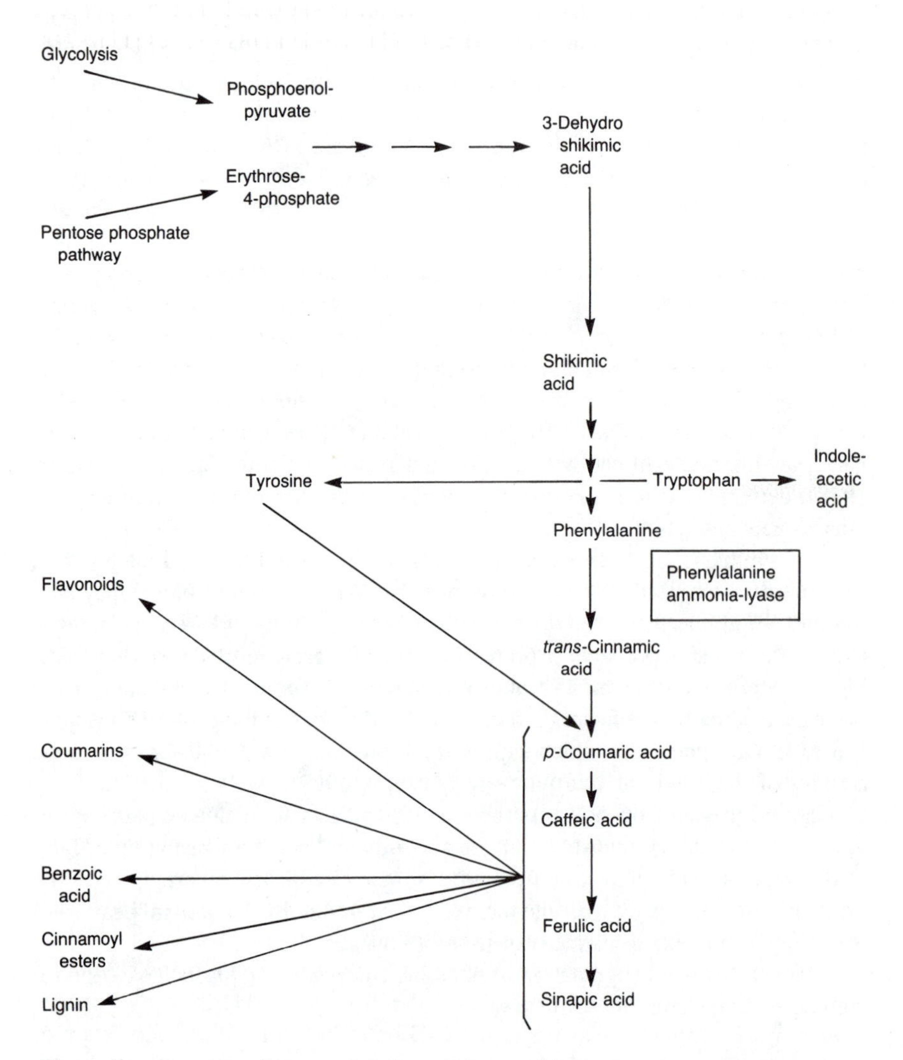


**Figure 28.** Some phenolic compounds formed via the shikimic acid pathway. Only the key intermediates are shown.

Artsikhovakaya, 1964; Patil and Dimond, 1967; Rohringer and Samborski, 1967; Metlitskii and Ozeretskovskaya, 1968; Kosuge, 1969). Many of these compounds and their derivatives (certain phytoalexins) exhibit antibiotic properties and therefore, are considered by some to play a role in disease resistance. In tobacco plants inoculated with TMV and kept at 20°C, the induction of hypersensitve response was associated with increased activity of PAL, accumulation of phenolic and flavonoid compounds, and repression of virus multiplication. PAL activity and the level of phenolic compounds were reduced, and multiplication of the virus resumed, following transfer of plants to 32°C (Paynot and Martin, 1977). Carrasco *et al.* (1978) reported that treatment of susceptible tomato plants with quinic acid and phenylalanine increased both phenolic content of the tissue and resistance to *Fusarium oxysporum*. On the other hand, in near-isogenic lines of wheat carrying the $Sr_6$ gene for resistance to *Puccinia graminis* f. sp. *tritici*, resistance was not related to changes in the levels of phenolic substances (Seevers and Daly, 1970a).

Infection in certain diseases is characterized by increased synthesis of certain precursors of phenolic compounds. In tobacco plants infected with *Pseudomonas solanacearum* increases in scopoletin and indoleacetic acid (IAA) content (Sequeira and Kelman, 1962) were associated with increases in the levels of phenylalanine and tryptophan, the respective precursors of scopoletin and IAA, 24 hr after inoculation (Pegg and Sequeira, 1968). The increases in the levels of the two amino acids were probably due to accelerated synthesis and not to protein breakdown, because the levels of most other amino acids dropped during the first 24 hr after infection. The rate of phenylalanine and tyrosine synthesis was also increased in wheat in response to infection by wheat rust (Rohringer *et al.*, 1967).

Oxidation products of phenolics, such as quinones, exhibit more toxicity to microorganisms than their reduced forms. In addition to their antimicrobial activities, quinones are also involved in both plant and fungal cell wall formation. Oxidized products of phenolics may also participate in the development of sexual structures of certain fungi and in the regulation of fungal metabolism (see Leatham *et al.*, 1980, for references).

The diverse biological activities of certain oxidized phenolics, such as quinones and free radicals, have been attributed to the ability of these compounds to react with, precipitate, and thus inactivate certain biologically important substances such as proteins, enzymes, and nucleic acids (Mason and Peterson, 1965; Kosuge, 1969; Pierpoint, 1971; Swain, 1977; Stahmann *et al.*, 1977; Leatham *et al.*, 1980). Quinones may also interfere with enzyme activity by binding metal cofactors, by reacting with sulphydryl groups, substrates, or cofactors, by binding nonspecifically to enzymes, and by competing with substrates (Kosuge, 1969; Webb, 1966). Stahmann and his associates (Leatham *et al.*, 1980) demonstrated, in an *in vitro* system, the ability of quinones and free radicals, formed from certain phenolic substrates by plant and fungal oxidative enzymes, to crosslink and precipitate certain soluble proteins. The phenolic substrates tested were representative members of phenols, tannins, hydroxycoumarins, flavonoids, and aromatic amino acids and their derivatives. Most of these phenolic substrates were able to crosslink soluble proteins (ribonuclease, lysozyme, and cytochrome *c*) only after they were oxidized by three oxidative systems: peroxidase–$H_2O_2$, polyphenol oxidase–$O_2$, and periodate. The rate and intensity of crosslinking were found to be influenced by factors such as the nature and concentration of the enzymes and reactants. Among the phenolic substrates tested, benzidine and tannic acid were the most effective crosslinking and precipitating agents, respectively.

Considering their diverse biological activities it is not surprising that quinones and other reactive oxidized phenolics have been frequently implicated in resistance. Quinones can influence disease reaction by precipitating certain pathogen-produced enzymes (Stahmann, 1965), by localizing pathogens in infected sites through formation of large polymers from proteins and other materials, and by increasing structural strength of the plant tissue (Leatham *et al.*, 1980) (see the discussion of lignification in Chapter 12).

Quinones can be reduced and consequently detoxified both enzymatically (i.e., quinone reductase) and nonenzymatically (i.e., ascorbic acid). While the role of quinone reductase in diseased tissue is not known, a pathogen may be able to avoid toxic effects of quinones by reducing them with quinone reductase.

Certain observations do not support the view that accumulation of phenolic substances and the increased activity of phenol oxidizing enzymes have a role in disease reaction. First, certain phenolic compounds do not exhibit effective toxicity to some pathogens *in vitro* at concentrations present in diseased tissue. Second, in some cases, differences in the levels of total phenolic substances among healthy, inoculated susceptible, and inoculated resistant plants are not significant (Seevers and Daly, 1970a). Third, since changes in the levels of phenolic compounds are generally observed in both diseased and injured tissues, accumulation of phenolics in diseased tissue may be due to nonspecific responses to injury caused by infection.

The literature on the role of phenolic compounds in pathogenesis has been reviewed by Kosuge (1969), Friend (1979), and Schlösser (1980).

## B. Changes in the Activities of Enzymes Involved in Phenol Metabolism

Inoculation of plants with both compatible and incompatible pathogens causes changes in the activity of a number of enzymes involved in phenol metabolism. The enzymes most often studied include phenylalanine ammonia-lyase (PAL), peroxidases, catalase, and phenoloxidases.

*1. Phenylalanine Ammonia-Lyase (PAL)*

PAL activity has been reported to change in plants in response to injury and to infection by fungi, bacteria, viruses, and nematodes (Rahe *et al.*, 1969a,b; Fritig *et al.*, 1973; Friend *et al.*, 1973; Rathmell, 1973; Giebel, 1973; Chylinska and Knypl, 1975; Paynot and Martin, 1977; Partridge and Keen, 1977; Brueske, 1980).

The increase in PAL activity in diseased tissue can be prevented by a number of inhibitors of protein synthesis such as actinomycin D, blasticidin S, puromycin, *P*-fluorophenylalanine, and 2,4-dinitrophenol, an uncoupler of oxidative phosphorylation (Minamikawa and Uritani, 1965; Zucker, 1968). This suggests that the increase in PAL activity in response to injury is due to *de novo* synthesis and that the synthesis of this enzyme in uninjured tissue is strongly repressed and is derepressed by injury (Kosuge, 1969).

In some cases PAL activity has been correlated with production of phytoalexins and with increased lignification. Hadwiger *et al.* (1970) have shown that production of the phytoalexins, pisatin and phaseollin, in pea and bean plants was dependent on PAL activity. Activities of PAL as well as 4-coumarate:CoA ligase increased in cell suspension culture of parsley in response to treatment with an elicitor of the phytoalexin, glyceollin (Hahlbrock *et al.*, 1981). The increased PAL activity was due to increased synthesis of the enzyme because increased activity of PAL paralleled that of mRNA of the enzyme. Friend *et al.* (1973) have suggested that in the *Phytophthora infestans*–potato system the rate and localization of lignification associated with hypersensitive response were controlled by PAL activity. These results are the basis for the suggestion by some that PAL activity may be associated with resistance. Brueske (1980) found significant increases in PAL activity in resistant but not in susceptible tomato roots following infection with the root-knot nematode, *Meloidogyne incognita*. Corsini and Pavek (1980) monitored changes in PAL activity and in the levels of certain phenolics following wounding or inoculation of potato tubers with *Fusarium roseum sambucinum*. They concluded that the differences in PAL activity and in the level of total glycoalkaloids and rishitin among four cultivars of potato did not appear to be great enough to account for the differences in their resistance to the fungus. It is apparent that the involvement of PAL in disease resistance has not been clearly established.

*2. Peroxidases*

Peroxidases catalyze the oxidation of a number of compounds such as aliphatic and aromatic amines and phenolics. Moreover, the enzymes are capable of catalyzing the hydroxylation of different aromatic substances in the presence of dihydroxyfumarate and reducing nitrate in the presence of appropriate electron donors. They also participate in the biosynthesis of ethylene and the polymerization of phenylpropane substances oxidatively into ligninlike substances, and can function as catalase by reducing $H_2O_2$ into $H_2O$ and $O_2$. Additionally, peroxidases catalyze conversion of IAA to 3-hydroxymethyl oxindole, which is then converted to 3-methylene oxindole. This is an important reaction because in addition to its growth-regulating activity, methylene oxindole is capable of inhibiting sulphydryl enzymes and interfering with feedback inhibition of certain regulatory enzymes (Kosuge, 1969).

Several isoenzymes of peroxidase are known. The isoenzymes differ with respect to certain properties including substrate-binding affinity and specific activity. Peroxidases are present in various cellular components including nuclei, mitochondria, ribosomes, cell walls, and membranes (Frič, 1976).

Peroxidases have also been implicated by some as having a role in resistance. Resistance of sugarbeets infected with *Cercospora beticola* was correlated with increased peroxidase and *o*-diphenol oxidase activities (Rautela and Payne, 1970). Induction of resistance to *Fusarium* wilt in susceptible tomato plants by treatment with Ethephon (an ethylene-releasing substance) was associated with increases in peroxidase and polyphenoloxidase activities (Retig, 1974). Peroxidases may also contribute to resistance by increasing the rate of lignification. Lignification of cell walls and papillae may interfere with pathogen growth and development and may contribute to the localization of viruses in local lesion hosts (Chapter 12). It should be pointed out, however, that peroxidase activity increases in both resistant and susceptible tissues following infection (Kosuge, 1969; Johnson and Cunningham, 1972; Vance *et al.*, 1976). The role of peroxidases in disease resistance is also discussed in Chapter 12.

### 3. *Phenoloxidases*

Phenoloxidases (phenolase, polyphenoloxidase, laccase) which catalyze oxidation of *o*-diphenols and *p*-phenols with molecular oxygen are known to be present in soluble forms in cytoplasm or bound to mitochondria, chloroplasts, and certain other subcellular organelles. The purported role of phenoloxidases in resistance is based on the observations that the activity of these enzymes is increased in infected tissue and that the oxidized phenols such as quinones are highly reactive and more toxic to microorganisms compared to their nonoxidized forms. However, the precise roles of these enzymes in resistance have not been clearly demonstrated and in many cases it is not clear if the increased activity of the enzymes is due to the host or pathogen.

Phenolase occurs in multiple forms mainly bound to cell components. The increased activity of the enzyme in infected and injured tissue is due either to increased solubilization of the enzyme or to its *de novo* synthesis (see Kosuge, 1969).

*4. Catalase*

The role of catalase in diseased reaction is not clearly understood. The enzyme has frequently been suggested to help reduce the level of $H_2O_2$, which may accumulate to toxic levels in diseased tissue.

While results of studies on phenol metabolism in infected tissue show that the activities of certain enzymes involved in synthesis and metabolism of phenolic compounds increase in response to infection, the overall results have failed to establish a clear correlation between disease resistance and activities of PAL, peroxidases, and phenoloxidase. Since wounding of plants causes changes in the activity of some of the enzymes involved in phenol metabolism (Uritani, 1971, 1976; Daly, 1972), it has been suggested that part or all of the observed changes in the activity of these enzymes may be nonspecific responses to injury caused by infection. In healthy plants certain enzymes, including those participating in oxidation of phenolic compounds, are separated from their substrates through compartmentation. In some cases increased activity of the compartmentalized enzymes in response to injury or infection is the result of decompartmentation that allows contact between the enzymes and their substrates.

The biochemistry of phenolic compounds in wounded plant tissues has been discussed by Rhodes and Wooltorton (1978).

*Chapter 10*

# Growth Regulator Imbalance in Plant Disease



## I. INTRODUCTION

Plants infected with pathogens exhibit various growth abnormalities, including excessive growth, stunting, tumors, epinasty, and formation of adventitious roots. Although in many cases the mechanisms responsible for the growth abnormalities are not known, they are commonly associated with alterations in the levels of growth regulators.

Very little is known regarding the nature of hormonal changes in diseased plants and the significance of such alterations in pathogenesis. Moreover, in many cases it is not known whether hormonal imbalances in diseased tissues are the cause or the result of infection.

In this chapter the involvement of growth regulators in pathogenesis is examined and the significance of their changes is discussed. The literature on the subject has been reviewed by Hislop *et al.* (1973), Sequeira (1973), Van Andel and Fuchs (1972), Daly and Knoche (1976), Pegg (1976a–d, 1981b), Dekhuijzen (1976), and Helgeson (1978).

## II. KNOWN GROWTH REGULATORS

### A. Indoleacetic Acid

Indoleacetic acid (IAA), a major growth regulator, serves various functions in plants. It is involved in cell enlargement and differentiation, leaf and fruit abscission, root formation and growth, and apical dominance.

A survey by Gruen (1959) has shown that about 80 species of pathogenic and nonpathogenic fungi are capable of producing substances with IAA activity. IAA also accumulates in plants infected with *Puccinia graminis* f. sp. *tritici* (Shaw and Hawkins, 1958), *P. carthami* (Daly and Inman, 1958), *Albugo candida* (Yu, 1971), *Fusarium oxysporum* f. sp. *lycopersici* (Matta and Gentile, 1964), *Verticillium albo-atrum* (Pegg and Selman, 1959; Wiese and DeVay, 1970), *Pseudomonas savastanoi* (Hutzinger and Kosuge, 1967; Kuo and Kosuge, 1969), and *Taphrina deformans* (Sziráki *et al.*, 1975). On the other hand, potatoes, beans, and sugarbeets infected with the curly top virus contain lower levels of IAA (Smith *et al.*, 1968).

Certain lines of evidence implicate IAA as having a role in pathogenesis in a number of diseases. First, symptoms of certain diseases caused by fungi and bacteria resemble closely those caused by IAA imbalance. The symptoms include galls and overgrowth, epinasty, and formation of adventitious roots. Second, some of these symptoms can be induced in healthy plants by application of IAA.

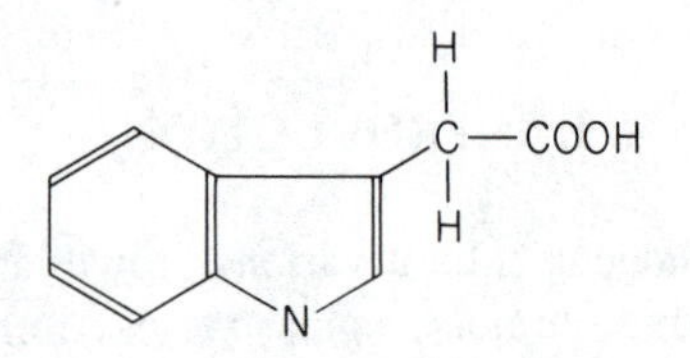


**Structure 1.** Indoleacetic acid.

The involvement of IAA in pathogenesis is discussed in Sections IV, V, and VI of this chapter.

### B. Ethylene

Ethylene is a ubiquitous hormone capable of eliciting a wide range of responses similar to symptoms of certain plant diseases. It is produced by plants and by a number of fungal as well as bacterial pathogens. Of the 228 species of fungi tested, 58 produced ethylene in culture (Ilag and Curtis, 1968).

Responses of plants to ethylene include growth inhibition, chlorosis, premature defoliation, epinasty, stimulation of adventitious roots, and early ripening of fruits (Abeles, 1972, 1973). Ethylene also causes an increase in membrane permeability, which is a common response to infection (Abeles *et al.*, 1971).

Many plants are known to produce increased levels of ethylene in response to stress and injury induced by biotic and abiotic agents (Abeles, 1973; Yang and Pratt, 1978). These observations are the basis for the assumption that increases in the levels of ethylene in certain diseased tissues are of plant rather than parasite origin.

A number of different ethylene-generating systems have been described in fungi and bacteria (Yang, 1967; Primrose, 1976; Chalutz *et al.*, 1977; Axelrood-McCarthy and Linderman, 1981). Most of these systems involve conversion of methionine to ethylene in the presence of flavinlike compounds and light. In some cases, a single microorganism may possess more than one ethylene-generating system (Chalutz *et al.*, 1977; Axelrood-McCarthy and Linderman, 1981). For more details see the review by Lieberman (1979).

The possible involvement of ethylene in pathogenesis is discussed in Sections IV and VI of this chapter.

### C. Cytokinins

A number of cytokinins with multiple regulatory functions are produced in plants. Many of the known cytokinins are $N^6$-substituted purines (Table 2). They stimulate cell division and enlargement, enhance DNA synthesis, delay senescence, counteract abscisic acid inhibition, and promote mobilization.

Like other growth regulators, cytokinins have been implicated in certain plant diseases because they are produced by few plant pathogens and undergo quantitative changes in diseased tissue (Dekhuijzen, 1976). Cytokinins are thought to contribute to the development of a number of disease conditions including fasciation of peas, gall and overgrowth, green-island formation, and altered translocation (see Sections IV and VI of this chapter).

**Table 2. The Free Bases of Natural Cytokinins[a]**

(General structure: purine with $NH-R_1$ at position 6 and $R_2$ at position 2, N-9 bearing H)

| | $R_1$ | $R_2$ | Chemical name | Common name or abbreviation | Concentration (M) for half-maximal response |
|---|---|---|---|---|---|
| I | $-CH_2-CH=C(CH_3)(CH_2OH)$ | H | 6-(4-Hydroxy-3-methyl-*trans*-2-butenyl)aminopurine | *trans*-Zeatin | $5 \times 10^{-9}$ |
| II | $-CH_2-CH=C(CH_2OH)(CH_3)$ | H | 6-(4-Hydroxy-3-methyl-*cis*-2-butenyl)aminopurine | *cis*-Zeatin | $10^{-7}$ |
| III | $-CH_2-CH_2-CH(CH_2OH)(CH_3)$ | H | 6-(4-Hydroxy-3-methylbutyl)aminopurine | Dihydrozeatin | $3 \times 10^{-8}$ |
| IV | $-CH_2-CH=C(CH_3)(CH_3)$ | H | 6-(3-Methyl-2-butenyl)aminopurine | IPA | $10^{-8}$ |
| V | $-CH_2-CH=C(CH_2OH)(CH_3)$ | $CH_3S-$ | 6-(4-Hydroxy-3-methyl-2-butenyl)2-methylthioaminopurine | $CH_3S$-Zeatin (cis or trans) | $10^{-8}$ |
| | $-CH_2-CH=C(CH_3)(CH_2OH)$ | $CH_3S-$ | 6-(4-Hydroxy-3-methyl-2-butenyl)2-methylthioaminopurine | $CH_3S$-Zeatin (cis or trans) | |
| VI | $-CH_2-CH=C(CH_3)(CH_3)$ | $CH_3S-$ | 6-(3-Methyl-2-butenyl)2-methylthioaminopurine | $CH_3S$-IPA | $6 \times 10^{-8}$ |
| VII | $-C(=O)-NH-CH(COOH)-CH(OH)-CH_3$ | H— | 6-(Threonylcarbamoyl)purine | — | ? |

[a]From Varner, J. E., and Ho, D. T. H. 1976. In *Plant Biochemistry* (J. Bonner and J. E. Varner, eds.), pp. 713–770. Academic Press, New York. Used with permission.

## D. Gibberellins

A growth regulator named gibberellin was first reported by Japanese investigators working with a disease of rice known as Bakanae disease (Yabuta and Hayashi, 1939). Today more than 40 different gibberellins with minor structural differences have been isolated from plants and fungi. They are individually referred to as GA1, GA2, and so on.

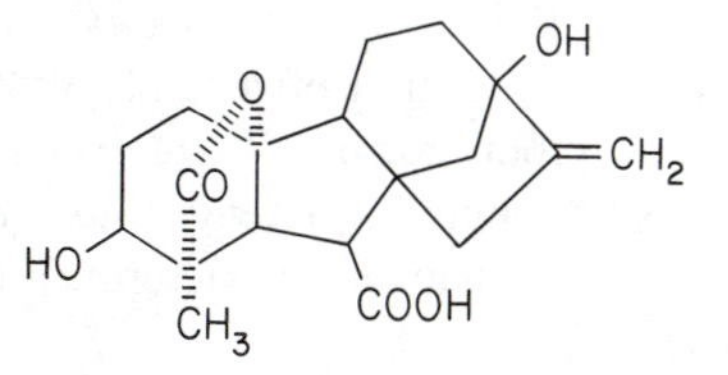


**Structure 2.** Gibberillin $A_1$.

Biological activities of gibberellins include internode elongation, reversal of dwarfism, and increases in the activities of α-amylase, protease, ribonuclease, and RNA polymerase (Varner and Ho, 1976).

Excessive elongation of internodes, a characteristic symptom of certain diseases is thought to be due to the accumulation of gibberellins in diseased tissue (see Section IV.C of this chapter).

## E. Abscisic Acid

Abscisic acid, which is produced by plants, serves several functions, including dormancy, inhibition of growth and seed germination, stimulation of fungal spore germination, and stomatal closure.

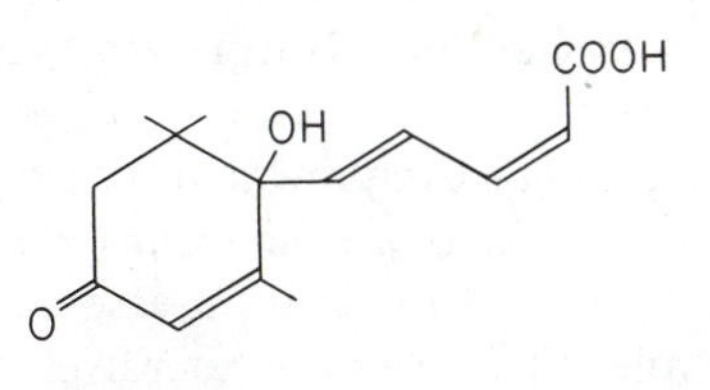


**Structure 3.** (+)-Absicisic acid.

Stunting, a characteristic response of plants to infection by vascular pathogens and certain viruses, has been attributed to increases in the levels of growth inhibitors, particularly abscisic acid (see Section IV.D of this chapter).

## III. PATHOGEN-PRODUCED TOXINS WITH GROWTH REGULATOR ACTIVITY

A few toxins produced by plant pathogenic fungi exhibit growth-regulating activities. Helminthosporal produced by *Helminthosporium sativum* can cause physiological changes in host tissue similar to those induced by gibberellins (Luke and Gracen, 1972b). Some of the effects of *Helminthosporium maydis* T toxin on the cytoplasm of Texas male-sterile maize are similar to those of abscisic acid (Arntzen *et al.,* 1973a). In some cases, fusicoccin produced by *Fusicoccum amygdali* mimics the action of natural growth regulators (Marré, 1979). Coronatin, a toxin produced by *Pseudomonas coronafaciens* var. *atropurpurea,* seems to possess auxinlike activity (Sakai *et al.,* 1979). Finally, IAA-mediated cell elongation is inhibited by victorin, a toxin produced by *H. victoriae* (Wheeler, 1969).

These observations reflect the complex nature of host–pathogen interaction and emphasize how metabolism of infected tissue can be affected by altered levels of known growth regulators as well as other metabolites in diseased tissue.

The mechanism of growth-regulator activities of pathogen-produced toxins is not known. These toxins may not act as growth regulators directly but may stimulate increased production of growth regulators by plants.

## IV. INVOLVEMENT OF GROWTH REGULATORS IN CERTAIN DISEASE-INDUCED ABNORMALITIES

Certain disease-induced abnormalities are thought to be mediated by changes in the levels of growth regulators in diseased tissue. A few examples are discussed below.

### A. Gall and Overgrowth

The development of galls and overgrowth in certain diseases is most likely the result of changes in the balance of growth regulators. One of the best examples of pathogen-induced galls is olive-knot disease caused by *Pseudomonas savastanoi,* in which the role of IAA has been extensively studied by Kosuge and his associates (Kuo and Kosuge, 1969, Smidt and Kosuge, 1978; Comai and Kosuge, 1980). Infection of the aerial parts of the susceptible hosts (olive, oleander, ash species, and swamp privet) by the bacterium results in the formation of galls due to excessive cell enlargement and cell division. The bacterium is capable of producing IAA *in vitro* (Magie *et al.*, 1963; Kuo and Kosuge, 1969). Moreover, galls contain higher levels of IAA than healthy tissue (Magie, 1963).

Evidence obtained by Smidt and Kosuge (1978) and Comai and Kosuge (1980) shows clearly that bacterium-produced IAA is the determinant of pathogenicity. First, mutants of the bacterium incapable of producing IAA do not produce galls. Second, production of IAA by the bacterium is coded by genes carried on a plasmid; the loss of it is associated with the loss of IAA production. Third, mutants incapable of producing IAA acquire the ability to produce the hormone upon acquisition of the missing plasmid. Finally, mutants without the plasmid are avirulent on oleander.

The pathway of IAA synthesis by the bacterium and its regulation have been defined by Kosuge and his associates. L-tryptophan is converted to indole-3-acetamide and $CO_2$ by the bacterial tryptophan oxidative decarboxylase, followed by hydrolysis of indole-3-acetamide to IAA. The enzyme is apparently inhibited by both IAA and indole-3-acetamide. Indoleacetyl-ϵ-L-lysine synthetase produced by the bacterium catalyzes the conversion of IAA to indoleacetyl-ϵ-L-lysine with reduced auxin activity (Hutzinger and Kosuge, 1967; Kuo and Kosuge, 1969).

Development of crown gall tumors induced by *Agrobacterium tumefaciens* also seems to be regulated by phytohormones. First, the bacterium is capable of producing IAA and a cytokinin in culture (Upper *et al.*, 1970). Second, a gene(s) required for IAA production is apparently located on the Ti-plasmid of the bacterium (Liu and Kado, 1979). A portion of the Ti-plasmid is now known to be stably incorporated into the plant genome following infection (Chapter 11). Third, unlike normal tissue, crown gall tumors synthesize all of their own growth regulators (Braun, 1969; Dekhuijzen, 1976). Fourth, tumor induction on bean leaves is promoted by the application of cytokinins, gibberellins, and IAA (El Khalifa and Lippincott, 1968).

Growth regulators may also play a role in the clubroot galls induced by *Plasmodiophora brassicae* because galls contain more cytokinins than healthy tissue (Matsubara and Nakahira, 1967; Katsura *et al.*, 1969; Dekhuijzen and Overeem, 1971; Ingram, 1969) and, unlike healthy tissue, diseased tissue containing the fungus can be cultured in the absence of cytokinins (Williams *et al.*, 1969).

The involvement of phytohormones in galls formed in response to infection by *Ustilago maydis* and by certain nematodes is less conclusive. A direct correlation was found between the size of galls on corn induced by *Ustilago maydis* and their cytokinin activity (Mills and Van Staden, 1978).

While cytokinin levels are generally increased in diseases involving overgrowth, the levels are decreased in *Verticillium*-infected tomato and cotton plants (Krikon *et al.*, 1971; Misaghi *et al.*, 1972; Patrick *et al.*, 1977).

## B. Fasciation of Peas

Evidence for the involvement of cytokinins in fasciation of peas caused by *Corynebacterium fascians* seems to be fairly strong. It includes production of a cytokinin [$N^6$-(3-methyl-2-butenylamino) purine] by the bacterium *in vitro* (Klämbt *et al.*, 1966; Helgeson and Leonard, 1966) and the development of disease symptoms similar to those caused by bacterial infection in healthy plants treated with a cytokinin, kinetin. Evidence for the involvement of cytokinins in this disease would be strengthened if cytokinin production by strains of the bacterium were correlated with their pathogenicity (Wheeler, 1975).

## C. Excessive Elongation

The excessive elongation of internodes, a characteristic symptom of certain diseases, has been correlated with the accumulation of gibberellins in diseased tissue. Such a correlation was first demonstrated in Bakanae disease of rice caused by *Gibberella fujikuroi* (*Fusarium moniliforme*). The fungus is capable of producing a number of gibberellins *in vitro,* and the symptoms of the disease, including accelerated internode elongation, can be reproduced by application of pathogen-produced gibberellins (Brian *et al.*, 1954). It is not clear if the increase in the level of gibberellins in diseased tissue is of fungus or host origin. In creeping thistle infected by *Pucinnia punctiformis,* as with *Gibberella*-infected rice, increased growth rate is also associated with high levels of gibberellins and auxins (Bailiss and Wilson, 1967). *Sphaceloma manihoticola,* the causal agent of cassava superelongation disease, produces a gibberellin in culture identified as gibberellin $A_4$. Some of the symptoms of the disease, including elongation of the internodes and asymmetrical growth around leaf spots, could be reproduced by application of known gibberellin $A_4$ or that purified from culture filtrate (Zeigler *et al.*, 1980). Finally, Daly and Inman (1958) have shown substantial increases in the level of IAA in safflower hypocotyls infected with *Puccinia carthami*. Increases in the level of IAA occurred during a period of hypocotyl growth.

## D. Stunting

Growth regulator imbalances have been implicated as the cause of stunting in certain diseases, particularly in those caused by viruses. While the gibberellin levels of diseased plants exhibiting excessive elongation are above the levels found in healthy plants, the levels are reduced in certain virus diseases characterized by growth retardation (Russell and Kimmins, 1971; Bailiss, 1974; Aharoni *et al.*, 1977; Ben-Tal and Marco, 1980). Moreover, stunting associated with certain virus infections has been partially or totally annulled by application of gibberellic acid (Maramorosch, 1957; Stein, 1962).

Virus-induced stunting has also been associated with a reduction in the level of IAA in diseased plants. Growth inhibition in cowpea seedlings infected with cowpea mosaic virus was attributed to either enzymatic breakdown of IAA or to the disruption of auxin transport or metabolism (Lockhart and Semanick, 1970).

Disease-induced stunting may also be due to abscisic acid imbalance. The reduced growth of tobacco leaves infected with systemic and local-lesion-forming strains of tobacco mosaic virus was associated with increased level of free abscisic acid. The level of bound abscisic acid was likewise increased in leaves infected with the local-lesion-forming strain of the virus (Whenham and Fraser, 1981). Leaf growth could also be reduced in healthy plants by application of abscisic acid. In tobacco plants infected with *Pseudomonas solanacearum* the level of a growth inhibitor, identified as abscisic acid, bacterial multiplication, and reduced internode elongation were correlated (Steadman and Sequeira, 1970). An increased level of abscisic acid in diseased tissue was attributed to the activity of the host rather than the pathogen. The level of abscisic acid was also increased in tomato plants infected with *Verticillium albo-atrum* (Pegg and Selman, 1959).

When considering the role of growth regulators in disease-induced stunting a distinction should be made between stunting resulting from reduced cell division and that caused by reduced cell elongation. The stunting of leaves of barley plants infected with yellow dwarf virus is caused by a reduction in cell division and not by changes in cell size. The virus-induced dwarfing can be overcome partially by application of gibberellin $A_3$. Reversal of virus-induced stunting by gibberellin, however, is due to increased elongation and not to counteraction of virus-induced reduction in cell division (Russell and Kimmins, 1971).

Growth retardation induced by virus infection is mostly due to changes in the balance of growth regulators rather than changes in the level of a single phytohormone which makes it difficult to establish a cause-and-effect relationship (Matthews, 1980). For example, a reduction in the rate of hypocotyl elongation in cucumber plants infected with cucumber mosaic virus was associated with increases in the levels of ethylene and abscisic acid and a decrease in the level of gibberellin-like substances (Marco *et al.*, 1976; Aharoni *et al.*, 1977).

## E. Water Stress and Wilting

Phytohormone imbalance in certain diseases may be the consequence of water stress induced by infection because the levels of growth regulators are known to change in water-stressed plants (Wright and Hiron, 1969; Wright, 1977). This seems to be the case in cotton plants infected with *Verticillium dahliae* because a drop in the level of cytokinins occurs in plants subjected to wilting, induced either by infection or by withholding water (Misaghi *et al.*, 1972). The increase in abscisic acid concentration in necrotic tissue in tobacco plants infected with tobacco mosaic virus was also thought to be due to water stress (Whenham and Fraser, 1981).

Wilting in oak trees caused by *Ceratosistis fagacearum* is associated with tylose formation and the reduction of water flow through twig cuttings. Beckman *et al.* (1953) have suggested that tylose formation in diseased plants contributes to the reduced water flow. They postulated that growth regulators may contribute to tylose formation by causing an increase in the plasticity of the cell wall, enabling plant pectolytic enzymes to weaken plant membranes. This allows protoplasts of ray cells to protrude into the vessels forming tyloses. This hypothesis is supported by the results of a study by Pegg (1959), who induced tyloses in tomato shoots with IAA.

Since some of the ethylene-induced symptoms such as epinasty, chlorosis, and defoliation are also associated with diseases caused by wilt pathogens, ethylene has been suspected of having a causal role in the pathogenesis of vascular wilt diseases. For example, the epinastic symptoms of *Fusarium* wilt of tomato have been attributed to pathogen-produced ethylene (Dimond and Waggoner, 1953b).

## F. Leaf Abscission

Defoliation, a characteristic response of plants to infection by certain pathogens, may partially or totally be due to changes in the balance of growth regulators. Premature leaf abscission in coffee and coleus plants infected with *Omphalia flavida* was attributed by Sequeira and Seevers (1954) to IAA degradation by a fungal oxidase. Culture filtrate of the fungus contained an IAA-inactivating factor. Moreover, the rate of IAA decarboxylation, presumably due to the activity of IAA oxidase, was higher in leaf discs of infected plants than in those of noninfected plants (Rodrigues and Arny, 1966) and increased substantially with the progress of the disease. While IAA may play a role in leaf abscission in *Omphalia*-infected plants, the possibility of the involvement of other growth regulators such as ethylene cannot be ruled out (Sequeira, 1973).

A fungus-produced IAA oxidase might also be involved in leaf abscission in roses infected with *Diplodia rosae*. The culture filtrate of the fungus, like that of *Omphalia flavida,* contained a factor capable of inactivating IAA (Kazmaier, 1960). Since IAA oxidase has not been found in diseased tissue, the involvement of this enzyme in leaf abscission in infected plants remains speculative. For a detailed discussion of the role of IAA oxidase in disease see Sequeira (1973).

Wiese and DeVay (1970) conducted a comprehensive study of the role of growth regulators in *Verticillium* wilt of cotton. Abscisic acid content of leaves increased in plants inoculated with a defoliating strain of *V. dahliae* but not in those inoculated with a nondefoliating strain. Since abscisic acid is known to accelerate leaf abscission in cotton plants (Addicot and Lyon, 1969) it was concluded that changes in the level of abscisic acid may, at least partly, account for disease-induced defoliation. However, the possible involvement of other regulators in leaf abscission could not be ruled out because defoliation in diseased plants was also associated with changes in the levels of ethylene and in the rate of IAA decarboxylation. In cotton plants infected with defoliating strains of the fungus, ethylene production was twice as high as in those infected with nondefoliating strains of the fungus (Wiese and DeVay, 1970).

Defoliation in *Verticillium*-infected hop (Talboys, 1972), *Xanthomonas citri*-infected citrus (Goto *et al.*, 1980), and *Cylindrocladium*-infected azalea (Linderman, 1974), is associated with altered levels of ethylene in diseased tissue.

Since abscisic acid stimulates ethylene production by leaf tissue, some of the activities attributed to abscisic acid, particularly leaf and fruit abscission, could be due to ethylene (F. B. Abeles, personal communication).

### G. Green-Island Formation

In certain diseases caused by biotrophic parasites the areas around infection sites remain green following leaf chlorosis, a phenomenon known as green-island formation. A number of hypotheses have been advanced to account for this phenomenon including changes in the levels of cytokinins. This is mainly because cytokinins are known to retard senescence in detached leaves (Fletcher, 1969) and to promote chlorophyll production (Fletcher *et al.*, 1973). However, the overall results of a number of studies have not provided incontrovertible evidence that cytokinins play a major role in green-island formation. Dekhuijzen (1976), who has reviewed the subject, concluded that green-island formation may be the result of changes in the balance of growth hormones including cytokinins brought about by infection.

### H. Altered Translocation

Although cytokinins seem to be involved in mobilization of metabolites in healthy plants (Kende, 1971) the proposed involvement of cytokinins in disease-induced alterations in translocation patterns has not been verified experimentally.

## V. MECHANISMS OF CHANGE IN INDOLEACETIC ACID CONTENT OF DISEASED TISSUE

Four mechanisms have been suggested as being responsible for changes in IAA content and/or IAA activity in diseased tissue.

### A. Changes in the Level of Indoleacetic Acid Precursors

Fluctuations in the level of IAA in diseased tissue may reflect changes in the level of IAA precursors. The level of DL-tryptophan (a precursor of IAA), which was found to be fairly constant in healthy tobacco plants, doubled in plants infected with *Pseudomonas solanacearum* and continued to increase with the rise in the level of IAA (Pegg and Sequeira, 1968). Accumulation of tryptophan and other aromatic amino acids in diseased tissue was attributed to the increased rates of their synthesis resulting from accelerated activity of the shikimic acid pathway and not to the breakdown of protein.

### B. Indoleacetic Acid Oxidation

The level of IAA in plant tissue may be altered as a result of changes in the activity of a peroxidase, active as oxidase, capable of oxidizing IAA to 3-methylene oxindole and $CO_2$. As was indicated earlier, premature defoliation in coffee and coleus plants infected with *Omphalia flavida* (Sequeira and Seevers, 1954; Rodrigues and Arny, 1966) and in roses infected with *Diplodia rosae* (Kazmaier, 1960) has been attributed to IAA oxidation by fungal oxidases.

Studies designed to establish a correlation between IAA oxidase activity and IAA level are complicated by the presence of activators and inhibitors of IAA oxidase in the host tissue (Table 3). These substances, which are mostly phenolics, accumulate in plants in response to infection and injury (Uritani, 1976). Moreover, plants contain enzymes capable of converting certain inactive phenolic substances to activators or inhibitors of IAA oxidase (Van Andel and Fuchs, 1972; Pegg, 1976a). In tobacco plants infected with *Pseudomonas solanacearum* a drop in IAA oxidase activity was correlated with a rise in the levels of scopoletin and chlorogenic acid, inhibitors of IAA oxidase (Sequeira, 1964). The activity of a peroxidase, active as oxidase, is also affected by stress, injury, and low water potential (Sequeira, 1973). The nature of the enzymatic breakdown of IAA in diseased tissue has been discussed by Van Andel and Fuchs (1972), Sequeira (1973), and Pegg (1976a).

**Table 3. Inhibitors and Activators of Indoleacetic Acid Oxidase**

| Inhibitor | Activator |
|---|---|
| Caffeic acid | Ferulic acid (at low concentration) |
| Chlorogenic acid | *p*-Hydroxybenzoic acid (at high concentration) |
| Catechol | *p*-Coumaric acid |
| Vanillyl alcohol | *m*-Coumaric acid |
| Ferulic acid (at high concentration) | *o*-Coumaric acid |
| Di-hydroxyphenylalanine DOPA | *p*-Hydroxybenzyl alcohol |
| Scopoletin | Phlarctic acid |
| Ascorbic acid | Tyrosine |

[a]Taken with permission from Pegg, G. F. 1976. In *Encyclopedia of Plant Physiology, New Series, Physiological Plant Pathology* (R. Heitefuss and P. H. Williams, eds.), Vol. 4, pp. 560–581. Springer-Verlag, Heidelberg.

## C. Conjugation of Indoleacetic Acid with Inactivator and Protector Molecules

IAA is known to be inactivated upon conjugating with a variety of chemicals such as aspartic acid, amino acids, certain sugars, polysaccharides, and proteins. Many IAA conjugates are considered to lack growth-promoting activities but can be readily converted to nonconjugated active forms. Auxin protectors (protein molecules) capable of causing a delay in peroxidase-catalyzed IAA oxidation have also been isolated from infected tissues including stems of *Agrobacterium tumefaciens*-infected sunflowers (Stonier, 1969) and galls induced by *Synchytrium endobioticum* on potato (Haard, 1978). IAA is also known to be protected by inhibitors of IAA oxidase such as orthodiphenols.

## D. Interference with Indoleacetic Acid Transport

The availability of IAA in diseased tissue may be influenced by factors which interfere with its transport in the plant (Van Andel and Fuchs, 1972).

In studies of the involvement of growth regulators in pathogenesis it is important to determine whether accumulation of phytohormones in diseased tissue is of host or of pathogen origin. This question was addressed by Phelps and Sequeira (1968) in their study of the role of IAA in the tobacco–*Pseudomonas solanacearum* system, in which infected plants contained higher levels of IAA compared to those of the noninfected plants. Both pathogenic and nonpathogenic isolates of the bacterium are capable of synthesizing IAA from tryptophan, a precursor of IAA. Phelps and Sequeira (1968) found that tobacco plants and *P. solanacearum* used different pathways for synthesis of IAA. They took advantage of this difference to determine the degree of host and pathogen contribution to the overall accumulation of IAA in diseased tissue. Judging from the results of a study with labeled tryptamine, a precursor of IAA for the plant but not for the bacterium, they concluded that accumulation of IAA in diseased tissue during the early stages of the disease is mostly of host origin.

## VI. INFLUENCE OF GROWTH REGULATORS ON THE OUTCOME OF DISEASE REACTION

An important, unresolved question is whether or not infection-induced alterations in the levels of growth regulators in plants have any influence on the outcome of disease reaction. Daly (1976b) has speculated that in certain compatible interactions disease-induced changes in the balance of growth regulators may redirect metabolic activity of the host in a way that favors the pathogen (Daly, 1976b). Whenham and Fraser (1981) pointed out that growth inhibition in tobacco leaves infected with TMV, which coincides with a rise in abscisic acid content of the leaves, might create a favorable condition for virus multiplication by reducing the competition for nucleoside triphosphates and amino acids needed for virus synthesis.

Growth regulators may also contribute to the establishment of incompatibility by altering a plant's biochemical environment in such a way that it becomes less conducive to the development of the pathogen.

In the following sections, the possible influence of IAA, cytokinins, and ethylene on the outcome of disease reaction is discussed.

### A. Indoleacetic Acid

The hypothesis that changes in the levels of IAA in infected tissue may play a role in disease reaction was originally put forth by Shaw and Hawkins (1958). This suggestion was based on the observation that in the wheat–*Puccinia graminis* system the rate of release of $^{14}CO_2$ from labeled IAA-fed tissue varied in compatible and in incompatible combinations. The rate of IAA decarboxylation (assumed to be due to the activity of IAA oxidase) initially increased in both compatible and incompatible combinations, during the early stages of infection. However, later the decarboxylation rate dropped in compatible combinations while it remained high in incompatible combinations. The IAA content of the susceptible tissue was very high during the late stages of disease development when decarboxylation rate was very low. Differences in the rate of decarboxylation of exogenous IAA were also demonstrated by Daly and Deverall (1963) and Antonelli and Daly (1966) in near-isogenic lines of wheat resistant and susceptible to the stem rust fungus.

To compare the pattern of IAA decarboxylation in resistant and susceptible tissue, Antonelli and Daly (1966) used near-isogenic lines of wheat resistant or susceptible to race 56 of the stem rust fungus. One set of lines carried the temperature-sensitive *Sr6* allele which shows incompatibility (resistance) at 20°C and compatibility (susceptibility) at 25°C. The second set of lines carried the sr6 allele which exhibits susceptibility of either temperature. With inoculated plants exhibiting compatibility at 20°C and 25°C there was an initial increase in the rate of IAA decarboxylation relative to noninoculated controls followed by a drop in the rate once the sporulation was underway. The percentage of increase in the rate of IAA decarboxylation in inoculated plants exhibiting incompatibility was essentially the same as that in plants showing compatibility during the first 2–3 days following infection. However, the rate in incompatible combination continued to increase to values approximately 8 times that of noninoculated controls, despite the absence of visible infection. It was concluded that while IAA and its metabolism might not be correlated directly with disease reaction, degradation of exogenous IAA might be influenced by mechanisms that control disease reaction.

To resolve the question of the involvement of IAA decarboxylation in disease resistance Seevers and Daly (1970b) compared the levels of peroxidases which seem to catalyze IAA decarboxylation in leaves of near-iosogenic lines of wheat carrying *Sr6* and *sr6* alleles to race 56 of the stem rust fungus. Judging from the results of the study (described in Chapter 12) it was concluded that in the above host–parasite system peroxidase activity does not appear to be related to resistance or susceptibility.

## B. Cytokinins

Treatment of plants with cytokinins is known to alter disease reaction in certain host–parasite combinations. Resistance of tomato plants to the root knot nematode, *Meloidogyne incognita,* was reduced following application of cytokinins to the roots. The change in disease reaction in cytokinin-treated infected roots was associated with increased production of giant cells (Dropkin *et al.*, 1969). Novacky (1972) showed that the development of the hypersensitive response in tobacco by *Pseudomonas pisi* is suppressed by pretreatment of leaves with cytokinins.

## C. Ethylene

The hypothesis that ethylene plays a role in resistance is based on the reported ability of this gas to induce phytoalexin formation in certain tissues and to stimulate synthesis and/or activity of a number of enzymes which are thought to participate in resistance response, including peroxidases and polyphenol oxidases, as well as β-1,3-glucanase and chitinase. The last two enzymes are capable of degrading β-1,3-glucans and chitin components of the fungal cell walls (Chapter 12).

Pegg (1976e, 1981a) has suggested a multiple role for ethylene in *Verticillium* wilt of tomato. The hormone acts as an inducer of resistance and also as a toxin synergist through interaction with the pathogen metabolites. Exposure of tulip bulbs to ethylene inhibited the synthesis of tuliposide A (a lactone with fungistatic activity) in the outer living bulb scales (Beijersbergen and Lemmers, 1978). Susceptibility of tobacco to TMV was increased by treatment with abscisic acid, ethylene, chloramphenicol, and actinomycin (Balázs *et al.*, 1973).

In some cases the involvement of ethylene in the disease may be indirect. Ethylene, which is produced in plants subjected to stress, may increase susceptibility of plants to certain pathogens by accelerating aging and senescence processes (F. B. Abeles, personal communication).

## VII. LIMITATIONS OF GROWTH REGULATOR STUDIES

While studies of growth regulators in healthy and diseased plants have been informative, some, nevertheless, have suffered from certain limitations. Recognition of these limitations is essential for a sound interpretation of the results and for designing future studies. Some of the most frequently cited limitations are discussed below:

1. In some cases the reported drop in the level of certain growth regulators in infected plants might have been due to their enzymatic breakdown and/or their inactivation during tissue extraction. As was pointed out earlier, IAA is known to be inactivated upon conjugation with aspartic acid, amino acids, and certain sugars, polysaccharides, and proteins.
2. Quantitative chemical methods employed prior to the advent of highly sensitive analytical techniques were not sensitive enough to detect minute quantities of growth regulators in plant tissues and culture filtrates.
3. In some cases, bioassay methods used to quantify growth regulators have not been specific. For example, the chlorophyll retention bioassay used by some early investigators to measure cytokinin activity was later found to be nonspecific compared to the highly specific and sensitive soybean and tobacco callus tissue bioassay (Helgeson *et al.*, 1969). Plant callus tissue bioassay tests also should be used with caution because the amount of growth is influenced by the level of sugars,

gibberellic acid, and the impurity of the test compounds (Helgeson, 1978).

4. The levels of growth regulators in diseased tissue have generally been determined at one or a few times during the course of disease development, not frequently enough to show the real pattern of changes in the level of the phytohormones. The problem associated with insufficient sampling has been discussed by Kosuge and Gilchrist (1976). They have pointed out that the results of a single analysis do not reveal metabolic changes that occur during the course of disease. Drastic temporal fluctuations in the level of growth regulators in diseased tissue have been observed in a few diseases. For example, increases in the level of IAA, indol-3-yl-acetonitrile (IAN), and myrosinase, which presumably liberates IAN from thioglucoside glucobrassicin in cotyledons of *Sinapis alba* infected with *Albuga candida* at early stages of infection were followed by a decline to the levels in noninoculated controls after 12 days (Yu, 1971).
5. In some early studies, consideration has been given to the changes in the level of a single growth regulator in diseased tissue with no regard to the changes in the levels of other growth regulators. Also, in some cases, the possibility of the interactions among different growth regulators and inhibitors has not been considered.

The limitations discussed above and a few others justify the suggestion that most of the early studies on the role of growth regulators in diseased and healthy plants should be reevaluated in the light of new knowledge. The limitations of the studies on the roles of growth regulators in plant disease have been discussed by Van Andel and Fuchs (1972), Sequeira (1973), Pegg (1976a–d), and Helgeson (1978).

## VIII. DISCUSSION

The overall results of studies on the role of growth regulators in both healthy and diseased plants have shown that hormone-mediated responses are generally not due to the activity of a single phytohormone but are the result of the interaction of a number of growth regulators. Moreover, since certain phytohormones are capable of stimulating production of other growth regulators, a change in level of a single phytohormone in response to infection may result in alterations in the level of other hormones. For example, since IAA is known to stimulate production of ethylene, some of the activities attributed to IAA might be caused by ethylene (Van Andel and Fuchs, 1972). Likewise, leaf and fruit abscission which has been attributed to abscisic acid might be triggered by ethylene which is produced in response to abscisic acid.

It is evident from the above discussion that in the study of the action of phytohormones in pathogenesis the involvement of all growth regulators and the nature of their interactions should be taken into consideration. Compared to the traditional approach of concentrating on one or a few growth regulators at a time, the integrated approach has greater potential of elucidating the role of growth regulators in pathogenesis. The need for an integrated approach to the question of hormone imbalance in diseased plants has been emphasized by Van Andel and Fuchs (1972) and by Helgeson (1978).

Studies on growth regulators have been mutually beneficial to students of the mode of action of hormones in both healthy and in diseased plants. Plant physiologists learned of the presence of gibberellins through the efforts of plant pathologists, and students of plant disease have used the knowledge obtained by plant physiologists of growth regulators to further their understanding of the role of phytohormones in disease.

Our knowledge of the role of growth regulators in pathogenesis is fragmentary and badly in need of expansion and consolidation. A better understanding of their roles can be derived only through innovative studies designed to provide answers to a number of questions. The following questions are generally considered to be particularly important:

1. Are hormonal imbalances in diseased plant essential for, or the consequence of, disease development?
2. Do growth regulators exert an influence on the biochemical activities associated with compatibility and incompatibility?
3. Are growth regulators involved in diseases not chararacterized by growth abnormalities?
4. What is the basis of disease-induced phytohormone imbalance?
5. Are changes in the levels of certain growth regulators in diseased plants due to altered levels of activators or inhibitors, to the formation of inactive complexes, or to altered biosynthesis of hormones?
6. What are the functions, if any, of certain metabolites (i.e., pathogen-produced toxins) with growth-regulator activities?
7. How do growth regulators interact with each other?
8. Do the activities of certain key enzymes change in diseased tissue as the result of growth-regulator imbalances?

The need for additional studies on the involvement of hormones in plant disease has frequently been suggested (Van Andel and Fuchs, 1972; Sequeira, 1973; Wheeler, 1975; Pegg, 1976a; Hegelson, 1978; and others).

*Chapter 11*

# Crown Gall Tumor Formation



## I. INTRODUCTION

*Agrobacterium tumefaciens* causes a disease known as crown gall in many species of dicotyledons. The disease is characterized by production of a non-self-limited, nondifferentiated neoplastic growth, to be distinguished from insect-induced galls, which are self-limiting (Braun, 1969). Two of the characteristics of crown gall tumors not shared by the normal cells are the abilities of the tumor to synthesize certain compounds known as opines and to proliferate in the axenic culture in the absence of IAA and cytokinins.

The nature of the interaction of the bacterium and the host has been the subject of intensive study in recent years, particularly during the past few years after the discovery of the involvement of bacterial plasmid DNA in the transformation process. Briefly, tumorigenic bacterial strains carry a large tumor-inducing (Ti) plasmid known as Ti-plasmid. A portion of its DNA (T-DNA) is incorporated stably into the plant DNA. Portions of T-DNA are apparently transcribed and translated. It is generally accepted that the oncogenicity is encoded by T-DNA.

The mechanism by which T-DNA is taken up by the plant cell and is integrated into plant DNA is not known. Moreover, it is not clear how and why introduction of T-DNA into the plant cell causes uncontrolled proliferation of the transformed cell in the absence of exogenously supplied phytohormones.

The intense interest in crown gall research is due to the uniqueness of the disease, its similarity to carcinogenesis in animals, and its potential for genetic engineering. In the following pages some of the features of this highly-evolved host–parasite system will be outlined briefly. Emphasis is placed on some of the recent findings regarding the nature of the tumor-inducing factor. The earlier literature on the subject of crown gall tumor formation has been reviewed by Beardsley (1972), Braun (1972), Lippincott and Lippincott (1975, 1976), and Kado (1976). For a discussion of the events during the transfer, replication, and expression of the bacterial plasmid DNA, see reviews by Merlo (1978, 1982), Drummond (1979), Van Montagu and Schell (1979), and Nester and Kosuge (1981).

## II. THE BACTERIUM

Four species of *Agrobacterium* have been recognized based on pathogenicity by Bergey's Manual (8th edition, 1974). They are *A. tumefaciens,* the causal agent of crown gall of many dicotyledons, *A. rhizogenes,* which causes hairy roots in many plants, *A. rubi,* which incites small overgrowth on a number of plants, and *A. radiobactor,* a nonpathogenic species. Pathogenicity in these species is, however, not a stable character because it is controlled by the Ti-plasmid, which can be transferred from one strain to another (see Nester and Kosuge, 1981).

Strains of *Agrobacterium* vary with respect to host range. That is, certain isolates have a narrow host range and some have a relatively broad host range. Certain lines of evidence suggest that in certain cases the host range trait is coded for by bacterial plasmids. Bacterial strains with a narrow host range acquired broad-host-range properties upon acquisition of plasmids from bac terial strains with broad host range (Loper and Kado, 1979; Thomashow *et al.*, 1980b). Conversely, narrow-host-range traits were acquired by a plasmidless bacterial strain derived from a broad-host-range strain upon acquisition of the Ti-plasmid from a narrow-host-range strain (Thomashow *et al.*, 1980b).

Pathogenicity of most of the Australian isolates of *A. tumefaciens* was initially correlated with their sensitivity to a bacteriocin produced by these isolates (Kerr and Htay, 1974; Roberts and Kerr, 1974). However, Kerr and Roberts (1976) later found many exceptions to this correlation.

## III. PHYSIOLOGY OF CROWN GALL TUMOR

### A. Phytohormone Independence

A feature which distinguishes crown gall tumors from normal tissue is the ability of the tumorous tissue to proliferate in the axenic culture in the absence of IAA and cytokinins. The exact role(s) of phytohormones in tumor development is not known. However, the following observations tend to support the hypothesis advanced by Kehr and Smith (1954) that growth regulators might be involved in tumor development. First, *A. tumefaciens* is capable of producing various hormones in cultures (Sukanya and Vaidyanathan, 1964; Upper *et al.*, 1970), and tumorous tissues contain increased levels of growth regulators (Wood, 1970; Ames and Mistretta, 1975; Dye *et al.*, 1962), auxin protectors (Stonier, 1971), and IAA conjugates (Feung *et al.*, 1976). Second, unlike normal tissue, fully transformed tumors are not stimulated by exogenously supplied cytokinins or auxin *in vitro* (Beardsley, 1972). Third, addition of auxins, cytokinins, and gibberellins to plant tissue causes an increase in tumor formation (El Khalifa and Lippincott, 1968). Finally, virulence of *A. tumefaciens* has been correlated with the amount of IAA production. Virulent strains of *A. tumefaciens* produced more IAA in the presence of tryptophan compared to avirulent Ti-plasmidless mutants (Liu and Kado, 1979). However, the mutants acquired both virulence and the ability to produce IAA at levels comparable to those of the wild type upon acquisition of the Ti-plasmid. The results show that the gene(s) necessary for the production of IAA by the bacterium is located on the Ti-plasmid, which also determines pathogenicity. Production of cytokinins, has also been correlated with virulence. The naturally occurring cytokinin, zeatin, and its derivatives, which were present in the cultures of a virulent strain (C 58) of *A. tumefaciens*, were not found in cultures of the avirulent strain of the bacterium which were cured of the plasmid (Kaiss–Chapman and Morris, 1977).

On the basis of the following observations Weiler and Spanier (1981) suggested that tumor growth cannot be due solely to increased levels of cytokinins and IAA: (1) several typical tumors formed on *Amaranthus caudatus* L. did not contain elevated levels of cytokinins; (2) cytokinin levels varied greatly in bacteria-free crown gall tumors from several species; (3) concentration of IAA did not increase in crown gall tumors on *Helianthus annus* L.; and (4) levels of gibberellins and abscisic acid in tumors and normal callus tissues were similar.

Amasino and Miller (1982) suggested that variations in tumor morphology in tobacco crown gall tumor lines may be due to the observed variations in the levels of auxin and cytokinins.

### B. Synthesis of Opines

Morel and his colleagues (Menage and Morel, 1964), and later others (Kemp, 1976, 1977, 1978; Hack and Kemp, 1977; Lippincott *et al.*, 1973; Petit and Tourneur, 1972), reported that crown gall tissues are capable of synthesizing certain compounds known collectively as opines, which are not present in normal plant tissue or in bacterial cultures. The list now includes octopine, noroctopine, octopinic acid, homooctopine, histopine, lysopine, nopoline, nopolinic acid (ornaline), agropine, and agrocinopine. Opines are placed into three families: octopines, nopalines, and agropines. The newly discovered agrocinopines may form the fourth family (Nester and Kosuge, 1981).

The number and species of opines synthesized by the crown gall cells seem to be determined by the bacterial strain, not the infected plant species. Petit *et al.* (1970) have shown that, in general, strains of *A. tumefaciens* capable of degrading certain opines, formed tumors that synthesized the same opines. This relationship was later examined by Kemp (1978), who found that crown gall tumors incited by *A. tumefaciens* strains that utilize octopine as a sole source of carbon or nitrogen for growth synthesize octopine, histopine, lysopine, and octopinic acid. Tumors incited by nopaline-utilizing bacterial strains synthesized nopaline and ornaline. A normal tissue culture and a crown gall culture incited by a strain of the bacterium unable to utilize either octopine or nopaline did not synthesize any of the opines. However, some exceptions have been found to the correlation between octopine and nopaline synthesis and utilization (Kerr and Roberts, 1976). For instance, some mutants of the bacterium which cannot degrade octopine or nopaline induced tumors capable of synthesizing normal levels of the two opines (Montoya *et al.*, 1977).

Opine synthesis does not seem to be initiated by wounding or infection. For example, octopine accumulated in crown gall tissue on Jerusalem artichoke and carrot 1–3 weeks after morphological appearance of tumors (Toothman, 1982).

The synthesis (Montoya *et al.*, 1977) and catabolism (Chilton *et al.*, 1976; Watson *et al.*, 1975) of octopine and nopaline are coded by bacterial plasmids.

## IV. TUMOR DEVELOPMENT

Tumor development has arbitrarily been divided into three distinct stages: (1) the wounding of host tissue and subsequent attachment of bacteria to wound sites (conditioning), (2) the transfer of bacterial plasmid DNA to conditioned host cells and its processing (transformation), and (3) the uncontrolled proliferation of the transferred host cells (tumor development).

### A. Attachment of the Bacterium to Plant Cell—Wound Requirement

The attachment of the bacterium to wound sites during the conditioning phase is a necessary prerequisite for transformation of plant cells to a tumorous state (Lippincott and Lippincott, 1969; Lippincott *et al.*, 1977; Glogowski and Galsky, 1978). Lippincott and Lippincott (1975) have suggested that wounding, which is required for tumor initiation, exposes specific sites on the host cell wall for such attachment. Attachment sites can be competitively occupied by avirulent bacterial strains (Lippincott *et al.*, 1969) and by bacterial cell envelopes (Beiderbeck, 1973).

In addition to the wound's providing attachment sites for bacterium, wound sap may contain certain compounds capable of stimulating cells adjacent to wound sites to divide, thus accelerating the wound-healing process. Klein (1965) found that tumor formation could be inhibited if wound sap was removed by washing prior to inoculation or during the first 6 hr thereafter. He suggested that wound sap may function primarily as a stimulator of cell division and cell conditioning. Wound sap may also contain substances which may serve as sources of nutrients for the bacterium. The greater ability of auxotrophic mutants of *A. tumefaciens* to form tumors in large wounds was attributed to the greater availability of nutrients for the bacteria in large wounds (Lippincott and Lippincott, 1969). Lippincott and Lippincott (1965) have suggested that tumor size and the rate of its development are directly correlated to wound size when other conditions are not limiting. Merlo (1978) has discussed the possibility that wound sap may stimulate formation of sex pili of the bacterium. Such sex pili, which are involved in transfer of DNA between gram-negative bacteria, may also function in plasmid DNA transfer from *Agrobacterium* to the host cells.

The healing process initiated following wounding involves resumption of meristematic growth. Wilson (1978) has suggested that wound-induced meristematic growth facilitates communication for the multiplication and dissemination of the tumor-inducing bacterial plasmid DNA.

In some cases, binding of bacteria to host cell wall has been found to be a plasmid-borne trait. For example, certain bacterial strains lacking the Ti-plasmid did not bind to plant cell walls. Binding in these strains was observed upon acquisition of a Ti-plasmid (Whatley *et al.*, 1978). On the other hand, two avirulent strains of the bacterium without the Ti-plasmid were capable of binding (Whatley *et al.*, 1976). It is therefore possible, depending on the bacterial strain, that genes which control binding are carried on either plasmid or nuclear DNA.

It has been suggested that the lipopolysaccharide portion of the bacterial cell envelope is involved in site attachment (Whatley *et al.*, 1976; Lippincott and Lippincott, 1977). The nature of this surface interaction will be discussed in Chapter 14.

### B. The Nature of the Tumor-Inducing Principle

Unlike the conditioning phase, the transformation phase does not require the presence of the bacterium. This observation was the basis for the hypothesis that tumor induction was mediated by a bacterial component called the tumor-inducing principle (TIP) (Braun and White, 1943). For several years, the identity of TIP was the most controversial subject in crown gall research. Since transformed plant tissue is very stable, it was logically deduced that the TIP must have self-replicating properties. During the early and mid-seventies, bacterial nuclear DNA was the prime suspect and became the subject of intense research by many investigators (Drlica and Kado, 1974; Chilton *et al.*, 1975; Kado and Lurquin, 1975; Gardner and Kado, 1977). However, the overall results did not support the hypothesis that bacterial nuclear DNA was involved in transformation.

Spurred by the reports of the consistent presence of large plasmids in all tumorigenic strains of *A. tumefaciens* reported by Zaenen *et al.* (1974), Schell *et al.* (1974), Van Larebeke *et al.* (1974), and others, attention was focused on the bacterial plasmid as a source of TIP. Evidence for the involvement of bacterial plasmid DNA in pathogenesis was later strengthened by two concurrent reports that tumorigenic strains of the bacterium lost their tumor-inducing ability when cured of their large plasmids (Van Larebeke *et al.*, 1975; Watson *et al.*, 1975). Moreover, avirulent strains became virulent upon acquisition of the missing plasmid through conjugation with virulent strains (Chilton *et al.*, 1976; Hooykaas *et al.*, 1977; Lin and Kado, 1977; Van Larebeke *et al.*, 1975; Watson *et al.*, 1975). Other lines of evidence for the involvement of plasmid DNA are the reported stable incorporation of a portion of DNA from a tumor inducing (Ti) plasmid (T-DNA) into crown gall tissue (Chilton *et al.*, 1976, 1977, 1980; Matthysse and Stump, 1976; Merlo *et al.*, 1980; Lemmers *et al.*, 1980; *Yang et al.*, 1980; Thomashow *et al.*, 1980a, c; Yadav *et al.*, 1980), and transcription of the tumor-inducing plasmid DNA of *A. tumefaciens* in sterile crown gall tumor cells (Ledeboer, 1978; Drummond *et al.*, 1977; Chilton *et al.*, 1977; Nester, 1979; Willmitzer *et al.*, 1980; Gelvin *et al.*, 1981).

It is now known that all pathogenic strains of *Agrobacterium* carry a large Ti-plasmid. Moreover, transformation of normal cells to tumor cells requires the stable incorporation of T-DNA into plant DNA.

## V. RELATIONS AMONG BACTERIAL PLASMIDS

Classification of Ti-plasmids is based on the ability of plasmids to code for catabolism of opines, unusual compounds synthesized only by tumorous tissue (Nestor and Kosuge, 1981). For example, octopine plasmids code for degradation of octopine and related substances and nopaline plasmids code for catabolism of nopaline and related compounds. Plasmids from *Agrobacterium* have been placed into incompatibility groups. Octopine and nopaline plasmids belong to the first group (Rh-1), the agropine plasmids are placed in the second group (Rh-2), and plasmids responsible for hairy root disease caused by *A. rhizogenes* form the third group (Rh-3). Despite the presence of these incompatibilty groups, results of studies on the relationships among plasmids point to the possibility that they have evolved from a common plasmid (Nester and Kosuge, 1981). For example, several wild-type strains of *A. rhizogenes* belonging to biotypes 1 and 2 were shown to carry at least one plasmid comparable in size to the Ti-plasmid of *A. tumefaciens*. Also the plasmids from various strains of *A. rhizogenes* shared an extensive sequence (Costantino *et al.*, 1981). However, the close relatedness originally suggested for the octopine plasmids of *A. tumefaciens* was not supported when additional strains of the bacterium were examined (Nester and Kosuge, 1981).

## VI. MECHANISM OF TRANSFORMATION

Because of its complexity, crown gall has been a constant source of challenge for scientists interested in unveiling its mystery. Although the mystery regarding the nature of the tumor-inducing factor has been solved through recognition of bacterial plasmid DNA, the very solution of the problem has created additional challenging questions regarding the nature of transfer of a portion of plasmid DNA to plant cells, its replication, and its translation. Does the bacterium enter the plant cell? If not, how does the bacterial plasmid DNA enter the plant cell? Does the plant cell receive only a single copy of the bacterial plasmid DNA? How is bacterial plasmid DNA maintained in plant cells? If bacterial DNA is translated, where in the cell do transcription and translation take place?

In the following section some of these questions will be discussed. These questions have also been addressed by a number of reviewers cited earlier.

### A. Acquisition of Bacterial Plasmid DNA

The mechanisms of acquisition, replication, and transcription of bacterial plasmid DNA are not well understood. The subject is currently being investigated vigorously in a number of laboratories. It is generally accepted, but not unequivocally proven, that intact bacterial cells do not enter the host cells following their attachment to the wound site. If this assumption is correct, bacterial plasmid DNA is somehow released and transferred to the plant cells by an unknown mechanism. A number of hypothetical mechanisms discussed by Merlo (1978) are considered here. First, the bacterial cells may be lysed, probably by lysogenic phages, following contact with plant cells, releasing their DNA. Lysogenic phages have been detected in homogenates of crown gall tissue by Parsons and Beardsley (1968). Second, pili which participate in DNA transfer between gram-negative bacteria may also participate in DNA transfer between *Agrobacterium* and plant cells. Merlo (1978) has speculated that the pili might be formed in response to a plant stimulus. Third, bacterial DNA may be transferred through conjugation. The conjugation hypothesis is supported by the observation that conjugal transfer of Ti-plasmids between two bacterial cells is promoted by genes located on Ti-plasmids (Genetello *et al.*, 1977; Kerr *et al.*, 1977). Moreover, both conjugation and the early steps in tumor initiation are inhibited at 30°C (Tempé *et al.*, 1977). Conjugation hypothesis for plasmid DNA transfer should, however, be balanced against reports that virulence of *Agrobacterium* strains does not appear to be correlated with their ability to conjugate (Holsters *et al.*, 1980; Koekman *et al.*, 1979). While the transfer of DNA between procaryotes and eucaryotes could be accomplished by the same process which mediates transfer of DNA between two procaryotes such as conjugation and pili formation, the DNA transfer from *A. tumefaciens* to plants might be mediated by a unique process not yet identified (see Merlo, 1978).

Marton *et al.* (1979) and Wullems *et al.* (1981a,b) have developed an *in vitro* transformation system in which Ti-plasmid DNA can be introduced into cell-wall-regenerating plant protoplasts by *A. tumefaciens*. This technique was said to facilitate biochemical studies of a large number of tumor lines obtained with different strains of the bacterium.

## B. Incorporation of T-DNA into Plant DNA

Results of a number of studies show that a portion of the T-DNA is incorporated stably into plant nuclear DNA (Chilton *et al.*, 1976, 1977, 1980; Matthysse and Stump, 1976; Merlo *et al.*, 1980; Thomashow *et al.*, 1980a,c; Lemmers *et al.*, 1980; Yadav *et al.*, 1980). However, integration of T-DNA into plant DNA does not occur at a specific site (Yadav *et al.*, 1980; Thomashow *et al.*, 1980a,c). While different tumor lines thus far studied contain different amounts of T-DNA, all of these lines contain a ''common DNA'' sequence (Lemmers *et al.*, 1980), which is highly conserved in different Ti-plasmids (Chilton *et al.*, 1978; Depicker *et al.*, 1978). The ''common DNA'' sequence has been incorporated into plant DNA in all tumors studied thus far and most likely functions in oncogenicity and maintenance of the tumorous state. Analysis of cloned DNA segments isolated from an octopine-producing crown gall tumor (A65/2) showed that T-DNA was covalently joined to plant nuclear DNA (Thomashow *et al.*, 1980a,c).

The involvement of T-DNA in oncogenicity is supported by the observation that the virulence of the bacterium is affected by additions and deletions within the T-DNA (Koekman *et al.*, 1979; Garfinkel and Nester, 1980). Evidence for the involvement of the "common DNA" sequence of T-DNA in oncogenicity and tumor maintenance also has been obtained by Yang and Simpson (1981). They found that the loss of tumorous traits (phytohormone independence and nopaline production) in revertant tobacco seedling tissue derived from crown gall tumors containing nopaline-type Ti-plasmid was associated with the loss of most, but not all, of the T-DNA sequence present in the parental tumors. "Southern blot" analysis showed that revertant tissues had lost the control region of the T-DNA-containing "common DNA" sequence.

It is not yet known whether plasmid or plant genes code for the incorporation of T-DNA into plant DNA.

### C. Transcription of Incorporated T-DNA in Tumors

The T-DNA portion of the Ti-plasmid which is stably incorporated into tumors is transcribed in a number of crown gall tumor lines thus far studied (Drummond *et al.*, 1977; Ledeboer, 1978; Gurley *et al.*, 1979; Gelvin *et al.*, 1981). To locate the regions of the Ti-plasmid which are transcribed in the bacteria and in tobacco crown gall tumors, Gelvin *et al.* (1981) compared patterns of RNA synthesis from the T-DNA in the bacterium and in tumors. They found that large portions of Ti-plasmid were transcribed when the bacterium was grown in minimal or rich medium. Additional regions of the plasmid were transcribed when the bacterium was exposed to octopine and agropine. In contrast to the bacterium, in which all portions of the T-DNA were weakly transcribed, only specific regions of T-DNA were transcribed in tumors. According to Gelvin *et al.* (1981) selective transcription of T-DNA in the tumor compared to that in the bacterium may indicate that T-DNA has evolved to function best in eucaryotic cells.

## VII. IDENTIFICATION OF PLASMID AND CHROMOSOMAL GENES INVOLVED IN ONCOGENICITY

Attempts have been made to identify different plasmid and chromosomal genes involved in oncogenicity. The approach has been to insert different transposons and the plasmid $RP_4$ into the test plasmids and to characterize the phenotypes of the resulting mutants at insertion sites (Holsters *et al.*, 1980; Klapwijk *et al.*, 1980; Garfinkel and Nester, 1980). Results of these studies have revealed that induction and maintenance of tumors require a number of regions in both plasmid and bacterial DNA. For example, Garfinkel and Nester (1980) isolated 37 mutants which differed in tumorigenicity following the introduction of Tn5 into the DNA of the bacterium carrying the octopine plasmids, A6 and $B_6$ 806. Of these mutants, 25 were mapped on the plasmid and 12 on the chromosome. Most of the plasmid mutants and some of the chromosomal mutants were avirulent, indicating that both plasmid and chromosomal genes function in tumorigenesis.

Some of the characters coded by Ti-plasmid genes include the ability of the tumors to synthesize opines (Klapwijk *et al.*, 1976; Montoya *et al.*, 1977), utilization of octopine and nopaline by *Agrobacterium* isolate as the sole source of carbon and nitrogen (Bomhoff *et al.*, 1976; Montoya *et al.*, 1977), host range (Loper and Kado, 1979; Thomashow *et al.*, 1980b), sensitivity of certain isolates of *Agrobacterium* to Agrocin-84, a bacteriocin produced by *A. radiobacter* (Engler *et al.*, 1975; Kerr and Roberts, 1976), and conjugal transfer of the Ti-plasmid between bacteria (Kerr *et al.*, 1977; Hooykaas *et al.*, 1977). One function which has been attributed to the chromosomal genes is bacterial adherence to the plant cell wall, which is presumably a prerequisite for tumor initiation (Whatley *et al.*, 1978).

## VIII. TUMOR REVERSAL

Tumor cells can partially or totally recover from the tumorous state. Normal-appearing reverted plants regenerated from teratoma-type tumors were earlier reported to retain both T-DNA (Yang *et al.*, 1980; Lemmers *et al.*, 1980) and tumorous traits (phytohormone independence and production of opines) (Braun and Wood, 1976) prior to undergoing meiosis. However, tumorous traits were reported to be lost completely following meiosis (Turgeon *et al.*, 1976) and the tissue contained no detectable T-DNA (Lemmers *et al.*, 1980; Yang *et al.*, 1980). More recently Yang and Simpson (1981) were able to induce normal revertant shoots from BT37 crown gall teratoma containing nopaline-type Ti-plasmid by treating the cloned tobacco tumor tissue with kinetin. The shoots could be induced to form roots and to produce viable seeds. Unlike revertants described by others, tissues derived from revertant seedlings lost tumorous traits (phytohormone independence and the ability to produce nopaline). The loss of the tumorous traits in the revertant was accompanied by the loss of most, but not all, of the T-DNA sequence present in the parental tumor tissue. However, both the reverted plants and plants derived from seeds contained sequences homologous to the ends of the T-DNA of the original tumor. Retention of DNA markers during meiosis was also demonstrated by Wullems *et al.* (1981a). They found that shoots from both octopine- and nopaline-type tumors grafted into normal tobacco plants produced flowers and seeds. Plants developed from these seeds retained morphological markers of their parental shoots. Moreover, the nopaline-synthesizing trait was retained by one seedling. Retention of a portion of bacterial T-DNA in reverted plants following meiosis and seed formation is significant since it makes it possible to use Ti-plasmid as a vector for introducing genes into plants that are propogated either sexually or vegetatively (Yang and Simpson, 1981).

The molecular basis for the reversion of tumorous tissue to normal tissue is not known. Since in many cases all or part of the T-DNA is present in reverted tissue, reversion may be the result of repression of tumor maintenance genes (Merlo, 1978).

*Chapter 12*

# Mechanisms of Disease Resistance



## I. INTRODUCTION

During the course of their coevolution, plants and pathogens have evolved a highly complex and intricate relationship. Pathogens have developed sophisticated offensive systems to parasitize plants and plants, rather than being an easy prey, have in general exhibited a remarkable potential to defend themselves against the onslaught of pathogens. Some plants are capable of preventing certain pathogens from penetrating their walls. Others seem to unleash their defensive weapons shortly after their walls are breached and a few of their cells are invaded.

While plant resistance to disease is the rule and not the exception, a few parasites are capable of killing their susceptible hosts. In general, the outcome of the host–parasite struggle seems to depend upon the efficacy of the plant's defense. A plant may become diseased when a parasite manages to invade it, unhindered by preexisting defense system(s) and/or without eliciting the plant's induced resistance response(s). Absence of disease may reflect the inability of the invading pathogen to overcome the plant's defense system(s).

Man has not remained indifferent to the outcome of the host–parasite struggle since his survival is largely dependent upon the health and productivity of crops. In an attempt to tip the balance in favor of the plant, man has selected, bred, and propagated disease-resistant lines of certain crops and tried to eradicate certain potential pathogens by chemical and cultural means. The current surge of interest in the study of the nature of plant disease resistance is also based on a firm conviction that effective management of plant diseases is, for the most part, contingent upon a clear understanding of the nature of plant disease resistance. The knowledge thus obtained would be used to manipulate plants and pathogens in such a way as to reduce crop losses to disease.

Results of numerous studies on resistance mechanisms show that plants defend themselves against infectious agents both actively and passively, and that active defense involves mobilization of a number of defensive strategies in a highly coordinated manner. To facilitate appreciation of the nature of defense systems, resistance factors are commonly divided into preformed (passive) and infection-induced (active) categories. Preformed resistance factors include those which are present in plants prior to their contact with the pathogens. Infection-induced resistance factors comprise those which are either absent or present at low levels before infection and are produced or activated upon infection. Separating resistance factors into the above two categories is rather arbitrary. For example, flavonols and glycosides could be placed in either category.

## II. PREFORMED RESISTANCE FACTORS—STRUCTURAL FEATURES

Structural features have been implicated in resistance but their importance has not been established. The purported roles of cuticles, stomates, and root hairs in resistance will be discussed briefly.

### A. Cuticles

A direct correlation between cuticle thickness and disease resistance has not been clearly demonstrated. This is probably because small cuts and fissures on the cuticular layers may provide points of entry for parasites. Moreover, some fungi seem to be capable of degrading cuticular layers by their cutin-degrading enzymes (Chapter 2).

Certain properties of the cuticle, such as wettability, may alter the rate of infection. For example, because of the tendency of leaf surfaces to repel water, wettable spores of certain fungi carried in water droplets are not retained on the leaf surfaces. Apple fruits inoculated with *Venturia inaequalis* showed more scab lesions on the red side than on the green side (Fisher and Corke, 1971). The difference in the amount of infection between the green and red sides was attributed to differences in wettability of the two surfaces. The thicker cuticle of the green side could not easily be wetted because of the presence of more surface wax and, therefore, could not readily retain drops of spore suspension.

The cuticles of certain plants contain materials capable of stimulating or inhibiting the growth of certain pathogens (Royle, 1976; Campbell *et al.*, 1980). While antifungal substances of cuticular origin are generally nonselective, selectivity has also been demonstrated. Germination of fungal spores of a nonpathogen and a weak pathogen of chrysanthemum was found to be inhibited more strongly than germination of a pathogen by materials extracted from the cuticle of the plant (Blakeman and Atkinson, 1976). Sporangia of *Peronospora hyoscyami* f. sp. *tabacina* germinated more readily on leaves of *Nicotiana debneyi* than on leaves of *N. tabacum*. The difference was attributed to the presence of an inhibitor of spore germination in the cuticle of *N. tabacum* but not in that of *N. debneyi* (Cruickshank *et al.*, 1977). See the discussion by Royle (1976) for additional examples.

## B. Stomates

The structure and behavior of stomates have been implicated as factors contributing to resistance to certain diseases (Rich, 1963). Daub and Hagedorn (1979), who studied the role of stomates in the resistance of bean plants to *Pseudomonas syringae,* concluded that stomatal behavior in the above host–parasite system could not contribute to resistance. It was pointed out that stomatal apertures in this host were so large, compared to the size of the bacterium, that even small openings were sufficient to allow bacterial ingress (Chapter 2).

## C. Plant Hairs

The possible contribution of plant hairs to resistance has also been considered. Plant hairs contain certain inhibitors (Bailey *et al.*, 1974) which may play a role in plant defense. Presently, however, there is no direct evidence for the involvement of plant hairs in resistance.

The literature on the role of structural features in resistance has been reviewed by Royle (1976), Campbell *et al.* (1980), and Akai and Fukutomi (1980). Royle (1976) found little evidence that structure alone can protect plants against pathogen invasion. However, he added that the results of the most careful studies have provided strong indications that structural features may sometimes confer at least a background level of resistance. He therefore emphasized the need for additional research before drawing firm conclusions.

Structural features of plant surfaces may contribute to resistance in a more sophisticated fashion than those that have been visualized thus far. For example, penetration of a pathogen may be affected by factors such as configuration of surface cuticular wax, chemical nature of structural barriers, characteristics of guard cells, or topography of leaf surfaces, rather than by the thickness of the cuticle and the degree of stomatal opening (Royle, 1976).

In many cases, the contribution of structural features to plant defense may be limited to causing a delay in parasite penetration. The delay may have no or only a little influence on the outcome of disease in certain host–parasite systems. In certain other systems the delay may contribute significantly to plant defense by allowing time for the induction of a more sophisticated and effective resistance response.

## III. PREFORMED RESISTANCE FACTORS—PREFORMED ANTIMICROBIAL SUBSTANCES

A number of preformed substances with antimicrobial properties have been implicated in resistance. These materials are generally present at relatively high concentrations in healthy plants, and in some cases are converted into more potent toxins as a result of infection. The roles of preformed substances in resistance have been discussed by Schönbeck and Schlösser (1976), Overreem (1976), Schlösser (1980), and Weinhold and Hancock (1980).

### A. Catechol and Protocatechuic Acid

Resistance of colored onions to *Colletotrichum circinans* has been attributed to the presence of catechol and protocatechuic acid (Fig. 29) (Walker and Stahmann, 1955).

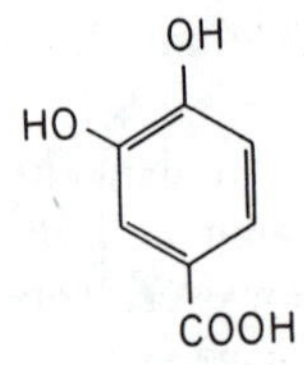


**Figure 29.** Protocatechuic acid.

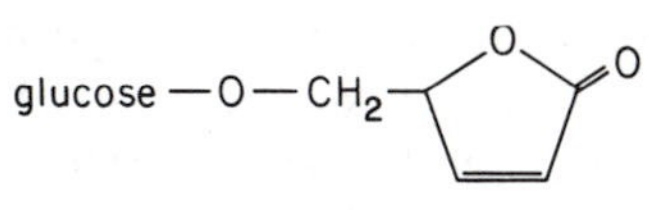


**Figure 30.** Proposed structure for ranunculin. [Hill and Van Heyningen. 1951. *Biochem.* **49:**332–335.]

## B. α-Tomatine

Tomato pathogens generally have been reported to be less sensitive to α-tomatine, a glycosidic alkaloid found in tomatoes and potatoes, than nonpathogenic fungi (Arneson and Durbin, 1968). Tomatine has been implicated in resistance of tomatoes to *Corticium rolfsii* (Schlösser, 1975). Moreover, *Septoria lycopersici,* a pathogen of tomato, was capable of detoxifying α-tomatine (Arneson and Durbin, 1968). However, Drysdale and Langcake (1973) consider it unlikely that tomatine is involved in resistance of tomato to *Fusarium oxysporum* f. sp. *lycopersici*.

## C. Unsaturated Lactones

Unsaturated lactones, known to be toxic to both plants and microorganisms, occur in plants principally as glucosides and are released by the action of β-glucosidase after wounding. Fungitoxic lactones have been isolated from tulips (Bergman and Beijersbergen, 1968), members of the Ranunculaceae family (Schönbeck and Schlösser, 1976), and cultivated avocados (Zaki *et al.*, 1980). Lactones with antimicrobial and phytotoxic properties were also isolated from two plant-pathogenic fungi, *Cephalosporium gregatum* and *C. gramineum* (Kobayashi and Ui, 1979). The best-known lactones from plants are ranunculin (Fig. 30) and tuliposides (Fig. 31). Tuliposides have been implicated in resistance of tulip pistils to *Botrytis cinerea* (Schönbeck and Schröeder, 1972) and of tulip bulbs to *Fusarium oxysporum* f. sp. *tulipae* (Bergman, 1966).

## D. Sulfur Compounds (Leek Oils and Mustard Oils)

Glucoside esters of isothiocyanic acid (mustard oils) are found in certain plants, particularly in the cabbage family. Enzymatic hydrolysis of mustard oil

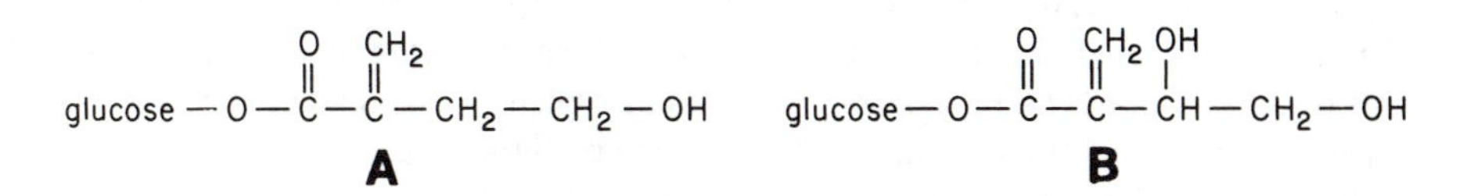


**Figure 31.** Proposed structures for tuliposide A (A) and tuliposide B (B) [Tschesche *et al.*, 1968. *Tetrahedron Lett.* **6:**701–706.]

glucosides by myrosinase yields isothiocyanates. The most important representative of leek oils is alliin (*S*-allyl-L-cysteine-sulfoxide), which is converted enzymatically first into allicin (diallyl-disulfide-oxide) and finally to diallyl-disulfide, the major component of garlic oil (Schönbeck and Schlösser, 1976). Isothiocyanates and allicin possess antibacterial and antifungal properties but their roles in resistance have not been firmly established (Schönbeck and Schlösser, 1976).

## E. Extruded Materials

Many plants extrude organic substances from their roots (Rovira and Davey, 1974; Hale *et al.*, 1978) and leaves (Tukey, 1970). Some of the extruded materials possess both stimulatory and inhibitory properties. Because these substances are available in virtually all infection courts, their potential antimicrobial activities and their role in resistance have been studied by many investigators (see the review by Weinhold and Hancock, 1980). These substances are considered to contribute to plant defense by acting directly against pathogens or by stimulating activities of organisms that are antagonistic to the pathogens. Moreover, the competitive ability of nonpathogens to utilize extruded materials as a source of food can limit the growth of certain pathogens that require these materials for the same purpose prior to infection. Despite the lack of direct evidence, the possibility that extruded materials may play a role in resistance cannot be ruled out.

Like plants, resident microflora also may extrude antimicrobial substances (Blakeman, 1971). Some nontoxic microbial products may interfere with the growth and development of pathogens by chelating, and thus removing, essential elements from the environment. For example, Kloepper *et al.* (1980) and Misaghi *et al.* (1980, 1982) showed that fluorescent pigments produced by fluorescent pseudomonads inhibited the growth of certain bacteria and fungi *in vitro* by chelating and removing iron from the media.

## F. Plant Glucanases and Chitinases

Glucanases and chitinases found in plants (Grassman *et al.*, 1934; Powning and Irzykiewicz, 1965; Abeles *et al.*, 1970) may play a role in resistance by degrading glucan and chitin components in the cell walls of fungal pathogens. Lysis of *Verticillium albo-atrum* observed in tomato plants (Dixon and Pegg, 1969) may be due to autolysis or the activity of tomato chitinase and β-1,3-glucanase (Pegg and Vessey, 1973). Nichols *et al.* (1980) also have shown that pea endocarp tissue contains a number of glycosidic enzymes including chitinase and chitosanase. Chitinase activity was higher in pea pods challenged with *Fusarium solani* f. sp. *phaseoli* and *F. solani* f. sp. *pisi* than in those of the noninoculated control group 0.5–6 hr after inoculation. In this connection it is interesting to note that the synthesis of β-1,3-glucanase and chitinase in bean leaves is stimulated by ethylene (Abeles *et al.*, 1970) produced in plants in response to infection and injury (Abeles, 1973). The literature on the role of glucohydrolases in resistance has been reviewed by Pegg (1977).

In summary, it is generally agreed that the preformed substances discussed above may contribute to general resistance. However, in many cases the evidence for their involvement in resistance is not very strong. Wood (1967) has proposed a set of criteria which must be met before a particular substance is implicated in resistance. The suspected chemical must be present in effective concentrations in those parts of the plant invaded in a form available to the pathogen. Moreover, the case for the involvement of a particular substance in resistance would be strengthened if it could be demonstrated that a change in the concentration of the chemical in the plant tissue parallels a change in disease reaction.

## IV. INFECTION-INDUCED RESISTANCE FACTORS—PHYSICAL FEATURES

In addition to the preformed defense mechanisms discussed above, plants possess a variety of other physical and chemical defense systems which are activated in response to infection. Physical changes in plants in response to injury and infection occur at both cellular and subcellular levels and reflect highly complex and coordinated events that begin soon after parasite encounter. Whether or not the physical changes contribute to resistance by preventing infection has been debated. It has been suggested that in some cases the physical barriers might only retard the development of the fungus to allow the time required for activation of nonphysical defense mechanisms such as hypersensitive response and phytoalexin accumulation. The nature and significance of some of the physical features which are considered to play a role in post-infectional resistance will be discussed briefly. This subject has recently been reviwed by Beckman (1980).

### A. Periderm and Callose

Resistance of papaya fruit to *Colletotrichum gloeosporiodes* has been attributed to periderm formation and callose deposition (Stanghellini and Aragaki, 1966). Callose and lignin-type materials have also been detected in plant tissues showing virus-induced hypersensitive necrosis (Kimmins, 1977). Formation of callose in plants in response to injury by biotic and abiotic agents may constitute a defense and/or a repair response. Callose deposits are impregnated with phenolic substances and thus may act as both chemical and physical barriers against the invading microorganisms.

## B. Gels and Tyloses

Resistance of banana to *Fusarium* wilt was attributed to the ability of resistant plants to produce gels and tyloses that prevent translocation of the fungus spores throughout the plant (Beckman, 1964). The rapidity of tylosis formation has been related to the varietal resistance of many plants to vascular wilt (Beckman, 1964). Resistance of cotton to *Verticillium albo-atrum* was attributed to the reduced systemic spread of fungus spores in plants due to a rapid occlusion of xylem vessels by tyloses followed by the synthesis of fungitoxic terpenoid aldehydes (Mace, 1978). Despite these reports, the role of gels and tyloses in resistance to vascular wilt pathogens remains undefined.

## C. Cell Wall Modifications

Cell wall modification is common in plants infected by fungi, bacteria, and viruses (Akai, 1959; Akai *et al.*, 1968, 1971; Goodman *et al.*, 1977; Sequeira *et al.*, 1977; Kimmins, 1977; Politis and Goodman, 1978; Sequeria, 1978). While the nature and the significance of cell wall modifications are not clearly understood, they generally are considered to be part of a complex series of defense mechanisms against pathogens. The literature on the roles of papillae, lignification, and haloes in resistance is reviewed briefly.

### *1. Papillae*

Apposition of substances on the inner surfaces of the plant cell wall is a common response of living plant cells to stress. The process involves the aggregation of host cytoplasm around the infection site with continuing deposition of wall materials and other unknown substances. Wall appositions formed directly beneath fungal penetration pegs are called papillae. Papilla formation is very common, particularly in several species of gramineae (Lupton, 1956; Bushnell and Bergquist, 1975; Aist, 1976a,b). Papillae, which generally appear as massive, dome-shaped structures, are considered by some to interfere with fungal penetration. However, the results of the studies on this subject are debatable and subject to different interpretations by different workers. Aist (1977) found that the wall appositions (papillae) induced in kohlrabi root hairs by mechanical wounding (bending) prior to inoculation with zoospores of *Olipidium brassicae* were effective in preventing infection by the fungus. However, in other studies (Aist and Israel, 1977a,b) certain penetration failures of the fungus could not be attributed to fungus-induced wall appositions, even though such failures were associated with large hemispherical appositions formed in advance of penetration tubes. Differences between fungus- and mechanically induced wall appositions were thought to be due to differences in the chemical composition of the wall appositions. In reed canary grass inoculated with *Helminthosporium catenarium,* penetration occurred through less than 2% of the papillae (Vance and Sherwood, 1976), while in barley and wheat inoculated with *Cochliobolus sativus* all penetrations were associated with papillae (Huang and Tinline, 1976). Sherwood and Vance (1980) studied the epidermal resistance of nonhost grasses to fungi that are pathogens of other species and found an association between cycloheximide-induced inhibition of wall thickening beneath the infection sites and cycloheximide-induced promotion of wall penetration. However, in these and similar studies the possible effect of cycloheximide on processes other than cell wall modifications cannot be ruled out. To clarify the role of papilla formation in the compatible barley–*Erysiphe graminis* f. sp. *hordei* interaction, Waterman *et al.* (1978) increased papilla formation in barley coleoptiles by centrifugation. They inferred that papillae are not responsible for the penetration failures that occur in the above host–pathogen system, but the enhancement of a cytoplasmic response such as papilla formation can potentially prevent penetration. In the same plant–pathogen combination, Aist *et al.* (1979) found that, unlike normal papillae, preformed and oversized papillae produced experimentally could prevent entry of the fungus into the host. To determine whether the functional difference between the two types of the papillae is due to variations in their physical properties, Israel *et al.* (1980) determined the elastic properties of the effective papillae, ineffective papillae, and contiguous cell wall *in vitro,* using acoustic microscopy. Acoustic microscopy allows visualization of certain properties (elasticity, density, and viscosity) of selected structures *in situ* during exposure to repeated stress waves (Maugh, 1978). Results of the study showed that papillae that were effective in preventing fungal penetration were acoustically more active (had greater elastic strength) than the ineffective papillae and contiguous area of the cell wall. The instrument used could not generate quantitative data. Judging from the results of the study, Israel *et al.* (1980) concluded that certain wall appositions could function in disease resistance by providing a viscoelastic barrier to the physical forces employed by the penetration structures of fungal pathogens.

The overall results of studies on papillae show that, except for a few cases where barriers to penetration may impede or even prevent fungal penetration, there is no unequivocal evidence that these structures play a major role in resistance. However, according to J. R. Aist (personal communication) a great deal of suggestive evidence is available, indicating that papillae are important.

Despite numerous investigations (Hirata, 1967; Edwards, 1970a; Stanbridge *et al.*, 1971; Kunoh and Ishizaki, 1976; Mayama and Shishiyama, 1976a, b), the exact chemical nature of papillae is not known. Mayama and Shishiyama (1978) studied localized accumulation of fluorescent- and ultraviolet-absorbing compounds at penetration sites in barley leaves infected with *Erysiphe graminis hordei* and concluded that the accumulated compounds might be polyphenols. In general, papillae are known to contain callose and possibly lignin, silicon, suberin, and cellulose (Aist, 1976b; Heath, 1980b). Further studies are needed to clarify the physical and chemical properties of papillae.

To verify the role of papillae in resistance it is worthwhile to establish whether or not the physical and chemical characteristics of papillae that are effective in preventing successful penetration are different from those of papillae that are ineffective.

The literature on the role of papillae in resistance has been reviewed by Aist (1976) and more briefly by Wheeler (1975) and Heath (1980b). The significance of cell wall modifications, induced in plants in response to infection by incompatible bacteria, is discussed later in this chapter.

### 2. *Lignification*

Lignification, which occurs in degenerating cytoplasm, in extracellular deposits, and in cell walls, may interfere with fungal growth in a number of ways (Ride, 1978; Ride and Pearce, 1979; Vance *et al.*, 1980). It may interfere with the translocation of water and nutrients from plant tissue to the fungus and with the movement of toxins and enzymes from the fungus to the plant. It may render walls more resistant to fungal penetration and to dissolution by fungal enzymes. Fungal penetration may be stopped due to lignification of hyphal tips, and low-molecular-weight phenolic precursors of lignin may inactivate certain fungal metabolites. Finally, lignification and other cell wall modifications may decelerate fungal development allowing phytoalexins to accumulate to effective levels. Phytoalexins, on the other hand, may retard fungal growth, providing enough time for cell wall modifications to become effective. Ligninlike polymers are synthesized in cell walls of young tomato fruits infected with *Botrytis cinerea* and may be responsible for the restriction of fungus development to a few epidermal cells after penetration (Glazener, 1982). Lignification has been suggested as a possible mechanism of virus localization in local lesion hosts (Kimmins and Wuddah, 1977). The literature on lignification as a mechanism of disease resistance has recently been reviewed by Friend (1976, 1977), Grisebach (1977), Ride (1978), Asada *et al.* (1979), and Vance *et al.* (1980).

Lignification frequently is more pronounced in plants infected with nonpathogens than in those infected with pathogens. Friend (1973) showed more rapid lignification in tuber disks and in detached leaves of potato plants resistant to *Phytophthora infestans* than in those of plants susceptible to the fungus. Ride (1973, 1975) studied lignification in wheat leaves inoculated with fungi and found that wound inoculations with nonpathogenic fungi resulted in a rapid production of lignin and confinement of fungi to wounds. Lignin was produced at a slower rate following wound inoculation with *Septoria* spp. and was not induced by wounding alone.

Lignification has been shown to be an inducible response elicited by certain compounds. Pearce and Ride (1982) reported that lignification could be induced in wounded wheat leaves by fungal cell wall preparations, chitin of both crab and fungal origin, chitosan, and ethyl alcohol chitin. According to Pearce and Ride (1980) induction of lignification seems to be more specific than phytoalexin induction. With the exception of mercuric ions, several known phytoalexin elicitors tested failed to induce lignification in wounded primary wheat leaves. Moreover, unlike mycelial fungi, yeasts and bacteria were not effective inducers of lignification.

It is reasonable to assume that parasites which are capable of developing through lignified host tissue can do so by degrading lignin. Ride (1980) studied the ability of 14 fungi including wheat pathogens to degrade lightly lignified walls from *Botrytis cinerea*-infected wheat leaves. Visual observations and quantitative measurements of the released carbohydrates from lignified and nonlignified walls in culture filtrates showed that the lignified papillae and haloes from these leaves were highly resistant to degradation by both pathogens and nonpathogens of wheat. Moreover, the activities of laccase and *P.* coumaryl esterase, often associated with lignin degradation (Christman and Oglesby, 1971) in culture filtrates of the fungi, were not correlated with their pathogenicity or their lignin-degrading ability. However, wheat pathogens were not capable of degrading lignin *in vitro*. Ride (1980) has pointed out that the production and activities of lignin-degrading enzymes *in vitro* might be different from those *in vivo*. Moreover, some fungal pathogens of wheat and other hosts might be able to grow and proliferate in lignified tissues particularly during the initial phase of the disease without having to degrade lignin. The literature on lignification as a mechanism of disease resistance has recently been reviewed by Vance *et al.* (1980).

### *3. Formation of Halo Spots*

Fungal penetration of leaves of the members of the family Gramineae is frequently accompanied by the formation of "haloes" as well as papillae (Lupton, 1956; Kunoh and Akai, 1969; McKeen *et al.*, 1969; Sherwood and Vance, 1976). The nature of halo spots, which are considered to result from an alteration of upper epidermal wall and sometimes of adjacent lateral walls, is not known.

Haloes are thought to be caused by wall degradation brought about by the activity of the penetrating fungi (Lupton, 1956; Akai *et al.*, 1968; McKeen *et al.*, 1969). However, alterations in wall structure in halo spots were not detected in electron micrographs (Bracker and Littlefield, 1973). Recent histochemical tests on reed canary grass haloes formed in response to nonpathogenic fungi showed the presence of cellulose, callose, and a small amount of lignin (Sherwood and Vance, 1976). Ride and Pearce (1979) also found small but possibly significant quantities of lignin in papillae and haloes on unwounded wheat leaves inoculated with nonpathogenic fungi. They used autoradiography with [$^{3}$H]phenylalanine or [$^{14}$C]cinnamic acid and found that a phenolic polymer was deposited locally in the upper epidermal wall during the early stages of interaction. They also found that papillae and cell walls in the halo regions were extremely resistant to chemical and enzymatic attack. Based on these results, they postulated that papillae and haloes which are formed shortly after appressorial production may have a role in the resistance of wheat plants to nonpathogenic fungi.

The overall results of these studies have not provided unequivocal evidence that cell wall modifications (lignification, papilla and halo formation) play major roles in resistance by interfering with fungal progress through physical and/or chemical means but they may contribute to the overall resistance. However, resolution of this question requires additional studies. The temporal relation of lignification to formation of papillae and haloes as well as penetration pegs is an important consideration when the role of cell wall modifications in resistance is considered.

## V. INFECTION-INDUCED RESISTANCE FACTORS—CHEMICAL FEATURES

### A. Hypersensitive Reaction

The hypersensitive reaction (HR) (defined in Chapter 1) is a wide-spread response of plants to infection by viruses, bacteria, fungi, and nematodes as well as to many nonpathogenic stimuli (Klement, 1963; Matta, 1971). The reaction is considered to be a resistance response culminating in the cessation of the growth and development of the invading pathogen in the tissue. The resistance (incompatible) response of potato tubers carrying R genes to *Phytophthora infestans* is associated with the HR, and is characterized by the rapid death of a limited number of host cells, tissue browning, and accumulation of fungitoxic substances (Kuć, 1976a). The literature on the nature of the HR and its possible role in resistance has recently been reviewed by Tomiyama *et al.* (1976), and Király (1980).

### *1. Mechanism of HR Induction by Fungi*

The nature of the metabolic events leading to fungus-induced HR is not clearly understood. Results of a number of studies seem to show that induction of the HR in incompatible combinations requires active metabolism of the interacting plant cells (Tomiyama, 1967; Kitazawa and Tomiyama, 1969; Sato *et al.*, 1971; Kitazawa *et al.*, 1973).

High-molecular-weight cell wall components of *Phytophthora infestans* have been reported to serve as nonspecific inducers of the HR in potato plants (Varns and Kuć, 1971; Varns *et al.*, 1971a; Kitazawa *et al.*, 1973; Lisker and Kuć, 1977). Hyphal wall components of *P. infestans* have also been shown to elicit hypersensitivelike responses in protoplasts from potato tubers (Doke and Tomiyama, 1980a,b). Protoplasts from potato cultivars with high degree of field resistance responded more actively to the fungal components than protoplasts from less resistant cultivars.

In addition to the HR-inducers, *Phytophthora infestans* also possesses substances that specifically suppress the initiation and intensity of the HR, tissue browning, and terpenoid accumulation in potato plants inoculated with incompatible races of the fungus or treated with elicitors of the HR (Varns and Kuć, 1971; Currier, 1974; Doke, 1975; Garas *et al.*, 1979; Doke and Tomiyama, 1980b). Suppressors of the HR have been found in germination fluids of the zoospores (Doke and Tomiyama, 1977) and cytospores (Doke *et al.*, 1980) of *P. infestans*. The HR suppressors in germination fluids of *P. infestans* cytospores were found to consist principally of water-soluble nonionic glucans containing $\beta(1 \rightarrow 3)$ linkages (Doke *et al.*, 1979). No significant differences were found in the quantities of glucan suppressors in the germination fluid of compatible and incompatible races of *P. infestans* (Doke *et al.*, 1980). However, the glucan from a compatible race of the fungus was a more effective suppressor than that from an incompatible race, indicating that qualitative differences in suppressors might be responsible for differences in their activities.

### *2. The Role of Fungus-Induced HR in Resistance*

It is generally, though not universally, accepted that in incompatible combinations characterized by the HR, resistance is due to death or cessation of the growth of the parasite following rapid collapse of the host cells. Based on the results of their studies on the wheat–*Puccinia graminis* f. sp. *tritici* system, Skipp and Samborski (1974) concluded that death of plant cells is closely associated with and may precede fungal death. Interaction of potato with an incompatible race of *Phytophthora infestans* resulted in rapid death of host cells while death of the fungus occurred 12 hr after inoculation (Shimony and Friend, 1975). Maclean and Tommerup (1979) also showed that the death of lettuce cells in incompatible interactions preceded the death of the downy mildew fungus. Brown *et al.* (1966), on the other hand, suggested that the rapid death of plant cells is the consequence and not the cause of death or inhibition of the invading fungus. Király *et al.* (1972) found that the inhibition by heat or antibiotics, of *Phytophthora infestans* growing in compatible potato tubers resulted in the appearance of dead host cells together with biochemical changes associated with resistance, including accumulation of the phytoalexin rishitin. They concluded, therefore, that the death or cessation of the growth of the fungus in plant tissue occurs prior to the collapse of the host cells and that the HR might be the consequence and not the cause of resistance. Similar results have been obtained by Barna *et al.* (1974) and Tani *et al.* (1975). In contrast, infiltration of soybean leaves with streptomycin 24 hr after inoculation with *Pseudomonas glycinea,* while causing tissue necrosis, did not result in significant accumulation of the phytoalexin glyceollin (Keen *et al.,* 1981). The conclusion derived by Király *et al.* (1972) was challenged by Maclean and Tommerup (1979) on the grounds that the sequence of events leading to the death of the host cell in Király's system was different from that operating in naturally induced resistance. Keen and Bruegger (1977) attributed the observation made by Király *et al.* (1972) to the release of nonspecific phytoalexin elicitors from *P. infestans* in response to antibiotic treatment causing rishitin accumulation.

Prusky *et al.* (1980) studied microscopically the sequence of events leading to cell death in the oat–*Puccinia coronata* var. *avenae* system and concluded that in this system haustorial damage preceded host cell death in both naturally and artifically induced hypersensitive reactions. They suggested that, when the sequence of events leading to the hypersensitive reaction is evaluated, a distinction should be made among hyphal growth, haustorial development, and viability of the fungus.

Jones and Deverall (1978) suggested that in wheat carrying the temperature-sensitive *Lr 20* gene for resistance to avirulent races of *Puccinia recondita,* inhibition of fungal growth was probably due to starvation of hyphae in the presence of dead or dying host tissue. They therefore suggested that in the wheat–*P. recondita* system necrosis is not essential for the expression of resistance to rust and that a distinction must be made between development of necrosis and processes leading to fungal growth inhibition. Campbell and Deverall (1980) were able to prevent necrosis in the above plant–parasite system by filtering out photosynthetically active wavelengths of light and by application of an inhibitor of noncyclic photophosphorylation. An avirulent race of the fungus grew slightly and produced very small colonies in resistant tissue in which necrosis had been prevented by the inhibitor. Daly (1976b) has discussed some of the drawbacks of the view that death or cessation of the growth of the invading pathogen in incompatible interactions occurs as a result of the death of a few host cells at the penetrated site.

The validity of the rationale that the HR occurs rapidly and therefore must constitute a resistance response also has been questioned. Daly (1976b) has pointed out that in certain cases the HR and its associated biochemical changes are not apparent immediately after spores of the biotrophic fungi come into contact with the host.

Some workers have adopted a compromising view and consider it unlikely that resistance to biotrophic fungi can be explained by a single hypothesis, such as hypersensitive resistance. Ingram (1978) suggested that, depending on the particular host–parasite system under study, host cell death may be either the cause or the consequence of events leading to resistance or may even be a totally unrelated phenomenon. This view has also been supported by Tani *et al.* (1975). Heath (1974) proposed that the ultimate expression of resistance in a host–pathogen system may be changed by a series of "switching points" during the infection affecting the course of host–parasite interactions.

### *3. HR Induced by Bacteria*

Infiltration of tobacco leaves with a number of phytopathogenic incompatible bacteria is known to result in a rapid collapse of the host cells within 9 hr, followed by bleaching and desiccation after approximately 18 hr. On the other hand leaves inoculated with *Pseudomonas tabaci,* a tobacco pathogen, show typical disease symptoms within 3–5 days.

Certain lines of evidence indicate that metabolic activity of bacteria is required for the development of the HR. First, at moderate inoculum levels the HR is induced only by living bacteria (Klement and Goodman, 1967). Second, auxotrophic mutants of *P. pisi* do not induce the HR unless the required amino acid is added (Sasser, 1978). Third, streptomycin prevents HR development when added to the inoculum or applied as an injection shortly after bacterial inoculation (Klement, 1971; Keen *et al.,* 1981). In the soybean–*P. glycinea* system, streptomycin not only prevented HR development but also blocked accumulation of the phytoalexin glyceollin (Keen *et al.,* 1981). The effect of streptomycin was considered to be on the bacteria because streptomycin treatment did not affect glyceollin accumulation elicited by iodacetate, an abiotic elicitor of glyceollin. Moreover, glyceollin accumulation and induction of the HR were not totally blocked if streptomycin was applied 2 hr or more after bacterial inoculation. This finding was interpreted by Keen *et al.* (1981) to indicate that recognition of the incompatible bacteria by soybean cells requires living bacterial cells and takes place fairly rapidly after inoculation. Rapid recognition is also a feature of bacterially induced HR in other host–parasite systems (Klement, 1971; Lyon and Wood, 1976).

It has been suggested that in some cases, HR-inducing incompatible and saprophytic bacteria attach to plant cell walls and are subsequently immobilized (encapsulated) following their introduction into intercellular spaces (Goodman, 1976; Goodman *et al.*, 1976b; Sequeira *et al.*, 1977; Sing and Schroth, 1977; Politis and Goodman, 1978; Cason *et al.*, 1978). Compatible bacteria are not attached to plant cell walls; thus they multiply freely in intercellular spaces.

Immobilization of bacteria near the cell walls is followed by a series of physical changes in the plant cell wall and in membranes near the site of bacterial attachment. Cells of an avirulent strain of *Pseudomonas solanacearum* were attached to the mesophyll cell walls and were surrounded by a thin pellicle within 3 hr after infiltration into tobacco leaves. Four hours later, a dense granular material accumulated in the space between the pellicle and the plant cell wall. Changes in the plasmalemma were also observed in the vicinity of the site of attachment of the bacterial cells to the host cell wall. The plasmalemma separated from the cell wall and the space thus formed was filled with numerous vesicles. Within 7 hr after infiltration cell water was lost, cell organelles were damaged, and the cells collapsed (Sequeira *et al.*, 1977).

Immobilization of the incompatible bacteria does not seem to be a universal response. Király *et al.* (1977) did not find a decrease in multiplication of incompatible bacteria, *P. pisi* and *P. syringae,* in tobacco at the onset of hypersensitive necrosis and for several hours later. In bean plants infiltrated with different pseudomonads, Hildebrand *et al.* (1980) did not detect any difference in type and degree of encapsulation between compatible and incompatible pathogens or saprophytes during the first 3 hr after inoculation. Moreover, they suggested that the structures observed around bacteria are formed as a result of physical forces involved in the drying cycle where dissolved cell wall materials encrust bacterial cells giving the impression they have been immobilized by the modified cell walls. In the *P. syringae*–bean system, Daub and Hagedorn (1979, 1980) also found no evidence that envelopment and immobilization processes are significant defense mechanisms. It is possible that in certain plants, but not in all, immobilization of the bacterial cells is required for the induction of the HR. The importance of the contact of the bacterium with the plant cell walls in the HR also has not been resolved. Stall and Cook (1979) suggested that contact of *Xanothomonas vesicatoria* with the pepper cell walls is necessry for the HR.

Multiplication patterns of compatible bacteria inside the leaves are also different from those of incompatible bacteria. For example, following their introduction into the tobacco leaves, compatible strains of *P. solanacearum* exhibit a rapid exponential growth for several hours after a short lag of about 8 hr. In contrast, the population of incompatible strains remains constant for 6 hr following injection into the leaves and then declines quickly with the onset of the HR (Sequeira and Hill, 1974).

The entrapment of certain incompatible bacteria following their introduction into leaf tissue may be associated with certain biochemical changes such as the accumulation of antibacterial products which may help reduce bacterial multiplication. Obukowicz and Kennedy (1981) suggested that tannins might contribute to immobilization of incompatible bacteria during the HR. Tannins, which are reported to have antibiotic activity (Swain, 1977), are formed through oxidative polymerization of phenolic substances by polyphenoloxidase. Results of an ultracytochemical study by Obukowicz and Kennedy (1981) showed that the deposition of phenolics increased in the vacuoles and vesiculating cytoplasm 10 hr after injection of an avirulent strain (B1) of *P. solanacearum* into tobacco leaves. The phenolic deposits diminished 20 hr after injection of B1, concomitant with the appearance of electron-opaque globules (possibly oxidized phenols or tannins) along the exterior side of the mesophyll cell walls. Reduction in the phenolic deposits in B1-injected leaves 20 hr after bacterial injection also coincided with a widespread appearance of polyphenoloxidase (PPO). PPO was mainly present in chloroplast grana but also in the cytoplasm and in vesicles in the space between the plasma membrane and the cell wall close to the entrapped bacteria. Phenolic deposits were not present in leaves infiltrated with a compatible strain (K60) of bacteria 10 or 20 hr after infiltration. Judging from these results, Obukowicz and Kennedy (1981) speculated that the disappearance of phenolics at 20 and 30 hr after injection of incompatible bacteria may have resulted from activation of polyphenoloxidase capable of converting the phenolics into antimicrobial quinones and tannins around the entrapped bacteria, promoting their attachment to the plant cell walls. An alternative possibility discussed by Obukowicz and Kennedy (1981) is that increase in PPO and polymerized phenolics prior to the development of the HR might be due to the formation of vesicles at the site of bacterial attachment as a consequence of the loss of compartmentation by membrane disruption. In contrast to the above study, Nemeth *et al.* (1969) did not detect changes in PPO activity in tobacco leaves injected with an HR-inducing incompatible strain of *P. syringae*. However, in their study the activity of the enzyme was detemined only once after HR development.

In certain cases, treatments that suppress induction of the HR also cause increased multiplication of bacteria in incompatible hosts. Gnanamanickam and Patil (1977a,b) reported that inoculation of resistant bean plants with *P. phaseolicola,* following their treatment with a toxin isolated from *P. phaseolicola*-inoculated susceptible bean plants, resulted in suppression of the HR and increased multiplication of bacteria. However, inoculation of toxin-treated bean plants with *P. tabaci, P. lachrymans,* and *P. tomato,* which are nonpathogenic to beans, led to the HR development and restriction of bacterial multiplication. Keen *et al.* (1981) showed that rifampicin-resistant (rif) mutant strains of incompatible race 1 of *P. glycinea* as well as *P. pisi,* nonpathogenic to soybeans, multiplied at rates similar to those of compatible races of *P. glycinea* in leaves treated with blasticidin S (an inhibitor of protein synthesis), which is capable of blocking the HR and glyceollin accumulation. Although these results tend to support the view that the HR and phytoalexin accumulation are responsible for restricting bacterial multiplication, other findings by Keen *et al.* (1981) were less supportive of this view. For example, blasticidin S also allowed multiplication of *P. fluorescens* rif, a saprophytic bacterium incapable of eliciting the visible HR and glyceollin accumulation in soybeans. Since treatment of noninoculated soybean leaves with blasticidin S resulted in increased loss of water and nutrients, the observed multiplication of the bacteria in blasticidin S-treated tissue might have resulted from the greater availability of nutrients (Keen *et al.,* 1981).

Phytoalexins, which are thought to contribute to resistance in certain plant–parasite systems, may also be involved in the expression of bacterially induced HR. The role of the HR and the phytoalexin glyceollin in resistance in the soybean–*Pseudomonas glycinea* system was recently investigated by Keen *et al.* (1981). Infiltration of soybean leaves with incompatible *Pseudomonas* spp. (pathogenic to plants other than soybeans) and with incompatible isolates of *P. glycinea* resulted in induction of a visible HR and accumulation of glyceollin and related isoflavonoids. Both the HR and glyceollin accumulation were blocked in leaves infiltrated with blasticidin S (an inhibitor of protein synthesis) within 0–9 hr after inoculation with the incompatible bacteria. Blasticidin S also inhibited glyceollin accumulation in leaves treated with sodium iodoacetate, an abiotic glyceollin elicitor. The induction of a visible HR and glyceollin accumulation was also blocked by streptomycin in leaves infiltrated with incompatible *Pseudomonas* spp., but not in those infiltrated with sodium iodoacetate. Moreover, the HR was induced and glyceollin was accumulated in leaves infiltrated with a temperature-sensitive, incompatible *P. glycinea* (race 1) at 22°C but not at 31°C. A visible HR developed and glyceollin was accumulated at both 22° and 31°C in response to sodium iodoacetate and to *P. glycinea* (race 6) and *P. pisi,* which are presumably temperature insensitive. The above results were interpreted by Keen *et al.* (1981) to indicate that the induction of the HR and accumulation of glyceollin are associated with the expression of resistance.

### *4. HR Induced by Viruses*

Virus-induced hypersensitive reaction associated with formation of local lesions is considered by some to be a resistant response because the formation of local lesions is associated with the suppression of virus multiplication (Harrison, 1955; Milne, 1966; Otsuki *et al.,* 1972) and restriction of virus movement through the adjacent tissue. Localization of the virus is not always associated with local lesion formation. For example, localization of TMV in cucumber cotyledons was not accompanied by necrosis or other visible symptoms (Cohen and Loebenstein, 1975). Metabolic changes associated with the formation of virus-induced local lesions include changes in cell permeability (Weststeijn, 1978); increases in the levels of ethylene (Balázs *et al.*, 1969; Nakagaki *et al.*, 1970; De Laat and Van Loon, 1982), phenolics, flavanoid compounds (Loebenstein, 1972), and formaldehyde (Tyihák *et al.*, 1978); and increases in the activity of peroxidase (Bates and Chant, 1970). The activity of phenylalanine ammonia-lyase was also increased in living cells adjacent to necrotic lesions (Farkas and Szirmai, 1969; Legrand *et al.*, 1976; Paynot and Martin, 1977). The significance of the observed changes in the initiation, development, and termination of local lesions is not well understood. Treatment of tobacco leaves with α-amino-oxyacetate, a competitive inhibitor of phenylalanine ammonia-lyase, resulted in a two- to fourfold increase in the size of TMV-induced local lesions. However, virus content of expanded lesions did not increase proportionately. Judging from these results Massala *et al.* (1980) suggested that increased synthesis of phenylalanine-derived metabolites (such as lignin) might contribute to virus-induced hypersensitive resistance.

Speculative mechanisms other than the HR which may contribute to plant resistance to viruses include plant tolerance of virus multiplication, failure of the virus to enter a living cell, failure of the virus RNA to release from the capsid, inability of the virus to spread from an initially infected cell to adjacent cells, and failure of the virus to spread systemically (Siegel, 1979).

## B. Peroxidases

Peroxidases, which are present in plant tissues, are known to participate in many important physiological processes, including oxidation of indoleacetic acid and phenolic compounds and the biosynthesis of ethylene. The hypothesis that peroxidases are involved in disease resistance is based on the reported increases in peroxidase activity in certain tissues in which resistance has been induced (Lovrekovich *et al.*, 1968; Ross, 1961; Rathmell and Sequeira, 1975). Moreover, phytotoxic hydrogen peroxide liberated through the activities of some oxidases in diseased tissue is thought to be removed by peroxidases, thus protecting the plant tissue from cellular damage and necrosis. Peroxidases are also known to inhibit fungal growth (Lehrer, 1969).

The possible role of peroxidases in resistance has been alluded to in certain diseases caused by bacteria, fungi, and viruses. Lovrekovich *et al.* (1968) showed that injection of heat-killed cells of *Pseudomonas tabaci* into tobacco leaves caused an increase in the activity of peroxidase and resistance to subsequent infection. They also protected susceptible tobacco plants from the bacteria by infiltrating the leaves with a commercial preparation of horseradish peroxidase. Sweet potato slices exposed to ethylene became resistant to *Ceratocystis fimbriata* and showed increased activities of peroxidase and polyphenol oxidase (Stahmann and Demorest, 1973). The activity of high- and low-molecular-weight peroxidases also is increased in wheat leaves inoculated with incompatible races of leaf rust fungus (Johnson and Lee, 1978). In virus-infected plants, formation of necrotic lesions is accompanied by an increase in peroxidase activity (Bates and Chant, 1970).

While peroxidase activity in some incompatible host–parasite combinations was found to be considerably higher than that in compatible combinations (Simons and Ross, 1970), opposite results have also been reported (Frič, 1969; Wood and Barbara, 1971). Moreover, increased resistance of older cotton seedlings to *Rhizoctonia solani* was not correlated with increased activity of peroxidase (Veech, 1976).

To verify the role of peroxidases, including IAA oxidase, in resistance, Daly and his associates (Seevers and Daly, 1970b; Daly *et al.*, 1971) used near-isogenic lines of wheat carrying the *Sr6* allele for resistance or the *sr6* allele for susceptibility to race 56 of *Puccinia graminis* f. sp. *tritici*. This system is unique because it exhibits incompatibility at 20–21°C and compatibility at 25–26°C. Therefore, a host response associated with resistance should be present at 20–21°C and not at 25–26°C. Similar increases in peroxidase activity were observed in both susceptible and resistant combinations at 20°C during the first and second days of infection. However, peroxidase activity continued to increase substantially in leaves exhibiting resistance while the rate of activity increased only slightly in leaves showing susceptible reaction. Peroxidase activity detected in wheat plants kept at 20°C for 6 days remained high after transfer to 26°C, despite a change in disease reaction from incompatibility to compatibility. Moreover, in the above system, increased peroxidase activity induced by ethylene treatment did not cause an increase in resistance. It was therefore concluded that in the above host–parasite system disease-induced peroxidase activity did not appear to be related to resistance or susceptibility.

Peroxidase activity increases in tobacco leaves following infiltration with heat-killed cells of an avirulent isolate of *Pseudmonas solanacearum* (Rathmell and Sequeira, 1975). Nadolny and Sequeira (1980) showed that in the above host–parasite system, the induction of resistance and the increase in peroxidase activity were apparent 8 hr after infiltration of heat-killed cells of an avirulent form (B1) of the bacterium, which rules out the possibility that peroxidase triggers the metabolic sequences leading to resistance. The following treatments also did not induce resistance but resulted in changes in peroxidase activity comparable to those induced by infiltration of B1: shading of leaves following infiltration with B1, infiltration of leaves with *Bacillus subtilis,* and injection of asbestos fibers into leaves. Nadolny and Sequeira (1980) therefore concluded that the increased peroxidase activity in the above system is not directly involved in disease resistance and is probably triggered by injury caused by toxic compounds produced by bacteria.

Interpretation of the data in the above studies is complicated by many factors, including the presence of several electrophoretically distinct peroxidase forms in diseased tissue. For example, 14 peroxidase isozymes were detected in wheat but the activity of only one (isozyme 9) consistently increased in the incompatible interaction (Daly, 1972). Also, in many cases the concentration of hydrogen donors (e.g., polyphenols) for peroxidase reaction in tissues under study has not been determined.

Plant resistance is often a function of a number of mechanisms operating in concert, each contributing somewhat to the overall defense. While in certain host–parasite systems peroxidases may not be a major factor in resistance, they may nevertheless contribute somewhat to the outcome of the host–parasite interaction. For example, peroxidases may contribute to resistance by participating in the synthesis of lignin and antimicrobial oxidized phenols. The literature on the role of peroxidases in resistance has been reviewed by Sequeira (1973), Stahmann and Demorest (1973), and Hislop *et al.* (1973).

### C. Phenolic Compounds

Many of the phenolic substances and their oxidation products, such as quinones that accumulate in diseased tissue, exhibit antimicrobial properties and therefore are considered by some to contribute to plant resistance. The involvement of phenolic substances in resistance is discussed in Chapter 9.

### D. Catalase

The possible involvement of catalase, which presumably causes a reduction in the level of hydrogen peroxide in cells, in resistance is unsettled (Farkas and Király, 1958; Frič, 1969).

### E. Glycosides

Many plants are known to contain nontoxic glycosides (mostly glucosides) which can be hydrolyzed to toxic substances by glycosidases present in wounded and infected tissues. Quinone, which is formed through oxidation of arbutin (hydroquinone glucoside), has been implicated in resistance of pears to fireblight (Hildebrand and Schroth, 1964; Powell and Hildebrand, 1970). Similarly, oxidation products of Phloridzin (dihydrochalcone glucoside) have been thought to be involved in the resistance of apples to scab (Kuć, 1967; Williams and Kuć, 1969).

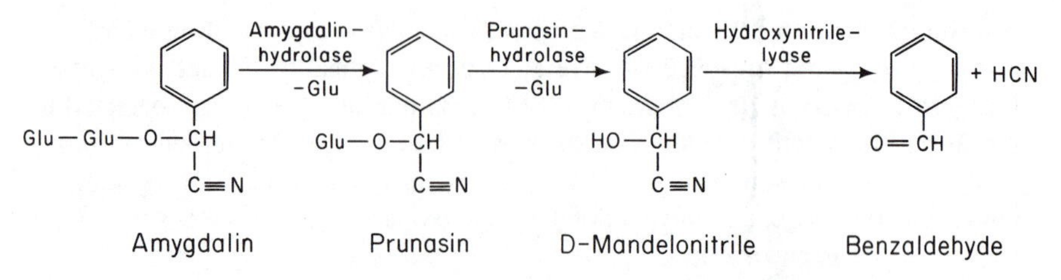


**Figure 32.** Enzyme-catalyzed formation of hydrocyanic acid (HCN) from a cyanogneic glucoside, amygdalin. [Reproduced with permission from Schönbeck and Schlösser, 1976. In *Encyclopedia of Plant Physiology. New Series, Physiological Plant Pathology* (R. Heitefuss and P. H. Williams, eds.), Vol. 4, pp. 653–678. Springer-Verlag, Heidelberg.]

More than one thousand species of plants contain cyanogenic glycosides, which are nontoxic in their glycosidic forms but release toxic hydrocyanic acid (HCN) when their tissues are subjected to physical (Wattenbarger *et al.*, 1968), chemical (Eyjolfsson, 1970), and biotic (Millar and Higgins, 1970; Millar and Hemphill, 1978) stimuli (Fig. 32). Because of their potential antimicrobial activity, cyanogenic glycosides have been implicated in resistance of cyanogenic plants to certain pathogens. Since no consistent correlation was found between the level of cyanogenic glycosides in plants and their resistance to specific pathogens (Snyder, 1950; Trione, 1960), the ability of certain pathogens to infect cyanogenic plants has been attributed to their potential to cope with HCN (Fry and Evans, 1977). The mechanism whereby the pathogens of cyanogenic plants cope with HCN may include enzymatic detoxification of HCN (Fry and Millar, 1972), conversion of HCN into certain nontoxic metabolites (Castric and Strobel, 1969; Oaks and Johnson, 1972), and induction of an alternate cyanide-insensitive mitochondrial respiration (Henry and Nyns, 1975; Rissler and Millar, 1977).

Of the enzymes capable of detoxifying HCN, formamide hydro-lyase (FHL), which converts HCN to nontoxic formamide, has been studied extensively. The overall results of a number of studies show a fairly good correlation between pathogenicity of fungal pathogens of cyanogenic plants and the ability of these pathogens to detoxify HCN by FHL (Fry and Millar, 1972; Fry and Munch, 1975). Fry and Evans (1977) showed that, while all 11 fungal pathogens of cyanogenic plants tested produced FHL, the enzyme was produced by only 1 of 6 fungi nonpathogenic to cyanogenic plants and by 9 of 14 fungi pathogenic to noncyanogenic plants. The specific activity of FHL was substantially greater in cell-free homogenates of pathogens of cyanogenic plants than in those of pathogens of noncyanogenic plants and certain nonpathogens. Myers and Fry (1978) determined the activities of a number of enzymes that may participate in production and metabolism of HCN in the extract of *Gloeocercospora sorghi*-infected sorghum leaf tissue. The activity of FHL in diseased tissue was more pronounced than that of any other enzyme tested. The enzyme activity was first detected 18 hr after inoculation and increased about 200-fold after 36 hr. Moreover, the increase in the activity of the enzyme corresponded with a drop in HCN potential (the amount of HCN potentially available in tissue). No FHL activity was detected in the extract of healthy tissue. A comparison of diseased and healthy tissue 24 hr after inoculation revealed a 20-fold increase in the activity of β glucosidase, which hydrolyzes dhurrin, a cyanogenic glucoside in sorghum, in the former. The activity of the enzyme, however, remained essentially constant between 24 and 72 hr after inoculation. The activities of the following enzymes in diseased tissue remained constant during pathogenesis: oxynitrilase, which releases HCN from *p.* hydroxymandelonitrile; β-cyanoalanine synthase, which catalyzes formation of β-cyanoalanine from either cysteine or serine and HCN; and rhodanase, which forms thiocyanate from HCN in the presence of an appropriate sulfur source. Judging from the results of the study, Myers and Fry (1978) concluded that in the sorghum–*G. sorghi* system, the fungus is capable of coping with HCN through detoxification by FHL.

In contrast to the fungal pathogens the tolerance of the bacterial pathogens of cyanogenic plants to HCN was reported to be only slightly higher than that of bacterial pathogens of noncyanogenic plants. The range of HCN sensitivity of the test bacteria was close to that of fungi, which are considered sensitive to HCN (Rust *et al.*, 1980).

## F. Ethylene

The question of the involvement of ethylene in resistance has been debated extensively. The idea was originally put forth by Stahmann *et al.* (1966) who reported that resistance of sweet potato to *Ceratocystis fimbriata* was increased by treatment of the tissue with ethylene. Moreover, evolution of ethylene was more pronounced in sweet potato slices inoculated with a nonpathogenic strain of the fungus than in those inoculated with a pathogenic strain. These observations, however, were not confirmed by Chalutz and DeVay (1969). Moreover, ethylene level increased in susceptible but not in resistant tomato plants in response to infection by *Fusarium oxysporum* f. sp. *lycopersici* (Gentile and Matta, 1975). However, treatment of susceptible tomato plants with ethephon (an ethylene-releasing substance) caused increases in resistance to *Fusarium* wilt and in the activity of peroxidase and polyphenol oxidase (Retig, 1974). The involvement of ethylene in disease resistance was not supported by the results of the study of Daly and his associates (Daly *et al.*, 1970; Daly, 1972). Ethylene treatment of wheat leaves carrying a temperature-sensitive gene for resistance to wheat stem rust rendered them susceptible even at 20°C, a temperature at which plants are normally resistant. Moreover, in susceptible plants ethylene-induced increases in peroxidase did not cause a change in disease reaction. Ethylene production by

*Penicillium digitatum,* which causes rot in orange and lemon fruits, is not required for pathogenicity because an isolate of the fungus incapable of producing ethylene was as pathogenic as an ethylene-producing isolate (Chalutz, 1979). In certain plant–parasite combinations ethylene has been reported to stimulate production of phytoalexins which are considered by some to play a role in resistance. Ethylene and ethrel (2-chloroethylphosphonic acid, an ethylene-releasing compound) induced production of phytoalexins such as 6-methoxymellein in carrot roots (Carlton *et al.*, 1961; Chalutz and DeVay, 1969; Chalutz *et al.*, 1969) and pisatin in pea pods (Chalutz and Stahmann, 1969). Henfling *et al.* (1978), who studied the effect of ethylene on accumulation of phytoalexins in potato tuber slices, concluded that ethylene is not an elicitor of phytuberin and phytuberol but can influence the quantities of these phytoalexins that accumulate in slices treated with cell-free sonicates of *Phytophthora infestans*.

Judging from the results of studies on the role of ethylene in plant disease it is unlikely that this hormone plays a major role in resistance. The possible roles in resistance of β-1,3-glucanases and chitinases, the synthesis of which is stimulated by ethylene, was discussed earlier in this chapter.

## G. Histones

Hadwiger *et al.* (1977) have shown that pea histones inhibit the growth of *Fusarium solani* f. sp. *phaseoli* and *pisi* and suggested that these basic proteins, which are rich in lysine and arginine, are potentially more important than the phytoalexin pisatin in resistance of pea tissues to plant pathogenic fungi.

## H. Phytoalexins

Phytoalexins as factors contributing to disease resistance have been studied for more than four decades by many workers, and their literature has been reviewed quite extensively. Recent reviews include Deverall (1976), Kuć, (1976a,b), Kuć *et al.* (1976b), Van Etten and Pueppke (1976), Albersheim (1977), Keen and Bruegger (1977), Stoessl *et al.* (1977), Van Etten (1979), Cruickshank (1980), and Keen (1981).

Phytoalexins are commonly defined as antimicrobial plant metabolites which are present at nondetectable levels in healthy plants and accumulate to high levels as a result of pathological, physical, and environmental stimuli. Phytoalexins are induced by fungi (Müller and Börger, 1940; Cruickshank, 1963), bacteria (Cruickshank and Perrin, 1971; Keen and Kennedy, 1974; Lyon and Wood, 1975; Gnanamanickam and Patil, 1975, 1976, 1977a,b; Weinstein *et al.*, 1981), viruses (Bailey and Ingham, 1971; Bailey and Burden, 1973), nematodes (Rich *et al.*, 1977; Veech and McClure, 1977; Veech, 1978), toxic chemicals (Cruickshank and Perrin, 1963; Bell, 1967; Hadwiger and Martin, 1971), and physical treatments (Hadwiger and Schwochau, 1971). There are about 60 different chemicals with phytoalexinlike properties for which chemical structures have been suggested. Phytoalexins are produced by more than 100 species of plants belonging to 21 families (Keen, 1981).

Several distinct but chemically related phytoalexins are produced by different plants. Moreover, various combinations of isoflavonoid and terpenoid phytoalexins are produced in response to specific pathogens in a single plant (Keen, 1975; Price *et al.*, 1976; Van Etten and Pueppke, 1976; Rich *et al.*, 1977). Multiplicity of some of these phytoalexins may sometimes be due to chemical modifications of parent molecules by microbial activities (Ward and Stoessl, 1977).

In addition to being antifungal, some phytoalexins are also toxic to bacteria (Lyon and Bayliss, 1975; Gnanamanickam and Patil, 1977a; Wyman and Van Etten, 1978; Gnanamanickam and Smith, 1980; Weinstein *et al.*, 1981), nematodes (Rich *et al.*, 1977; Kaplan *et al.*, 1980a,b), plants (Shiraishi *et al.*, 1975; Skipp *et al.*, 1978; Glazener and Van Etten, 1978), and animals (Van Etten and Bateman, 1971; Oku *et al.*, 1976a). However, phytoalexins are most effective against fungi and some are considered to be capable of limiting fungal colonization of plant tissues (Johnson *et al.*, 1976; Skipp and Bailey, 1976; Deverall, 1977; Smith, 1978; Yoshikawa *et al.*, 1978a).

Studies on phytoalexin production by fungi have generally been confined to saprophytes and nonbiotrophic parasites. However, biotrophic fungi are also known to elicit phytoalexin production (Oku *et al.*, 1976b; Shiraishi *et al.*, 1977).

### *1. Isoflavonoid Phytoalexins*

Isoflavonoid phytoalexins are produced primarily by genera of leguminoseae, and include pisatin in pea, phaseollin and related substances in bean, and glyceollin in soybean, alfalfa, and clover. However, some legumes also produce nonisoflavonoid phytoalexins. All of the known isoflavonoid phytoalexins are produced in plants in response to infection by fungi. However, a few bacteria, viruses and nematodes also are capable of eliciting production of isoflavonoid phytoalexins.

Phytoalexins are produced in bean plants inoculated with *Pseudomonas phaseolicola* (Lyon and Wood, 1975; Gnanamanickam and Patil, 1975, 1976, 1977a) soybeans challenged with *P. glycinea* (Keen and Kennedy, 1974), and soybeans inoculated with *Erwinia carotovora* (Weinstein *et al.*, 1981). In the *P. glycinea*–soybean system the incompatible interaction is characterized by the accumulation of high levels of the phytoalexin glyceollin and reduced bacterial

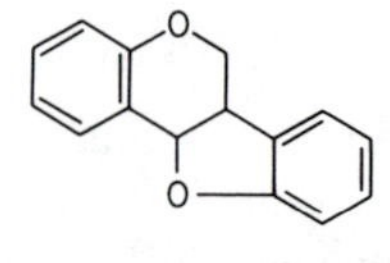

I - PTEROCARPAN

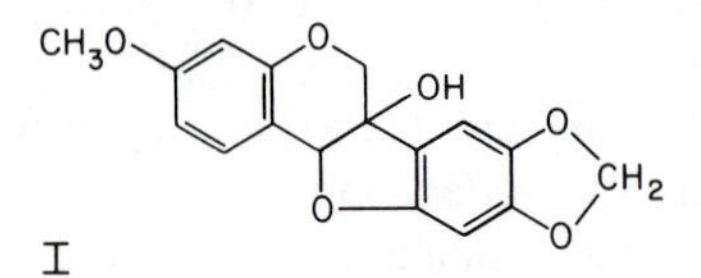


I

PISATIN
(Garden pea and 4 other species of pea)

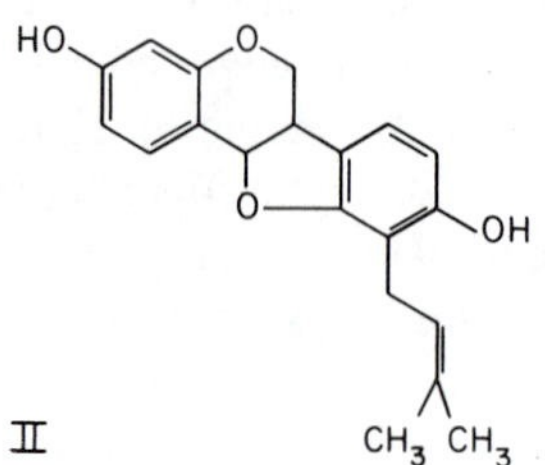


II

PHASEOLLIDIN
(French bean; cowpea)

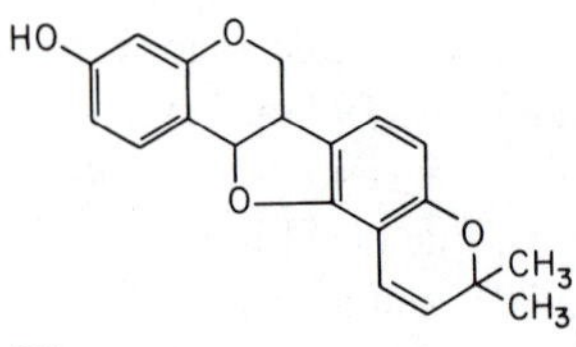


III

PHASEOLLIN
(bean)

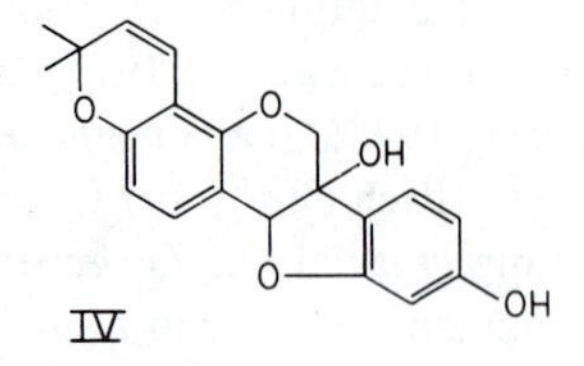


IV

GLYCEOLLIN I
(Soybean)

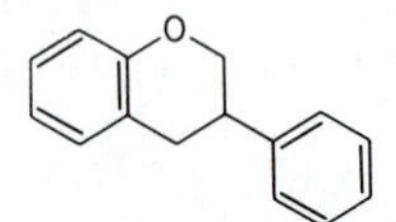


2 - ISOFLAVAN

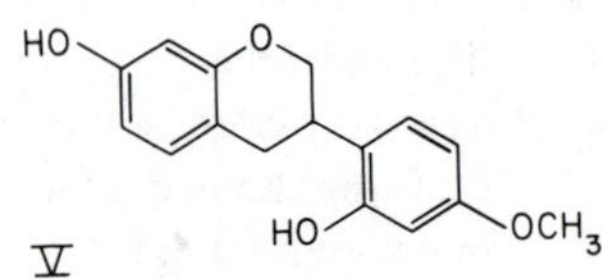


V

VESTITOL
(birdsfoot-trefoil)

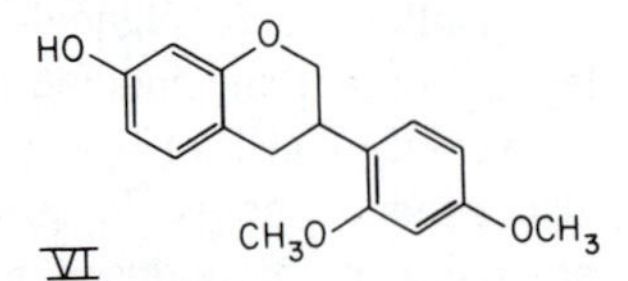


VI

SATIVAN (SATIVIN)
(Alfalfa)

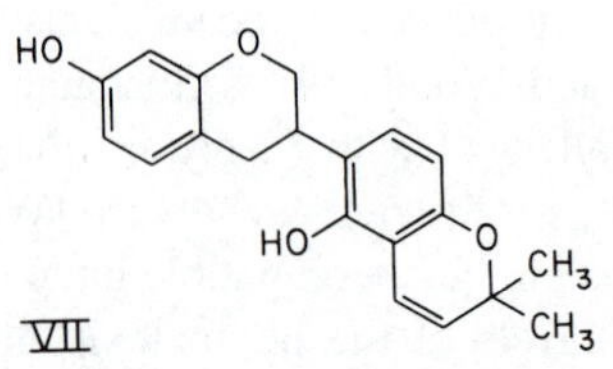


VII

PHASEOLLINISOFLAVAN
(bean)

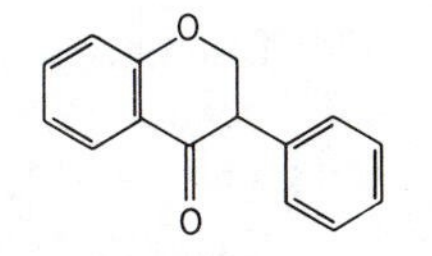


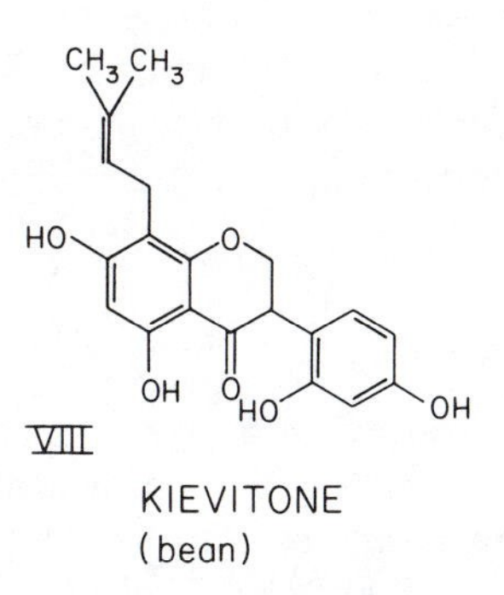


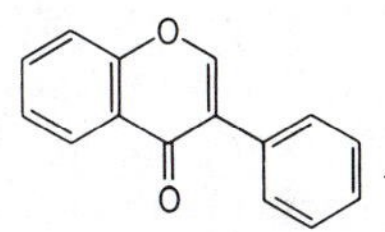


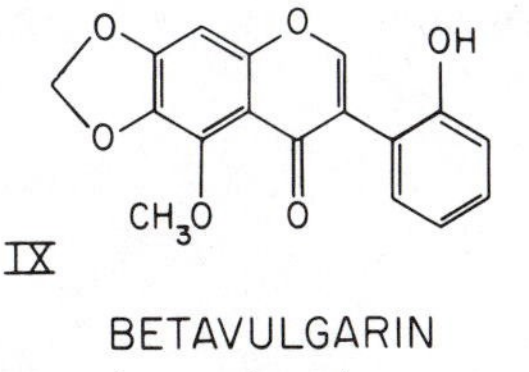


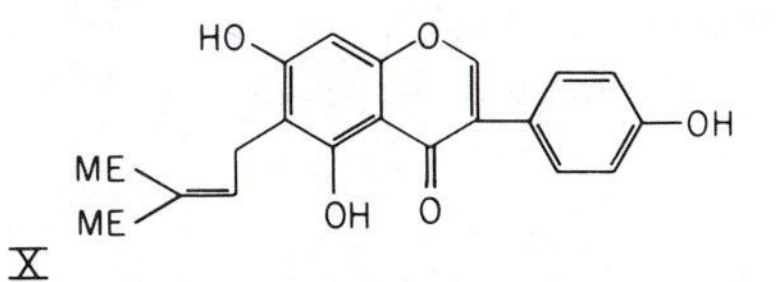


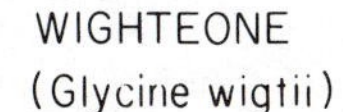


**Figure 33.** Structures of some phytoalexins belonging to four different classes of isoflavonoids taken from the following sources: (I) Perrin and Perrin (1962); (II) Burden *et al.* (1972), Bailey and Burden (1973); (III) Burden *et al.* (1972), Bailey and Burden (1973); (IV) Burden and Bailey (1975); (V) Bonde *et al.* (1973); (VI) Bonde *et al.* (1973), Ingham and Millar (1973); (VII) Burden *et al.* (1972), Bailey and Burden (1973); (VIII) Burden *et al.* (1972), Bailey and Burden (1973); (IX) Geigert *et al.* (1973); (X) Ingham *et al.* (1977). [See References for complete citations.]

multiplication. A compatible interaction, on the other hand, is featured by accumulation of a small amount of glyceollin and by a high rate of bacterial multiplication (Keen and Kennedy, 1974; Bruegger and Keen, 1979).

The phytoalexin, glyceollin, accumulated in an incompatible soybean cultivar but not in a compatible one, 2 or 3 days after inoculation with *Meloidogyne incognita* (Kaplan *et al.*, 1980a). Moreover, phytoalexins did not accumulate to significant levels in the above two cultivars following inoculation with compatible species of *M. javanica*. Mobility of *M. incognita* in vitro, but not that of *M. javanica,* was inhibited by glyceollin (Kaplan *et al.*, 1980b).

The structures of some isoflavonoid phytoalexins are shown in Fig. 33. Following Van Etten and Pueppke (1976), they have been placed under four different classes based on their ring structures.

### 2. *Terpenoid Phytoalexins*

Terpenoid phytoalexins (reviewed by Kuć *et al.*, 1976b) are accumulated in some plants in response to infection by pathogenic and nonpathogenic fungi (Tomiyama *et al.*, 1968a; Varns *et al.*, 1971a,b; Kuć, 1972; Stoessl *et al.*, 1976), bacteria (Beczner and Lund, 1975; Lyon *et al.*, 1975), and cell-free sonicates of compatible and incompatible races of *Phytophthora infestans* (Varns *et al.*, 1971a,b). The role of the fungitoxic terpenoids produced in potato tubers in resistance has been studied extensively (Allen and Kuć, 1968; Tomiyama *et al.*, 1968a,b; Sato *et al.*, 1971; Varns *et al.*, 1971b; Kuć, 1972, 1976a). The purported ability of sesquiterpenoids to limit the development of *P. infestans* in incompatible tissues is not supported by the results of recent studies (Henfling *et al.*, 1980b). The involvement of gossypol and related compounds in resistance of cotton plants to *Verticillium* wilt has been studied by Bell and Presley (1969a,b), Zaki *et al.* (1972a,b), and Mace (1978). Structures of some terpenoid phytoalexins are shown in Fig. 34.

### 3. *Metabolism of Phytoalexins by Plant Cells*

Certain plants are known to metabolize phytoalexins. For example, rishitin and lubimin are metabolized in aged, cut potato tubers in the presence or absence of an incompatible race of *Phytophthora infestans* (Horikawa *et al.*, 1976; Ishiguri *et al.*, 1978). Stoessl *et al.* (1976) also reported that the sweet pepper phytoalexin, capsidiol, is metabolized to an appreciable extent by healthy pepper tissue. Moreover, cell suspension cultures of kidney bean (a producer of phaseollin) and mung bean (a nonproducer of phaseollin) metabolized phaseollin (Glazener and Van Etten, 1978).

1. FROM SOLANACEAE:

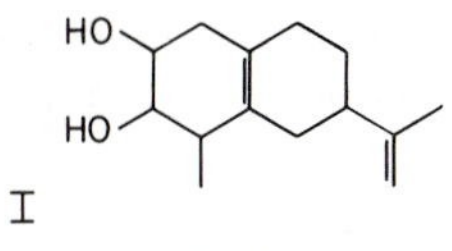


I RISHITIN

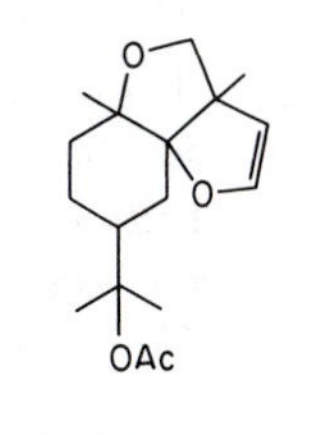


II PHYTUBERIN

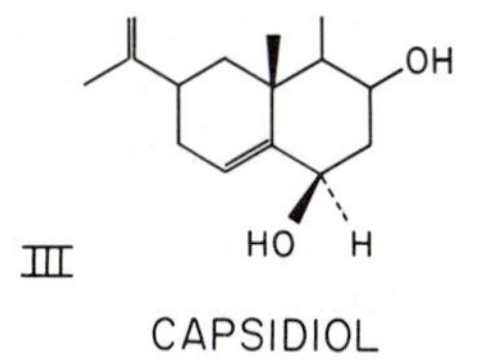


III CAPSIDIOL

2. FROM MALVACEAE:

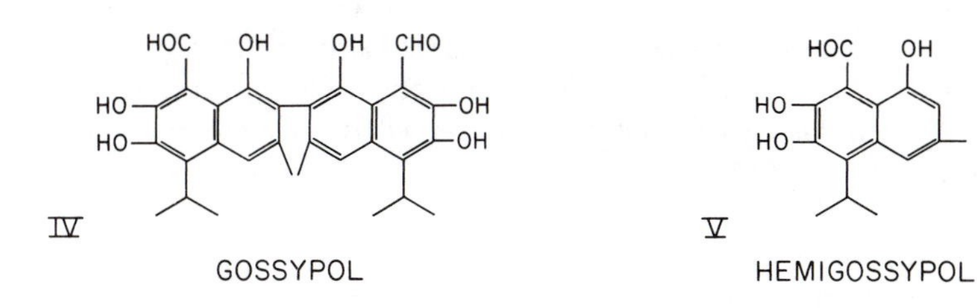


3. FROM LEGUMINOSEAE:

$CH_3.CH_2.CH{=}CH.C{\equiv}C.CO.C{=}CH.CH{=}C.CH{=}CH.COOR$ (O bridge between the C atoms of C=CH.CH=C)

VI WYERONE, R = $CH_3$

VII WYERONE ACID, R = H

**Figure 34.** Structures of some terpenoid phytoalexins from different plant families taken from the following sources: (I) Katsui *et al.* (1968); (II) Coxon *et al.* (1974); (III) Gordon *et al.* (1973); (IV) Adams *et al.* (1960), Edwards and Cashaw (1957); (V) Zaki *et al.* (1972a,b); (VI) Fawcett *et al.* (1971); (VII) Letcher *et al.* (1970). [See References for complete citations.]

*4. Mode of Action of Phytoalexins*

Little is known about the mode of action of phytoalexins. Phaseollin was reported to cause leakage of metabolites, loss of dry weight, reduction of nutrients, and alterations in respiration in treated mycelium of *Rhizoctonia solani* (Van Etten and Bateman, 1971). For further discussion see Van Etten and Pueppke (1976).

*5. Structure–Function Relationship of Phytoalexins*

The antifungal activity of petrocarpan isoflavonoids has been related to their molecular stereochemistry (Perrin and Cruickshank, 1969). However, this hypothesis was not supported by the results of other studies (Van Etten, 1976; Ravise and Kirkiacharian, 1976; Harborne *et al.*, 1976; Smith, 1978).

*6. Role of Phytoalexins in the Hypersensitive Response*

Metabolically active cells adjacent to cells undergoing hypersensitive response have been reported to synthesize phytoalexins and export them to the cells which are being colonized by the pathogen (Nakajima *et al.*, 1975; Tomiyama and Fukaya, 1975). Accumulation of large quantities of phytoalexins in tissues undergoing hypersensitive reaction (Keen, 1971; Bailey, 1974) is thought to contribute to the hypersensitive death of host cells (Shiraishi *et al.*, 1976). This view is speculative and requires experimental verification. For further discussion see Section V.A in this chapter and also Deverall (1976), Hargreaves and Bailey (1978), and Keen (1981).

*7. Mechanisms by Which Phytoalexins May Influence the Outcome of Disease Reaction*

Some investigators have attempted to explain compatibility in the context of the phytoalexin theory. Two hypotheses have been advanced to account for phytoalexin-mediated compatibility in plant–pathogen systems: (1) the pathogen does not allow phytoalexins to be accumulated to toxic levels and (2) the pathogen is tolerant of the accumulated phytoalexins. In their treatment of the subject, Van Etten and Pueppke (1976) used the terms *differential synthesis* and *differential sensitivity*, respectively, for these two hypotheses. This approach has been adopted here.

*a. Differential Synthesis.* According to this hypothesis, an incompatible interaction is characterized by a greater and more rapid accumulation of phytoalexins than that of a compatible interaction (Van Etten and Pueppke, 1976). In soybean plants infected with *Phytophthora magasperma*, the phytoalexin glyceollin accumulated sooner and reached higher concentrations in incompatible than in compatible interactions (Yoshikawa *et al.*, 1978; Keen, 1981). However, accumulation of phytoalexins is not always limited to nonhost plants or to resistant cultivars. For example, phaseollin accumulated in hypocotyls of resistant and susceptible bean cultivars in response to inoculation with *Colletotrichum lindemuthianum* (Bailey and Deverall, 1971). Moreover, in the *P. infestans*–potato system levels of terpenoid phytoalexins produced in tubers in compatible interactions were equal to or more than those produced in incompatible interactions (Kuć, 1976b). Similar results were obtained by Varns *et al.* (1971a) and Lisker and Kuć (1978), who studied terpenoid accumulation in potatoes inoculated with compatible or incompatible isolates of *P. infestans*. In some cases the greater accumulation of phytoalexins in incompatible host–parasite combinations compared to compatible combinations may be due to reduced rates of biodegradation rather than increased rates of biosynthesis of phytoalexins. This was suggested to be the case in the soybean– *P. megasperma* var. *glycinea* (Yoshikawa *et al.*, 1979).

*b. Differential Sensitivity.* According to this hypothesis, a compatible interaction occurs when the parasite is insensitive to the accumulated phytoalexins. Incompatibility, on the other hand, reflects the sensitivity of the parasite to phytoalexins (Van Etten and Pueppke, 1976). Support for this hypothesis was originally provided by the results of studies by Cruickshank (1962), who screened 50 fungal isolates for their sensitivity to phytoalexins and found that pathogens are commonly tolerant of their host's isoflavonoid phytoalexins while nonpathogens are sensitive. Pathogenicity of *Botrytis fabae* toward broad bean has been attributed to its tolerance of phytoalexins compared to the nonpathogen, *Botrytis cinerea* (Deverall *et al.*, 1968; Rossall *et al.*, 1980). High levels of phytoalexins accumulated in compatible interactions in the pea–*Fusarium solani* f. sp. *pisi* and the bean–*F. solani* f. sp. *phaseoli* systems; however, each organism showed tolerance of its host phytoalexin *in vitro* (Van Etten, 1979).

Van Etten and Stein (1978) found that sensitivity of *Fusarium solani* f. sp. *pisi* and *F. solani* f. sp. *phaseoli* to pisatin and phaseollin was influenced by *in vitro* bioassay conditions. Tolerance of these fungi to phaseollin was greatly increased following their exposure to a low concentration of the phytoalexin in shake culture suggesting the presence of an adaptive tolerance mechanism (Van Etten and Stein, 1978).

Some fungi are known to metabolize phytoalexins (Müller, 1958; Higgins and Millar, 1969a,b; Van Etten and Smith, 1975; Ward and Stoessl, 1976a; Lappe and Barz, 1978; Kuhn and Smith, 1979; Smith *et al.*, 1980, 1981). Therefore, in some cases, differential sensitivity of fungi to phytoalexins may be due to their ability to metabolize the accumulated phytoalexins into nontoxic forms. *Colletotrichum coccodes* (=*C. phomoides*), a pathogen of tomato, is capable of demethylating medicarpin to demethylmedicarpin, which is considerably less inhibitory than medicarpin to spore germination, germ tube elongation, and mycelial development of the fungus (Higgins and Ingham, 1981). While phytoalexin accumulation was induced in alfalfa leaves by *Stemphylium botryosum,* a pathogen, and *Helminthosporium turcicum,* a nonpathogen of alfalfa, only the pathogen could metabolize the phytoalexin into nonfungitoxic forms (Higgins and Millar, 1969a,b). Van Etten *et al.* (1980) screened 57 field isolates and two ascospore isolates of *Nectria haematococca* mating population (MP) VI for their sensitivity to pisatin, ability to demethylate pisatin, and their virulence on pea. No highly virulent isolates of the fungus were found which showed sensitivity to pisatin and/or were incapable of demethylating the phytoalexin. Many of the nonpathogenic isolates and those with low virulence were most sensitive to pisatin and/or incapable of demethylating it. However, an additional group of isolates with low virulence was found to be tolerant of pisatin and/or capable of metabolizing it. They concluded that high virulence of *N. haematococca* MP VI on pea appears to require the ability to tolerate pisatin and/or metabolize it to nontoxic forms.

In contrast to the above studies, there are numerous cases where successful pathogens were sensitive to and/or were unable to degrade the host's phytoalexins (Van den Heuvel and Glazener, 1975; Pueppke and Van Etten, 1976; Wyman and Van Etten, 1978). Differential sensitivity of twelve fungi to phaseollin *in vitro* also was not correlated with their pathogenicity to bean plants (Van den Heuvel and Glazener, 1975). Similar results were obtained by Smith *et al.* (1975). Moreover, many fungi which are capable of degrading and/or tolerating phytoalexins are unable to infect plants which produce the phytoalexins (Heath and Higgins, 1973; Stoessl *et al.,* 1973; Van Etten, 1973; Lyon, 1976; Duczek and Higgins, 1976; Bailey *et al.,* 1977; Skipp and Bailey, 1977).

Except for a few cases, the overall results do not show the presence of a clear correlation between pathogenicity of fungi and their ability to tolerate or degrade phytoalexins to nontoxic forms. The literature on the metabolism of isoflavonoid phytoalexins by fungi has been reviewed by Van Etten and Pueppke (1976).

The results of *in vitro* tests of the antibacterial properties of isoflavonoid phytoalexins have been controversial. For example, results of studies of Keen and Kennedy (1974), Lyon and Wood (1975), and Gnanamanickam and Patil (1977a) showed that *Pseudomonas phaseolicola* and *P. glycinea* were sensitive to isoflavonoid phytoalexins from French bean and soybean, respectively, while relatively little sensitivity was reported by Wyman and Van Etten (1978). The discrepancy is considered to be due to the use of different isolates by different investigators, to variations in the chemical and physical properties of the medium used, and to the length of incubation period.

*8. Nonspecific Elicitors and Phytoalexins*

Many fungi are known to contain high-molecular-weight substances capable of eliciting phytoalexins. These elicitors are nonspecific because they are present in both compatible and incompatible races of plant pathogens. Nonspecific elicitors have been isolated from *Rhizopus stolonifer* (Stekoll and West, 1978), *Phytophthora infestans* (Chalova *et al.*, 1976; Henfling *et al.*, 1980a), *Fusarium solani* (Daniels and Hadwiger, 1976), *Colletotrichum* spp. (Anderson, 1978), *Cladosporium fulvum* (DeWit and Roseboom, 1980), and *Phytophthora megasperma* f. sp. *glycinea* (Kuan and Erwin) (formerly *P. megasperma* var. *sojae*) (Ayers *et al.*, 1976a–c). A polygalacturonase from *Rhizopus stolonifer* was also found to elicit phytoalexin accumulation in castor beans (Stekoll and West, 1978; Lee and West, 1981a,b). The nonspecific elicitors from *P. megasperma* f. sp. *glycinea* (Pmg) were found to be efficient inducers of phytoalexins in the tissues of both susceptible and resistant cultivars of soybean plants (Keen *et al.*, 1975; Ayers *et al.*, 1976a; Albersheim *et al.*, 1977; Valent and Albersheim, 1977) and in potato and red kidney bean plants, which are not hosts for the fungus (Wade *et al.*, 1977). A glucan elicitor which appeared to be structurally similar to the Pmg elicitor capable of eliciting glyceollin in soybeans has also been isolated from brewer's yeast extract (Hahn and Albersheim, 1978). This suggests that soybean plants are capable of recognizing cell wall glucans of other fungi. Cline *et al.* (1978) showed that four of the phytoalexins which accumulated in red kidney bean cotyledons following exposure to the glucan elicitors from Pmg or yeast are the same phytoalexins which have been reported to accumulate in red kidney bean hypocotyls following inoculation with *Colletotrichum lindemuthianum*.

Many of the nonspecific elicitors isolated from fungi appear to be $\beta(1\rightarrow3)$ and $\beta(1\rightarrow6)$ glucans (Ayers *et al.*, 1976a,c; Albersheim and Valent, 1978). However, as we will see later in this chapter, chitosan, a derivative of chitin (Hadwiger and Beckman, 1980), and a fragment of cell wall pectic polysaccharide (Hahn *et al.*, 1981) are also capable of eliciting phytoalexins.

Mycelial cell walls of fungi have been shown to possess effective elicitors of phytoalexins (Albersheim and Valent, 1978; Ayers *et al.*, 1976 b,c; Keen, 1978; Yoshikawa, 1978). It has been aptly pointed out that the insoluble elicitors associated with fungal cell walls must be located on the surfaces of the walls if they are to be recognized by plant cells during the infection process. Alternatively, non-surface-borne insoluble elicitors may be solubilized upon contact with the host and thus become recognized by the plant cell walls. This was found to be the case in the soybean–*Phytophthora megasperma* f. sp. *glycinea* system by Yoshikawa *et al.* (1981). These workers reported that soluble elicitors of glyceollin accumulation were released from insoluble mycelial walls of the fungus following treatment of the walls with living soybean cotyledon tissue or cell-free extracts from soybean cotyledons or hypocotyls.

The literature on the structure and function of complex carbohydrates involved in host–parasite interaction has been reviewed by Albersheim *et al.* (1981).

### 9. *Specific Elicitors of Phytoalexins*

Keen (1975, 1978) and Keen and Bruegger (1977) have isolated factors from *Phytophthora megasperma,* f. sp. *glycinea* that appeared to function as specific glyceollin elicitors in soybeans. The elicitors exhibited the same specificity for glyceollin accumulation on certain compatible and incompatible soybean genotypes as the races of the pathogen themselves. Surface glycoproteins extracted from isolated fungus cell walls have also been reported to act as specific elicitors (Keen, 1978). However, as Bruegger and Keen (1979) pointed out, the soybean *P. megasperma* f. sp. *glycinea*–system is complicated by the presence in the fungus of nonspecific glyceollin elicitors. Specific elicitors of glyceollin have also been found in the cellular envelopes of incompatible races of *Pseudomonas glycinea* (Bruegger and Keen, 1979). With one exception, five races of the bacterium and the specific glyceollin elicitors from them exhibited the same specificity for glyceollin accumulation in cotyledons of two cultivars of soybeans. The presence of specific elicitors in certain pathogens is highly significant. However, there is presently no evidence that these specific elicitors function *in vivo*.

The search for phytoalexin elicitors has mostly been limited to preformed compounds (those present in the fungal and bacterial cells or in their cultures). Consideration has been given recently to the elicitors formed in plant tissues in response to infection. Hargreaves and Bailey (1978) have speculated that the production of phytoalexins in virus-infected necrotic bean tissues could be due to the release of a plant's constitutive elicitor from cells damaged by the virus. Nichols *et al.* (1980) showed that chitinase activity was higher in pea pods challenged with *Fusarium solani* f. sp. *phaseoli* and *F. solani* f. sp. *pisi* than those of the water-treated control, 0.5–6 hr after treatment. They suggested that chitinase activity in pea pods might assist in degrading *F. solani* cell walls, possibly releasing regulatory compounds such as phytoalexin inducers. Hadwiger and Beckman (1980) reported that chitosan (β-1,4-linked glucosamine, a deacetylated derivative of chitin) from *Fusarium solani* cells exhibited multiple activities: it elicited pisatin in pea pods, inhibited fungal germination and growth, and protected pea tissue from infection by *F. solani* f. sp. *pisi*. Because of its multiple roles, Hadwiger and Beckman (1980) suggested that chitosan may have a central role in disease resistance. Hahn *et al.* (1981), of Albersheim's group, reported that a fragment of the cell wall polysaccharide from soybean hypocotyls rich in galacturonic acid could elicit phytoalexin accumulation. They therefore speculated that certain pectic-degrading enzymes known to elicit phytoalexin accumulation (Stekoll and West, 1978; Lee and West, 1981a,b) might do so by releasing endogenous phytoalexin elicitors from plant cell walls.

The present intensive research on phytoalexin elicitors can be justified easily when their possible potential is considered. Studies of these substances seem to hold promise for unraveling unknown features of plant–pathogen interactions and for ellucidating the nature of the interactions of genes in pathogens with those in plants. Some have gone so far as to predict a role for them in plant disease protection.

Phytoalexin elicitors have been reveiwed by Albersheim and Anderson-Prouty (1975), Keen and Bruegger (1977), Albersheim and Valent (1978), and Keen (1981). See also Ward and Stoessl (1976b).

### *10. Phytoalexin Induction Hypothesis*

A number of hypotheses for induction of phytoalexins have been proposed. Some of these hypotheses, which have been reviewed recently by Cruickshank (1980), will be discussed briefly.

Hadwiger and Schwochau (1969) proposed that phytoalexin formation might involve derepression of enzyme synthesis. According to this hypothesis, microbial products and other substances at low concentrations could derepress segments of DNA that control the synthesis of regulatory enzymes involved in phytoalexin synthesis. Metlitskii (1976) has suggested that a phytoalexin elicitor(s) is produced by the parasite as a result of the interaction of a specific host plant metabolite with a specific receptor on the parasite's membrane. The elicitor then interacts with the host cells stimulating the synthesis of phytoalexins. According to Metlitskii (1976), the synthesis of the specific plant metabolites and the specific receptor on the parasite's membrane are controlled by genes for resistance and avirulence, respectively. However, the synthesis of phytoalexin elicitors and phytoalexins is nonspecific and is not directly related to resistance and avirulence genes. Based on the results of their studies on the soybean–*P. megasperma* f. sp. *glycinea* system, Ayers *et al.* (1976c) have considered that initiation, control, and biosynthesis of phytoalexins are controlled by a series of elicitors, inhibitors, and specificity factors. The initial rate of fungal growth, as well as the release of elicitors is controlled by inhibitors and specificity factors. The released elicitors then initiate and direct the synthesis of phytoalexins. A model proposed by Keen and Bruegger (1977), which also is based on the soybean–*P. megasperma* f. sp. *glycinea* system, holds that the synthesis and control of phytoalexins depend on the interaction between specific elicitors, produced by the parasite during the determinative phase of infection, and specific receptors on the host membranes. Specific elicitors are unique for dominant avirulence genes of each parasite and specific receptors are unique for dominant resistance genes of each plant. According to the "double induction" hypothesis proposed by Cruickshank (1980), plant metabolites regulate both the quality and the quantity of elicitors formed during infection. The elicitors in turn regulate the biosynthesis of phytoalexins by *de novo* synthesis of enzymes or by activating the existing enzymes.

In addition to phytoalexin elicitors, phytoalexin suppressors (blockers) have also been implicated as having a role in phytoalexin induction. Rahe and Arnold (1975) suggested that the synthesis of phytoalexins is regulated by specific repressors of phytoalexin formation present in compatible but not in incompatible combinations. Kuć *et al.* (1976a), Valent and Albersheim (1977), and Oku *et al.* (1977) found both elicitors and suppressors of phytoalexin synthesis in fungi and postulated that specificity might be determined by a balance between elicitors and suppressors of phytoalexins. Heath (1981c) and Bushnell and Rowell (1981) have presented models in which specificity at both race-cultivar and species levels is assumed to be controlled by the suppression of plant's defense responses (e.g., hypersensitive reaction, phytoalexin accumulation, lignification). In these models the outcome of disease reaction is considered to depend on the ability of pathogen-produced suppressors to bind to host's receptors. In the compatible interaction the defense system(s) of the plant is not elicited because the plant's receptor is attached to a suppressor and thus is incapable of binding to the pathogen-produced elicitor of defense system(s). In the incompatible interaction the inability of the suppressor to attach to the plant's receptor allows binding of the elicitor to the receptor, culminating in elicitation of a defense response such as phytoalexin accumulation. Suppressors have been implicated as determinant of specificity in certain host–parasite systems (Doke *et al.*, 1979, 1980; Doke and Tomiyama, 1980b; Heath, 1980a).

It should be pointed out that some of the phytoalexin induction hypotheses discussed previously have been formulated in conformity with Flor's gene-for-gene concept (Flor, 1955, 1971). The concept assumes that for each gene for resistance in the host there is a corresponding gene for avirulence in the pathogen and that the expression of resistance (incompatibility) requires an interaction between a gene for resistance in the host and a gene for avirulence in the pathogen. The concept has been used mostly to explain the genetic basis of host specificity exhibited by physiological races of the biotrophic fungi. Implicit in the gene-for-gene concept is induced resistance; that is, resistance is the result of activation of a defense mechanism in response to infection which serves to inhibit or suppress disease development. Susceptibility in the context of the "induced resistance" hypothesis may therefore be regarded as a passive (noninduced) response due to the absence of genes for resistance and avirulence (Wheeler, 1975). However, like resistance, susceptibility (compatibility) may also be an induced response. According to the so-called induced susceptibility hypothesis discussed by Daly (1972, 1976b,c,e), compatibility is the result of an induced mechanism which favors the growth and development of the pathogen, while incompatibility is the absence of such a mechanism. For example, pathogen-induced changes in the levels of plant hormones might trigger certain metabolic switches which favor the growth and development of the invading fungus, culminating in the establishment of a compatible interaction. The induction of a susceptible response might not necessarily involve the synthesis of enzymes, but might instead be the result of regulation of the activity of certain key regulatory enzymes in a manner that favors the parasite (Daly, 1972, 1976b,c). Daly (1972) has pointed out that depending on the plant–pathogen system, neither induced susceptibility nor induced resistance alone but a combination of both may be involved in the interaction.

Different aspects of the genetics of host–pathogen interactions have been reviewed extensively (Flor, 1971; Day, 1974, 1976, 1979; Wheeler, 1975; Ellingboe, 1976, 1981; Loegering, 1978; Bushnell, 1979; Leonard and Czochor, 1980).

### *11. Evaluation of the Phytoalexin Theory*

Evidence for and against the involvement of phytoalexins in resistance has been summarized by Wheeler (1975), Deverall (1976), Pueppke (1978), Keen (1981), and others. The following lines of evidence, compiled by Keen (1981), are considered by some to provide support for the involvement of phytoalexins in resistance:

1. There are many examples in which successful parasites are capable of degrading phytoalexins at a higher rate than their closely related nonpathogenic species (Cruickshank, 1963).
2. Treatment of susceptible soybean plants with ultraviolet (UV) light elicited glyceollin production and rendered the plants resistant to subsequent inoculation with *P. megasperma* f. sp. *glycinea* (Bridge and Klarman, 1973). In some cases, preinoculation treatments which reduced the levels of phytoalexin accumulation have altered a normally resistant reaction to a susceptible reaction (Cruickshank and Perrin, 1965; Bell and Presley, 1969a,b; Murch and Paxton, 1977; Yoshikawa *et al.*, 1978a).
3. In the flax–*Melampsora lini* and the soybean–*P. megasperma* f. sp. *glycinea* systems, substantial amounts of phytoalexins are produced only in incompatible interactions (Yoshikawa *et al.*, 1978; Keen and Littlefield, 1979). In the flax–*Melampsora lini* system, rapid cessation of parasite growth in the presence of resistance alleles was associated with a rapid accumulation of phytoalexins compared with alleles that restricted parasite development more slowly (Keen and Littlefield, 1979).
4. In the soybean–*P. megasperma* f. sp. *glycinea* system, growth of an incompatible race of the pathogen stopped at the time and cellular site in

which phytoalexins were present at highly toxic concentrations (Yoshikawa *et al.*, 1978a).

5. Co-inoculation of soybean plants with compatible and incompatible races of *Phytophthora* spp. resulted in accumulation of phytoalexins and restriction of the growth of both fungi (Paxton and Chamberlain, 1967).
6. In the soybean–*P. megasperma* f. sp. *glycinea* system, a normally compatible interaction was rendered incompatible by the application of purified phytoalexins to the infection site (Chamberlain and Paxton, 1968) while removal of phytoalexins led to increased compatibility (Klarman and Gerdemann,1963).
7. Limited studies by Keen (1978) and Bruegger and Keen (1979) suggest that in the soybean–*P. megasperma f. sp. glycinea* and the soybean–*Pseudomonas glycinea* systems elicitors isolated from certain races of the pathogens generally exhibited the same specificity for phytoalexin accumulation on certain compatible and incompatible genotypes of soybeans as the races of the pathogens themselves.

Following are some of the most frequently cited arguments against the involvement of phytoalexins in resistance:

1. Phytoalexin production can be induced by a number of nonspecific chemical and physical stimuli. However, judging from the results of the study of Yoshikawa (1978) this objection may not always be valid. Working with the soybean–*P. megasperma* f. s. *glycinea* system, Yoshikawa (1978) found differences in the mode of action of biotic elicitors (fungus cell wall) and abiotic elicitors (metal salts and detergents). The phytoalexin glyceollin was accumulated in response to both types of elicitors. However, accumulation of the phytoalexin in response to the biotic elicitor was due to a stimulation of biosynthesis of glyceollin while that in response to mercuric chloride was mainly due to an inhibition or inactivation of glyceollin degradation systems. These results are highly significant and show that in the above host–parasite system, and probably in other systems, a distinction should be made between biotic and abiotic elicitors. Results also seem to suggest that while phytoalexin elicitation might possibly be a nonspecific response to different elicitors, phytoalexin accumulation might be specific. Another relevant point is that certain abiotic phytoalexin elicitors are active in certain tissues but not in others. For example, sodium dodecyl sulfate elicited glyceollin accumulation in soybean cotyledons but not in soybean leaves. Moreover, $CuCl_2$, $HgCl_2$, mycolaminaran, and $K_2Cr_2O_9$ were active elicitors of glyceollin in soybean cotyledons but were relatively inactive in leaves (Yoshikawa, 1978; Keen *et al.*, 1981).

2. Most phytoalexins exhibit low or moderate toxicity to microorganisms.
3. Phytoalexins do not always show selective toxicity toward pathogens. Moreover, since many plants accumulate several phytoalexins, the insensitivity of a parasite to one phytoalexin may not contribute to its pathogenicity.
4. Phytolexins generally are not accumulated to inhibitory levels at the time resistance is expressed. Moreover, high levels of toxic substances in the plant tissue do not insure effective toxicity, because the actual contact between the toxic chemicals and the invading pathogens might be influenced by low water solubility of the compounds and their presence in certain organelles due to compartmentation (Daly, 1972).
5. It has been argued that in some cases very high concentrations of phytoalexins do not seem to impede the progress of the disease. For example, in pea plants infected with the common root fungus, *Aphanomyces euteiches,* lesions continued to increase up to 5 days following inoculation while the level of accumulated pisatin in the tissue, 36 hr after inoculation, was 8 times the level sufficient to inhibit the growth of the fungus (Pueppke and Van Etten, 1974, 1976).
6. Even in cases where the roles of certain antimicrobial chemicals in disease resistance have been accepted, a major question is whether or not the chemicals are present in sufficient concentrations at the proper time and site to account for resistance (Daly, 1972). It should be emphasized, however, that in some cases it is difficult or even impossible to obtain rigid experimental proof (Cruickshank, 1980).

The overall correlative evidence seems to show that phytoalexins may play a role in certain plant–pathogen combinations, notably, the soybean–*P. megasperma* f. sp. *glycinea* combination, where the nature of host–pathogen interactions has been carefully studied. In a number of other systems, however, phytoalexins have not been clearly linked to disease resistance. It is generally agreed that in this case, as in other cases of resistance, generalization should be avoided.

## VI. THE DYNAMIC AND COORDINATED NATURE OF RESISTANCE

The significance and relative contribution to plant defense of the mechanisms discussed in this chapter have been the subject of debate, and thus far none of these mechanisms has emerged as a universally accepted major factor in resistance.

It seems reasonable to assume that a single mechanism may rarely account for the resistance in a given host–parasite system and that plant defense is generally the function of a number of mechanisms operating in an integrated, coordinated manner. The level of resistance achieved in a given host–parasite combination may be the sum total of contributions of a number of resistance mechanisms in both constitutive and inducible categories. That is, the greater the number of participating resistance mechanisms in a given plant the higher the level of its tolerance. Multiplicity of plant's resistance mechanism has been emphasized by Kuć and Caruso (1977), Campbell *et al.* (1980), Bell (1980), Heath (1981b), and a few others.

Bell (1980) has pointed out that, in certain vascular wilt diseases, plants defend themselves against the spread of the pathogens by a series of sequential defense responses, which include entrapment of fungus conidia at end walls of xylem elements and formation of gels, tyloses, phytoalexins, tannins, and new xylem elements. To be effective, each of these defense components must occur in a specific sequence following infection. Accumulation of phytoalexins and tannins is aided by gels and tyloses while phytoalexins and tannins may prevent the pathogens from destroying gels and tyloses. The dynamic nature of defense in plants against vascular pathogens has been discussed recently by Beckman (1980), Bell (1980), and Bell and Mace (1980).

Spatial localization of certain chemicals in organs, tissues, and cells of plants may also be significant, since it allows deployment of a specific defense component at a designated site where it may be most effective (Bell, 1980). For example, in cotton roots, triterpenoids are formed in the epidermis, sesquiterpenoids in the cortex, and proanthocyanidins in the endodermis and hypodermis (Bell and Stipanovic, 1978; Bell and Mace, 1980). Plant defense also changes with time, with plant age, and in response to the environment. For further discussion see a review by Bell (1980).

*Chapter 13*

# Induced Resistance



## I. INTRODUCTION

Results of several studies have shown that pretreatment of plants with a number of physical, chemical, and biotic agents alters disease reaction to subsequent challenge inoculation. This phenomenon, which is known as induced resistance or so-called cross protection, has been observed in diseases caused by fungi, bacteria, viruses, and viroids. Sequeira (1979) believes that the term *interference* is a more appropriate term for describing phenomena known as cross protection.

In this chapter some of the features of induced resistance to fungi, bacteria, viruses, and viroids are discussed, and the proposed mechanisms of this interesting phenomenon are reported. The literature on this subject has been reviewed by Goodman *et al.* (1976a), Kuć and Caruso (1977), Sequeira (1979), Goodman (1980), Hamilton (1980), Matta (1980), Suzuki (1980), and Kuć (1981).

## II. INDUCED RESISTANCE TO FUNGI

There have been numerous reports of induced resistance to fungal pathogens (Yarwood, 1954; McLean, 1967; Deverall *et al.*, 1968; Sinha and Trivedi, 1969; Kuć *et al.*, 1975). Inoculation of leaves or etiolated hypocotyls of bean plants with an incompatible race of *Colletotrichum lindemuthianum* protected the plants against subsequent infection by normally compatible races of the fungus (Deverall *et al.*, 1968; Rahe *et al.*, 1969a; Skipp and Deverall, 1973). Heale and Sharman (1977) induced partial resistance to *Botrytis cinerea* in carrot tissue culture by pretreatment of the tissue with heat-killed conidia from the fungus. Systemic protection was elicited in cucumber against *C. lagenarium* by prior inoculation with the fungus (Kuć *et al.*, 1975). Four cultivars of watermelon and muskmelon were similarly protected against *C. lagenarium* by the same procedure in the greenhouse (Caruso and Kuč, 1977a) and in the field (Caruso and Kuć, 1977b). Hammerschmidt *et al.* (1976) protected cucumbers against *C. lagenarium* by prior inoculation with *Cladosporium cucumerinum* (a pathogen of cucumber) and against *C. cucumerinum* by prior inoculation with *C. lindemuthianum* (a pathogen of bean). Protection of cucumber against *C. lagenarium* also was elicited by *Pseudomonas lachrymans* (Caruso, 1977) and tobacco necrosis virus (Jenns and Kuć, 1977). Localized and systemic protection against *Phytophthora parasitica* var. *nicotianae* was induced by tobacco mosaic virus infection of tobacco hypersensitive to the virus (McIntyre and Dodds, 1979). Finally, the incidence of certain rust diseases on susceptible plants has been reduced by preinoculation with rust fungi pathogenic on other plant species (Jonhston and Huffman, 1958; Littlefield, 1969; Tani *et al.*, 1980).

In some cases infection of plants by certain pathogenic fungi has been shown to render plants somewhat susceptible to nonpathogenic fungal species or to avirulent strains of pathogens (Ouchi *et al.*, 1976; Heath, 1980a). Barley leaves became susceptible to incompatible races of *Erysiphe graminis hordei* and nonpathogenic powdery mildew fungi following their interaction with a compatible race of the fungus. Moreover, wheat and barley powdery mildews developed on leaves of melon, a nonhost, preinoculated with *Sphaerotheca fuliginea,* a pathogen of melon (Ouchi *et al.*, 1979).

Disease reaction has also been altered by exposure of plants to abiotic factors prior to inoculation. Andebrahan and Wood (1980) reported that UV irradiation of hypocotyls of beans susceptible to a race of *C. lindemuthianum,* 24–48 hr prior to inoculation, made them resistant to the fungus. However, UV irradiation of etiolated bean hypocotyls for short periods prior to inoculation resulted in a decrease in resistance to a virulent race of *C. lindemuthianum,* to *C. langenarium,* and to *C. coffeanum.*

The mechanism of fungus-induced protection is not clearly understood. Nonsystemic protection of potato tubers after exposure to avirulent isolates of *Phythophthora infestans* was attributed to the accumulation of antimicrobial substances (phytoalexins) in colonized tissue (Müller and Börger, 1940). Systemic protection of cucumbers against *C. langenarium* was suggested to be due to the reduced ability of the challenge fungus to penetrate the host tissue (Jenns and Kuć, 1977). This conclusion was confirmed by Richmond *et al.* (1979) who showed that removing a portion of leaf epidermal tissue of protected susceptible cucumber resulted in a marked reduction of protection. Jenns and Kuć (1980) showed that the level of protection (reduction of fungal penetration) against *C. lagenarium* in plants infected with tobacco necrosis virus was reduced when spores of *C. lagenarium* were injected into leaves, bypassing the epidermis.

Hammerschmidt and Kuć (1982) suggested that induced systemic resistance of cucumber against *C. lagenarium* by *C. lagenarium* might be associated with a rapid lignification of leaf epidermal tissue directly under fungus appressoria. The rate of lignin synthesis was found to be higher in muskmelon protected against *C. lagenarium* than in unprotected controls (Touze and Rossignol, 1977).

Kuć and Caruso (1977) have proposed that systemic protection in cucurbits against *C. lagenarium* may involve the following three mechanisms: (1) restriction of fungus penetration into the host tissue; (2) resistance to fungus spread brought about by agglutination of fungal hyphae by an agglutinating factor; and (3) production of phytoalexins in the infection site. In addition to agglutinating the developing hyphae, the agglutinating factor may inactivate pathogen-produced enzymes and toxins. Implicit in this hypothesis is the assumption that induction of resistance following contact with the inducer (the first inoculum) is associated with production of metabolites which can act against the challenger (the second inoculum). An alternative hypothesis advanced by McIntyre *et al.* (1981) assumes that infection of the host with the inducer results in the synthesis of a translocatable signal. While the signal is, by itself, inactive against the challengers, it may condition the plant to respond to challengers with an appropriate defensive response such as the production of antichallenger substances. McIntyre *et al.* (1981) reasoned that the signal scheme has an evolutionary advantage over direct synthesis of antichallenger substance(s) because of its greater economy. In plants with signal strategy genes responsible for defense (the formation of antichallenger substances) are switched off in the absence of the challenger and energy thus saved is used for increased production. This argument, however, is based on the assumption that the synthesis of a signal molecule(s) requires less energy than the synthesis of a metabolite(s) active against the challenger, an assumption which may or may not be correct depending on factors such as the nature of the hypothetical signal molecule and its effective concentration.

Jenns and Kuć (1980) reported that resistance to anthracnose induced systemically in cucumber plants by localized infection with tobacco necrosis virus, *Colletotrichum lagenarium,* and *Pseudomonas lachrymans* appeared dependent upon the presence of common mechanisms. They also suggested that the induced resistance was likely to be dependent upon a common function of the infectious agents and not upon the infectious agents themselves. The hypothetical mechanisms of induced protection by various biotic agents have been discussed by Kuć (1981), Matta (1980), and Kuč and Caruso (1977).

## III. INDUCED RESISTANCE TO BACTERIA

Inoculation of plants with certain bacteria has been shown to induce resistance to subsequent challenge inoculation by the same or other bacteria (Smith *et al.,* 1911; Brown, 1923; Lovrekovich and Farkas, 1965; Goodman, 1967;

Lippincott and Lippincott, 1969; Graham *et al.*, 1977). The nature of resistance induced by bacteria is not well understood. Sequeira *et al.* (1977) and Graham *et al.* (1977) have found that injection into tobacco leaves of heat-killed cells of avirulent *Pseudomonas solanacearum* as well as purified lipopolysaccharides (LPS) from both virulent (smooth) and avirulent (rough) strains of the bacterium induces resistance to the subsequent challenge by living virulent bacterial cells. Based on their electron microscopic studies, *Graham et al.* (1977) suggested that the protection phenomenon in the above system may be due to the inability of the challenge bacterial cells to attach to the binding sites on the host cell walls which are already occupied by dead cells or purified LPS (see Chapter 14). Mazzucchi *et al.* (1979) have also shown that injection of a protein–LPS complex from a compatible strain of *P. tabaci* and from two incompatible strains of *P. lachrymans* and *P. aptata* into tobacco leaves caused a delay or inhibition of development of disease symptoms by five compatible strains of *P. tabaci*. Protection of pear against *Erwinia amylovora,* the causal agent of fire blight, by *E. amylovora* has been attributed to DNA from virulent and avirulent isolates of the bacterium (McIntyre *et al.*, 1975). The literature on induced resistance to bacteria has recently been reviewed by Goodman *et al.* (1976a), Sequeira (1979), and Goodman (1980).

## IV. INDUCED RESISTANCE TO VIRUSES

Resistance to viruses is induced in certain plants following infection with the same or different viruses (Ross, 1961, 1965; Loebenstein, 1962, 1972; Balázs *et al.*, 1977; McIntyre *et al.*, 1981), with bacteria (Klement *et al.*, 1966; Loebenstein and Lovrekovich, 1966; Sziráki *et al.*, 1980), and with fungi (Hecht and Bateman, 1964; Mandryk, 1963).

The mechanism of induced resistance to viruses is not known (Zaitlin, 1976). Ross (1965) has suggested that the induction of necrotic local lesions in tobacco varieties that respond hypersensitively to TMV, results in formation of translocatable substances capable of activating mechanisms that limit local lesion size in noninfected areas. The nature of protecting factors is not known. Protection of upper leaves of tobacco against TMV following inoculation of lower leaves with the virus has been attributed to an RNA factor (Sela *et al.*, 1966). Loebenstein *et al.* (1966) have discussed the possibility that interferonlike substances (nonspecific antiviral proteins) may be involved in induction of resistance to viruses. Other proposed hypothetical mechanisms discussed by Hamilton (1980) include the inhibition of replication of RNA of the challenger virus (the second virus) by the proteins or by RNA-dependent RNA polymerase of the inducer virus (the first virus) and depletion of essential metabolites required for the synthesis of the challenger virus. The literature on the mechanism of induced resistance to viruses has recently been reviewed by Hamilton (1980), and more briefly by Sequeira (1979).

## V. INDUCED RESISTANCE TO VIROIDS

Preinoculation of tomato plants with a mild strain of potato spindle tuber viroid protected plants from a severe strain of the viroid (Fernow, 1967). Induced resistance also has been demonstrated with citrus exocortis, chrysanthemum stunt, and chrysanthemum chlorotic mottle viroids (Niblett *et al.*, 1979).

## VI. SOME FEATURES OF INDUCED RESISTANCE

While the diversity of responses and the presence of exceptions do not allow generalization, following are some of the features of the induced resistance phenomenon discussed by Sequeira (1979):

1. The induction of resistance seems to be dependent on light and temperature.
2. The induction of resistance is not immediate and requires a time interval between initial and challenge inoculation.
3. In many plant–pathogen systems the induced resistance is systemic; that is, it spreads from the site of initial inoculation. However, examples of localized induced resistance are also known.
4. Induced resistance seems to lack specificity with respect to the inducer and challenger. Systemic resistance was induced in a hypersensitive tobacco cultivar (*Nicotiana tabacum*) against TMV, *Phytophthora parasitica* var. *nicotianae, Pseudomonas tabaci,* and *Peronospora tabacina* following inoculation with TMV (McIntyre *et al.*, 1981). Reproduction of green peach aphid was also reduced on tobacco plants infected with TMV.
5. In some cases the acquired protection (resistance) has been found to be long lasting. For example, systemic resistance induced by TMV inoculation in tobacco varieties that are hypersensitive to the virus lasted for about 20 days (Ross, 1961). Similarly, the systemic protection elicited in cucumbers against *Colletotrichum lagenarium* with *C. lagenarium* lasted up to 5 weeks and could be extended by a booster inoculation (Kuć and Richmond, 1977).
6. The persistance of the systemic protection induced by fungi seems to depend on the survival of the initial inoculum.

*Chapter 14*

# Specificity in Plant–Pathogen Interactions



## I. INTRODUCTION

There is a remarkable degree of specificity in plant–pathogen interactions. That is, despite the presence of a diverse array of potential pathogens in the environment, only a few plant species or cultivars become infected. Moreover, only certain biotypes or strains of plant pathogens are capable of inducing disease in certain cultivars of a single plant species. Specificity is not confined only to plant–pathogen interactions, it is also a feature of animal–pathogen and plant–plant interactions. Brian (1976) has recognized five levels of specificity: First, the distinction between pathogens and absolute nonpathogens; second, the distinction between host and nonhost for a particular pathogen; third, the distinction among specialized races of some microorganisms in terms of their ability to attack a certain range of host species; fourth, specificity pertaining to the ability of certain pathogens to attack only a single plant species, or a few closely related species; and, finally, specificity related to the restriction of physiological races of a pathogen to a few genotypes of the host species.

In a discussion of host–parasite specificity, Heath (1981a) has emphasized the need to distinguish two types of specificity, one at species level (basic compatibility), and the other at race-cultivar level. Like Ward and Stoessl (1976), Heath (1981a) considers compatibility as a form of accommodation of the host or the pathogen. Host species specificity involves accommodation of the pathogen; race-cultivar specificity entails accommodation of the host. Compatibility at the host level may be achieved when a pathogen accommodates itself by avoiding or inactivating preformed or induced defense strategies of the plant, by killing host cells prior to the expression of resistance response, or by preventing elicitation of resistance mechanisms of the host. Incompatibility at the race-cultivar level may reflect the ability of the resistant cultivar to negate pathogen-produced factor(s) which contribute to susceptibility.

## II. MECHANISMS OF SPECIFICITY

The mechanism of specificity in plant–pathogen interactions is not well understood. However, in recent years, remarkable progress has been made toward understanding the nature of this phenomenon. The following mechanisms are considered to account for specificity in certain host–pathogen systems.

### A. Hypersensitive Response

It is generally, though not universally, accepted that in certain plant–pathogen systems resistance is mediated by the hypersensitive response (HR). HR-induced resistance is thought to be due to the cessation of the pathogen development following hypersensitive death of the infected tissue or the result of an accumulation of toxic quantities of certain antimicrobial substances in tissues undergoing the HR. Assuming that the HR is the determinant of resistance in certain host–parasite systems, specificity in such systems may conceivably be mediated by the ability of the pathogens to induce or suppress the HR. The involvement of the HR in resistance and particularly in specificity is not presently supported by compellingly strong evidence. The controversy surrounding the assumption that the HR is the determinant of resistance was discussed in Chapter 12.

### B. Phytoalexins

The involvement of phytoalexins in specificity is highly speculative. However, since phytoalexins may contribute to resistance in certain plant–pathogen systems, they may also play certain roles in specificity. A hypothetical phytoalexin-mediated specificity may be based on binding of pathogen-produced phytoalexin suppressors or elicitors to host receptors, on the rapidity of phytoalexin elicitation, on the rate of phytoalexin accumulation (the rate of phytoalexin synthesis and degradation), and on the sensitivity of the pathogen to the accumulated phytoalexins. For further discussion see Kuć (1976b) and Bushnell and Rowell (1981).

It has been speculated that specificity may be determined by the specific elicitors of phytoalexins, such as those isolated from *Phytophthora megasperma* f. sp. *glycinea* (Keen, 1975; Keen and Bruegger, 1977) and *Pseudomonas glycinea* (Bruegger and Keen, 1979). However, there is presently no evidence that these specific elicitors function *in vivo*. Moreover, the majority of the phytoalexin elicitors studied thus far proved to be nonspecific (Chapter 12). The role of phytoalexins in specificity has been discussed by Kuć (1976b) and Oku *et al.* (1979).

## C. Selective Plant Toxins

As already mentioned in Chapter 4, a few selective fungus-produced phytotoxins, particularly victorin produced by *Helminthosporium victoriae*, have provided the best example of pathogen specificity (Wheeler, 1976b; Scheffer, 1976). These selective toxins are produced only by pathogenic fungal isolates and in many cases are capable of producing all the known characteristic symptoms of the disease when applied to susceptible plants. The loss of pathogenicity of fungal isolates is correlated with the loss of toxin-producing ability. Moreover, plant cultivars which are susceptible to the toxin-producing pathogens are highly sensitive to their toxins, while resistant cultivars are insensitive. The controversy surrounding a number of hypotheses advanced to explain the mechanism of toxin-induced specificity is discussed in Chapter 4. The role of selective toxins in specificity has been discussed by Durbin and Steele (1979). For further discussion see also a recent book edited by Durbin (1981).

## D. Common Antigens

The basic tenet of the common antigen hypothesis, originated from the work on the flax–*Melampsora lini* system (Doubly *et al.*, 1960), is that tolerance (compatibility) is characterized by greater antigenic similarity between the host and the pathogen than resistance (incompatibility). Stated differently, in compatible interactions, hosts and parasites share certain common (cross-reactive) antigens not present in incompatible interaction. For example, in the flax–*M. lini* system, compatibility was found to be correlated with the presence of a serologically similar protein in the host and pathogen (Doubly *et al.*, 1960). Presence of common antigens only among parasites and their compatible plant hosts has been reported in a few other host–parasite systems such as cotton–*Xanthomonas malvacearum* (Schanthorst and DeVay, 1963), sweet potato–*Ceratocystis fimbriata* (DeVay *et al.*, 1967), potato–*Phytophthora infestans* (Palmerley and Callow, 1978), and possibly maize–*Ustilago maydis*

(Wimalajeewa and DeVay, 1971). However, in the following systems, antigen sharing between hosts and parasites was a feature of both compatible and incompatible combinations: cotton–*Fusarium oxysporum* f. sp. *vasinfectum,* cotton–*Verticillium albo-atrum* (Charudattan and DeVay, 1972), and cotton–*Meloidogyne incognita* (McClure *et al.*, 1973; Misaghi and McClure, 1974). Moreover, common antigens were not found between alfalfa and *Corynebacterium insidiosum* (Carroll *et al.,* 1972).

DeVay *et al.* (1981) have determined the cellular location of a major cross-reactive antigenic substance shared by cotton and a number of cotton pathogens. Results of indirect fluorescein isothiocyanate staining of antibodies against a purified common antigen from *Fusarium oxysporum* f. sp. *vasinfectum* in cotton roots showed that the common antigen was mainly present around xylem vessels, the endodermis, and the epidermis in cotton root sections cut below the root hair zone. The common antigen was largely present in hyphal tips and patchlike areas on the conidia of the fungus. The purified fungal cross-reactive antigen was partially characterized as mainly carbohydrate with less than 10% protein (Charudattan and DeVay, 1981).

While the presence of common antigens in certain plant–pathogen systems is highly significant, their influence on the outcome of host–pathogen interaction and their role in compatibility are not known. It has been postulated that in host–pathogen interactions where common antigens between host and parasite are present, compatibility may be due to the ability of the host to recognize a common protein in the pathogen. The host therefore responds to the pathogen in a favorable (compatible) manner allowing it to ramify through the tissue without responding defensively. In the absence of common antigens the resistance response is invoked, leading to incompatibility (DeVay, 1976). Certain polysaccharide and protein–polysaccharide complexes are known to function as suppressors of phytoalexin elicitation and the hypersensitive response (Doke *et al.*, 1979; Mazzucchi *et al.*, 1979). Charudattan and DeVay (1981) discussed the possibility that the cross-reactive antigen in *F. oxysporum* f. sp. *vasinfectum,* which is a protein–carbohydrate complex, may also have similar functions. The validity of this hypothesis remains to be established. The significance of antigen sharing in host–parasite interactions has been discussed by DeVay *et al.* (1972), DeVay (1976), and DeVay and Adler (1976).

## E. Recognition Phenomenon

Upon contact, two organisms may recognize each other through interaction of complimentary surface macromolecules. Recognition of two interacting organisms is considered to be of great importance because the outcome of their interaction may be determined by the nature of such recognition. For example, the type of responses that result in either compatibility or incompatibility in host–parasite interactions may be governed by the nature of the messages transmitted across the interacting surfaces of the two organisms. The participation of surface-borne components in plant–parasite recognition has been inferred by the presence on the plant surfaces of certain macromolecules capable of binding specifically to certain pathogens (Hinchi and Clarke, 1980). Moreover, cytoplasm from one species is usually compatible with that of another species (Bushnell, 1976), which suggests that compatibility might be determined more by physical and chemical properties of host–parasite surfaces than by cytoplasmic factors (Bushnell, 1979).

The importance of a direct physical contact between the host and pathogen for phytoalexin production has been clearly demonstrated by Nichols *et al.* (1980). They reported that molecular sieve barriers (with pores as large as 0.4 μm) which physically separated the host and the parasite in the pea–*Fusarium solani* f. sp. *phaseoli* system, prevented phytoalexin induction. Additionally, suppression of fungal growth after the initiation of resistance response required contact between the pathogen and the host cells.

In studies of recognition between the host and pathogen, it is important to distinguish between surface–surface contacts involving recognition and those which are purely physical such as adherence of microorganisms to the inert surfaces (Sequeira, 1980).

Of the macromolecules that participate in recognition, lectins have been studied the most. Lectins are proteins or glycoproteins found in many plant and animal cells (Liener, 1976) capable of binding to glycolipids and glycoproteins that cover the cell surfaces. Lectins are also known as *hemagglutinins* or *phytohemagglutinins* to denote their ability to agglutinate red blood cells. Some lectins can also agglutinate bacteria. The agglutinating ability of lectins is due to the presence of multiple binding sites on the lectins. Cell agglutination by lectins can be prevented by appropriate monomeric and small oligomeric sugars capable of binding to the specific binding sites on the lectins (Albersheim and Anderson–Prouty, 1975; Callow, 1975).

The specificity in surface–surface interactions may be determined by the ability of certain macromolecules, such as lectins, to recognize subtle differences in terminal monosaccharide residues that protrude from the glycolipids and glycoproteins that cover the cell surfaces. The differences among terminal monosaccharide residues are due to the type of linkages between individual monosaccharides and the location of hydroxyl groups on them. Albersheim and Anderson–Prouty (1975) have postulated that pathogen specificity may be mediated by glycosyl transferases which catalyze synthesis of the terminal monosaccharides in surface glycoproteins recognized by plant lectins. Sophisticated levels of specificity may be achieved by mutations that affect the synthesis of these enzymes.

Sequeira (1980) has pointed out that recognition mediated by the interaction of the surface components of host and pathogen is not unique and is reminiscent of several related phenomena in plant and animal systems such as graft rejection, pollen–pistil interaction, fertilization, and tissue differentiation, which are also determined by the interaction of complementary macromolecules.

Recognition of two organisms through interaction of complementary surface macromolecules has been the subject of intensive study in recent years. For reviews on this topic see Liener (1976), Callow (1977), Albersheim *et al.* (1977), Sequeira (1978, 1980), Albersheim and Valent (1978), Clarke and Knox (1978), Siegel (1979), Dazzo (1980), and Frazier and Glazer (1979).

In the following pages, the literature on the nature of surface–surface interaction between plants and a few microorganisms is briefly reviewed. We will see that, depending on the system, recognition is either for compatibility or incompatibility.

### *1. Rhizobium spp.*

Because of the presence of a remarkable degree of specificity among *Rhizobium* species in terms of their ability to nodulate legumes, plant–*Rhizobium* systems are amenable to studies on recognition and its possible role in determining the outcome of host–*Rhizobium* interactions.

Nodulation is initiated by the attachment of compatible bacterial cells to the root hairs of legumes. Recognition of compatible bacterial cells by the plant is thought to trigger certain biochemical and morphological changes which allow the bacterium to penetrate the host cell and to establish infection. While incompatible bacterial cells may sometimes attach to the root hairs, they presumably are not recognized by the host and thus are incapable of causing infection.

The extensive work on recognition between plants and rhizobia was spurred by an early report of Hamblin and Kent (1973) that *Rhizobium phaseoli* bound to blood group A erythrocytes following incubation with a lectin isolated from beans. Bohlool and Schmidt (1974) later showed that 22 of 25 nodulating strains of *R. japonicum* bound to a fluorescein-conjugated lectin from soybeans, the host of the bacterium. Rhizobia incapable of nodulating soybeans were not bound by the lectin. This general pattern was later confirmed by Bhuvaneswari *et al.* (1977), who also showed that 15 of the 22 nodulating strains of the bacterium tested were bound by the fluorescein-conjugated soybean lectin, while nine non-nodulating strains of the bacterium did not bind to the lectin. Despite reported exceptions (Chen and Phillips, 1976; Law and Strijdom, 1977), there seems to be a good correlation between the ability of *R. japonicum* strains to nodulate soybeans and to bind to the soybean lectin.

It has been speculated that specificity in the soybean–*R. japonicum* system depends upon recognition of carbohydrate components of the bacterial cell wall by a surface-borne soybean lectin. However, the evidence for the involvement of lectin in specificity would be strengthened if it could be shown that lectin is present on the surface of soybean root hairs.

A specific lectin named trifoliin, from clover root hairs, has been reported to be involved in recognition between white clover and *R. trifolii* (Dazzo and Hubbell, 1975; Dazzo *et al.*, 1978). Trifoliin was reported to bind to both the infective bacteria and clover roots presumably because of the presence of a trifoliin-binding common carbohydrate antigen on the surface of clover root cells and the infective strains of the bacterium. The infective bacterial strains are thought to be bound selectively to clover roots through trifoliin which forms a bridge between the common antigen on the surface of the bacterium and on the clover roots (Dazzo *et al.*, 1978). This bridging hypothesis is supported by a number of observations. Fluorescein-conjugated antiserum to trifoliin was absorbed mostly to the root hairs of clover and not to those of a number of nonhost legumes. Moreover, receptor sites for trifoliin have not been detected in noninfective *R. trifolii* mutants or other Rhizobium species incapable of nodulating clover (Dazzo *et al.*, 1978). Additional support for the bridging hypothesis would be provided if the presence of trifoliin on the clover root hairs were confirmed and if the binding of nodulating strains of *R. trifolii* to clover roots through trifoliin were directly demonstrated.

The identity of the cell wall components of the rhizobia which participate in recognition is unknown. Wolpert and Albersheim (1976) suggested that the lipopolysaccharide (LPS) portion of the cell wall of the bacteria might be a likely candidate. Carlson *et al.* (1978) compared the composition and immunochemistry of the purified LPS from strains of *R. leguminosarum*, *R. phaseoli*, and *R. trifolii* and found that these properties varied as much among strains of a single *Rhizobium* species as among the three species of *Rhizobium*. They concluded that there was no obvious correlation between the nodulating group to which a *Rhizobium* belongs and the chemical composition or immunochemistry of the bacterial LPS. However, they did not rule out the possibility that structural regions in the LPS, unrelated to immunoactive sites, could be involved in specificity of *Rhizobium*–legume interactions. Robertsen *et al.* (1981) have postulated that extracellular acidic polysaccharides secreted by *Rhizobium* species may play a role in recognition between *Rhizobium* and their host and subsequent entry of the bacteria into their hosts.

While the results of the studies on the soybean–*R. japonicum* and the clover–*R. trifolii* systems tend to show that lectins may play a role in specificity in both systems, the case is somewhat weakened by the presence of certain exceptions, namely the inability of certain nodulating strains of the bacteria to bind with their host lectin and binding of certain nonnodulating strains of the bacteria to lectins. These exceptions may, however, be explained in the light of new findings discussed below. Bhuvaneswari and Bauer (1978) made an interesting observation that the lectin-binding sites on the surface of rhizobia are transitory. That is, their nature and number change with the age and the condition of the medium. For example, certain strains of *R. japonicum* did not possess soybean lectin-binding sites unless they were grown in media containing soybean root exudate. Binding sites for the clover lectin were also present on the surface of *R. trifolii* only at certain periods during the growth of the bacterium (Dazzo *et al.*, 1979) (see also Bauer *et al.*, 1980).

The overall results of the studies on recognition in the host–*Rhizobium* systems do not provide unequivocal proof for the involvement of lectins in the specificity of rhizobia; they nevertheless suggest that lectins may play a role in recognition.

### 2. *Psuedomonas solanacearum*

Klement (1963) found that introduction of incompatible plant pathogenic bacteria into the intercellular spaces of tobacco leaves caused rapid collapse of leaf tissue characteristic of the hypersensitive response (Chapter 12). Sequeira and his colleagues (Sequeira and Graham, 1977) have suggested that the hypersensitive response in tobacco induced by avirulent strains of *P. solanacearum* involves interaction between the LPS portion of the bacterial wall and a lectin from plant cell wall. This interaction is avoided by virulent strains by their ability to secrete a readily soluble extracellular polysaccharide (EPS) which binds to the plant cell wall lectins. The binding of EPS to the host cell wall lectin may result in covering all its available binding sites, preventing attachment of the virulent bacteria to the lectin. The bacteria thus remain free in intercellular spaces to cause infection. This hypothesis is based on the following lines of evidence: First, virulent isolates of *P. solanacearum* which have both LPS and EPS do not attach to potato and tobacco cell walls while avirulent bacterial cells possessing only LPS were attached to these walls. Second, cell walls of potato and tobacco contain a lectin which is a hydroxyproline-rich glycoprotein similar to the glycoproteins present in the cell walls of other plants. Lectins from cell walls of both tobacco and potato agglutinated all avirulent strains of the bacterium tested. Virulent strains either were not agglutinated or were agglutinated only slightly. Third, potato cell wall lectin did not agglutinate avirulent cells in the presence of purified EPS. Fourth, while potato lectin bound LPS from both virulent and avirulent strains of the bacterium, interaction was stronger with LPS from avirulent forms. Finally, cells from a virulent strain of the bacterium were agglutinated by the potato lectin following removal of EPS by repeated washing. The above lines of evidence tend to support the hypothesis that the inability of the virulent strains of the bacterium to induce HR may be due to the inability of these strains to attach to tobacco cell walls by virtue of having EPS. The above hypothesis does not explain why certain isolates of *P. solanacearum* which produce copious amounts of EPS in culture but are not pathogenic to tobacco are capable of inducing a rapid HR in tobacco leaves (Sequeira, 1980).

Sequeira (1980) has pointed out that the ability of EPS to prevent binding of LPS to the lectin is analogous to the ability of a polysaccharide produced by certain strains of *Escherichia coli* to prevent phagocytosis (Rottini *et al.*, 1976).

Certain avirulent strains of *P. solanacearum* are not agglutinated by the lectin and are incapable of inducing the HR. The non-HR-inducing avirulent strains of the bacterium are thought to differ from the HR-inducing avirulent strains with respect to the chemical composition of bacterial LPS. For example, certain rough HR-inducing avirulent isolates of *P. solanacearum* lacking *O*-polysaccharide (*O*-antigen) in their LPS were readily agglutinated by potato lectin while certain non-HR-inducing smooth avirulent strains of the bacterium which appeared to have portions of the *O*-polysaccharide chain in their LPS were agglutinated only slightly by the lectin (Whatley *et al.*, 1979). Results of analyses of LPS components of a number of strains of *P. solanacearum* by gas chromatography showed that LPS from HR-inducing and non-HR-inducing strains of the bacterium were different in sugar composition. Moreover, LPS from the tested strains could be separated into two distinct size classes, corresponding to low-molecular-weight (rough) LPS and high-molecular-weight (smooth) LPS (Whatly *et al.*, 1980). It was concluded that LPS structure of *P. solanacearum* is correlated with the induction of the HR. So far, most of the HR-inducing strains (regardless of EPS production) have rough LPS (Sequeira, personal communication). Whatley *et al.* (1980) have speculated that the initial recognition, a prerequisite for induction of the HR, requires the presence of rough LPS. Smooth strains may not be recognized by the host because in these strains the portion of the LPS which participates in recognition is probably masked by the *O*-antigen.

While LPS by itself cannot induce the HR, both rough and smooth LPS are capable of preventing the HR induced by *P. solanacearum* when they are introduced into the leaves prior to bacterial inoculation (Graham *et al.*, 1977).

As was pointed out in Chapter 12, Hildebrand *et al.* (1980) believe that in bean plants the attachment of a number of pseudomonads to plant cell walls following their introduction into the intercellular spaces of leaves is a purely physical phenomenon unrelated to bacterial interactions with cell wall components.

### *3. Agrobacterium tumefaciens*

*A. tumefaciens* is capable of causing non-self-limited neoplastic growth in many species of dicotyledons (Chapter 11). Tumor initiation in plants requires attachment of the bacterium to a specific wound site. Lippincott and Lippincott (1969) have shown that tumor development could be reduced or inhibited in pinto beans when leaves were inoculated with a mixture of avirulent and virulent cells of the bacterium or with virulent bacteria following inoculation with avirulent bacteria. These results were interpreted to suggest that a bacterial component is involved in site attachment. The bacterial component participating in attachment appears to be a fraction of LPS of the outer envelope. This view is supported by the observation that tumor formation could be inhibited by LPS preparation from avirulent strains of *A. tumefaciens* capable of binding to plant cell walls (Whatley *et al.*, 1976; Whatley, 1977). LPS from bacterial strains incapable of binding to cell walls or from *A. radiobacter* were not effective (Whatley *et al.*, 1976). Moreover, no inhibition was observed if LPS preparations were added to the wounds 15 min after inoculation. It is not yet known which of the three components of LPS (lipid A, R core, or *O*-antigen) is involved in site attachment (Whatley, 1977). Whatley *et al.* (1978) have shown that the genetic information for host transformation during pathogenesis and that for the synthesis of LPS involved in bacterial attachment are carried on the same bacterial plasmid.

The identity of the binding site on the plant cell wall is not known. Based on the following lines of evidence, Lippincott and Lippincott (1977) have suggested that the binding site on the host cell wall may be a pectic polysaccharide: First, tumor formation can be inhibited by polygalacturonic acid at a very low concentration. Second, tumor inhibition properties of the host cell wall can be reduced by treatment with strong acids which remove most of the pectin and hemicelluloses. However, tumor inhibition activity was not significantly affected by treatment of the cell walls with pectinases or cellulases. Third, embryonic bean tissues do not inhibit tumor formation unless they are treated with pectinesterase. Lippincott and Lippincott (1977) have further speculated that the *Agrobacterium* binding site in plant cell walls might be the galacturonic acid residue of the polysaccharide portion of the middle lamellae which is probably exposed after wounding. The involvement of the pectic component of plant cell walls in binding of bacteria is not yet supported by direct evidence; therefore, the possibility of the participation of nonpectic components of the cell wall in binding cannot presently be ruled out. Judging from the results of their study on the interaction of *A. tumefaciens* with potato lectin and concanavalin A, Pueppke and Klueffel (1982) consider it highly unlikely that potato lectin plays a general role in bacterial attachment to wound sites on potato tubers.

Matthysse *et al.* (1982) found that *A. tumefaciens* strains bind specifically to intact cells and protoplasts from carrot and suggested that the specific receptor for the virulent bacterial isolates is located on the plant cell membrane.

It should be noted that recognition in the plant–*P. solanacearum* system is assumed to be for incompatibility while recognition in the legume–*Rhizobium* and plant–*A. tumefaciens* systems is for compatibility.

#### 4. *Erwinia amylovora*

Huang and Huang (1975) showed that an avirulent strain of the fire blight bacterium, *Erwinia amylovora,* lacking an extracellular slime layer, was agglutinated in the xylem vessels of apple petioles following inoculation. Agglutination of the avirulent strain was suggested to be a defense response which immobilizes bacterial cells in the host tissue. Hsu and Goodman (1978) also found that an avirulent strain of *E. amylovora* was agglutinated by filtrates from apple cell-suspension cultures inoculated with a virulent strain of *E. amylovora,* but not with an avirulent strain of the bacterium. The agglutinating factor in apple cell-suspension culture was thought to be largely from the outer envelope of the bacterial cells.

Sequeira (1980) has discussed the possibility that specific colonization of leaf surfaces by certain bacteria may involve recognition. While leaf surfaces covered by a cuticular layer do not seem to be ideal sites for recognition, materials capable of recognizing bacteria, including lectins, may be diffused out from epidermal cells. Lectin or lectinlike materials have indeed been detected on the surfaces of leaves of tobacco and potato by Sequeira and his associates (Sequeira, 1980). Whether lectin or lectinlike materials are present on the leaf surfaces of other plants is not known. Moreover, it is not yet known with certainty that colonization of leaf surfaces by pathogenic bacteria is a specific phenomenon. Bacterial attachment to plant cell walls has been reviewed by Whatley and Sequeira (1981).

### 5. *Fungi*

Interaction of fungi with plants may also involve recognition of surface molecules. Judging from the results of their histological studies on the wheat-stem rust system, Samborski *et al.* (1977) suggested that gene-specific recognition occurs between the cell wall of the fungus and host plasmalemma. Mendgen (1978) found that fluorescein isothiocyanate-labeled germ tubes of bean rust fungus bound only to tissue sections from susceptible plants but not to those from resistant and nonhost plants. Kojima *et al.* (1982) reported that an agglutinating factor from sweet potato roots agglutinated nongerminated spores of seven strains of *Ceratocystis fimbriata* including a strain pathogenic on sweet potato. The factor, however, exhibited agglutinating activity against germinated spores of five strains of the fungus pathogenic on hosts other than sweet potato but not against germinated spores of sweet potato and almond. The agglutinating factor required $Ca^{2+}$ for activity. The factor was a high-molecular-weight, poly- or oligosaccharide associated with membranes and was thought to function as a recognition factor in the above host–pathogen system.

Binding is not always specific. Nozue *et al.* (1979) found that infection hyphae of both compatible and incompatible races of *Phytophthora infestans* bound very firmly to the plasmalemma of potato cells *in vitro*. Furuichi *et al.* (1980) suggested that potato lectin plays an important role in the binding of cell wall surfaces of the infecting hyphae of both compatible and incompatible races of *P. infestans* to potato cell plasmalemma. Results of their electron microscopic observations showed that potato cell membrane vesicles bound to hyphal surfaces of both compatible and incompatible races of the fungus only in the presence of potato lectin. The cell wall surfaces of both compatible and incompatible races of the fungus were thought to possess the same amounts and/or locations of binding sites for the lectin. Furuichi *et al.* (1980) further showed that potato lectin-mediated binding was inhibited in the presence of *N,N'*-diacetylchitobiose, a specific potato lectin hapten, which is also known to inhibit induction of the hypersensitive response of potato cells infected by an incompatible race of *P. infestans* (Nozue *et al.*, 1980). These observations suggest that lectin-mediated binding of fungal hyphae to the host plasmalemma in this host–parasite system may be an essential factor for induction of the hypersensitive response (Furuichi *et al.*, 1980).

Evidently, the host components participating in recognition involving fungi are located on the host plasmalemma. With bacteria, on the other hand, the interacting components seem to be on the plant cell walls. This is to be expected because bacteria do not penetrate host cell walls forcibly and must, therefore, be recognized by plant cells at the cell wall level. With fungi, on the other hand, recognition does not take place until cell walls have been traversed by the fungus and a contact with the host plasmalemma has been made (Sequeira, 1980).

### 6. *Viruses*

The mechanism of specificity in plant–virus interaction is not known. Moreover, there is no direct evidence that interactions between plants and viruses involve recognition. Partridge *et al.* (1974) have found a correlation between seed transmissibility and the presence of glycoproteins in the capsid of plant viruses. A number of common glycoproteins were detected in the capsid of two seed-transmitted viruses but not in that of three non-seed-transmitted viruses. They suggested that the glycoproteins in the capsid of the seed-trasmitted viruses may be involved in attachment and recognition of the viruses to the gametophytic cell surfaces during viral invasion of the gametophytic tissue.

Protoplasts derived from certain plants which are not hosts of certain viruses can be infected by and can support multiplication of the viruses (Takebe and Otsuki, 1969; Takebe, 1975). Similar results were obtained by Beier *et al.* (1977, 1979) who showed that protoplasts of 54 field-immune lines of cowpeas supported the replication of cowpea mosaic virus. Finally, low levels of clover yellow mosaic virus were detected in cell suspension cultures of white clover, which cannot be infected with the virus, following inoculation of the cultures with the virus (Jones *et al.*, 1981). These results have been interpreted to mean that in virus–plant interactions specificity might be determined by the outcome of the events occurring prior to entry of virus particles into the host cytoplasm and not by cytoplasmic factors. Moreover, specificity in certain insect-borne viruses may reside more in vector–virus relations than in plant–virus relations. With plant viruses, specificity is maximal when a vector is involved, less so with mechanical inoculation, and least in cases where protoplasts are exposed to the viruses (Siegel, 1979).

The overall results of studies on recognition between host and pathogen suggest that specificity in a number of host–pathogen systems may be determined during the initial contact between interacting partners through recognition of surface molecules. It has been suggested that recognition involves transfer of messages across the interacting surface molecules to receptor sites where the messages are finally translated into the rejection or acceptance of the pathogen. However, we do not yet know how the messages, if any, are generated, transmitted back and forth, received, translated, and finally executed. The study of recognition between plant and pathogen is incomplete and results and conclusions are tentative. The speculative nature of the recognition between plant host and pathogen has been emphasized by Sequeira (1980).

# Some Useful Books


Abeles, F. B. 1973. *Ethylene in Plant Biology*. Academic Press, New York. 302 pp.

Akai, S., and Ouchi, S., eds. 1971. *Morphological and Biochemical Events in Plant–Parasite Interaction*. The Phythopathological Society of Japan, Tokyo. 415 pp.

Byrde, R. J. W., and Cutting, C. V., eds. 1973. *Fungal Pathogenicity and the Plant's Response*. Academic Press, New York. 499 pp.

Conn, E. E., ed. 1981. *Secondary Plant Products,* Vol. 7. Academic Press, New York. 800 pp.

Crawford, R. L. 1981. *Lignin Biodegradation and Transformation*. Wiley-Interscience, New York. 154 pp.

Daly, J. M., and Uritani, I., eds. 1979. *Recognition and Specificity in Plant Host–Parasite Interactions*. Japan Scientific Societies Press, Tokyo, and University Park Press, Baltimore. 355 pp.

Davies, D. D., ed. 1980. *Metabolism and Respiration,* Vol. 2. Academic Press, New York. 688 pp.

Day, P. R. 1974. *Genetics of Host–Parasite Interaction*. W. H. Freeman and Company, San Francisco. 238 pp.

Deverall, B. J. 1977. *Defense Mechanisms of Plants*. Cambridge University Press, London. 110 pp.

Durbin, R. D., ed. 1981. *Toxins in Plant Disease*. Academic Press, New York. 536 pp.

Friend, J., and Threlfall, D. R., eds. 1976. *Biochemical Aspects of Plant–Parasite Relationships*. Academic Press, London. 354 pp.

Goodman, R. N., Király, Z., and Zaitlin, M. 1967. *The Biochemistry and Physiology of Infectious Plant Disease*. D. Van Nostrand, Princeton, New Jersey. 354 pp.

Gunter, K., ed. 1978. *Biochemistry of Wounded Plant Tissues*. De Gruyter, Berlin, New York. 680 pp.

Hatch, M. D., and Boardman, N. K., eds. 1981. *Photosynthesis,* Vol. 8. Academic Press, New York. 500 pp.

Hedin, P. A., ed. 1977. *Host Plant Resistance to Pests*. ACS Symposium Series 62, American Chemical Society, Washington, D.C. 286 pp.

Heitefuss, R., and Williams, P. H., eds. 1976. *Physiological Plant Pathology, Encyclopedia of Plant Physiology (N.S.),* Vol. 4. Springer-Verlag, New York. 890 pp.

Horsfall, J. G., and Cowling, E. B., eds. 1978. *Plant Disease: An Advanced Treatise,* Vol. 3; *How Plants Suffer From Disease*. Academic Press, New York. 487 pp.

Horsfall, J. G., and Cowling, E. B., eds. 1979. *Plant Disease: An Advanced Treatise,* Vol. 4; *How Pathogens Induce Disease*. Academic Press, New York. 466 pp.

Horsfall, J. G., and Cowling, E. B., eds. 1980. *Plant Disease: An Advanced Treatise,* Vol. 5; *How Plants Defend Themselves*. Academic Press, New York. 534 pp.

Kadis, S., Ciegler, A., and Ajli, S. J., eds. 1972. *Microbial Toxins,* Vol. 8; *Fungal Toxins*. Academic Press, New York. 400 pp.

Király, Z., ed. 1977. *Current Topics in Plant Pathology*. Akadémiai Kiadó, Budapest. 443 pp.

Kozlowski, T. T., ed. 1976. *Water Deficits and Plant Growth,* Vol. 4. Academic Press, New York. 383 pp.

Kozlowski, T. T., ed. 1978. *Water Deficits and Plant Growth,* Vol. 5. Academic Press, New York. 323 pp.

Kozlowski, T. T., ed. 1981. *Water Deficits and Plant Growth,* Vol. 6. Academic Press, New York. 548 pp.

Littlefield, L. J., and Heath, M. C., 1979. *Ultrastructure of Rust Fungi.* Academic Press, New York. 277 pp.

Loewus, F. A., and Ryan, C. A., eds. 1981. *Recent Advances in Phytochemistry, The Phytochemistry of Cell Recognition and Cell Surface Interactions,* Vol. 15. Plenum Press, New York. 288 pp.

Mace, M. E., Bell, A. A., and Beckman, C. H., eds. 1981. *Fungal Wilt Diseases of Plants.* Academic Press, New York. 608 pp.

Matthews, R. E. F. 1981. *Plant Virology.* 2d ed. Academic Press, New York. 858 pp.

Metlitskii, L. V., and Ozeretskovskaya, O. L., 1968. *Plant Immunity: Biochemical Aspects of Plant Resistance to Parasitic Fungi.* Plenum Press, New York. 114 pp.

Mirocha, C. J., and Uritani, I., eds. 1967. *The Dynamic Role of Molecular Constituents in Plant–Parasite Interaction.* Bruce Publishing, St. Paul, MN. 372 pp.

Preiss, J., ed. 1980. *Carbohydrates: Structure and Function,* Vol. 3. Academic Press, New York. 656 pp.

Swain, T., Harborne, J. B., and Van Sumere, C. F., eds. 1979. *Recent Advances in Phytochemistry, Biochemistry of Plant Phenolics,* Vol. 12. Plenum Press, New York.

Tomiyama, K., Daly, J. M., Uritani, I., Oku, H., and Ouchi, S., eds. 1976. *Biochemistry and Cytology of Plant–Parasite Interaction.* Kodansha Ltd., Tokyo and Elsevier Sci. Pub. Co., Amsterdam. 256 pp.

Vanderplank, J. E. 1982. *Host–Pathogen Interactions in Plant Disease.* Academic Press, New York. 207 pp.

Wheeler, H. 1975. *Plant Pathogenesis.* Springer-Verlag, New York. 106 pp.

Wood, R. K. S. 1967. *Physiological Plant Pathology.* Bot. Monograph, Vol. 6. Blackwell Scientific Publications, Oxford, England. 570 pp.

Wood, R. K. S., ed. 1981. *Active Defense Mechanisms in Plants.* Plenum Press, New York. 381 pp.

Wood, R. K. S., and Graniti, A., eds. 1976. *Specificity in Plant Diseases.* Plenum Press, New York. 354 pp.

Wood, R. K. S., Ballio, A., and Graniti, A., eds. 1972. *Phytotoxins in Plant Diseases.* Academic Press, New York. 530 pp.

# References


Abeles, F. B. 1972. Biosynthesis and mechanism of action of ethylene. *Ann. Rev. Plant Physiol.* **23:**259–292.

Abeles, F. B. 1973. *Ethylene in Plant Biology*. Academic Press, New York and London.

Abeles, F. B., Bosshart, R. P., Forrence, L. E., and Habiq, Q. H. 1970. Preparation and purification of glucanase and chitinase from bean leaves. *Plant Physiol. (Bethesda)* **47:**129–134.

Abeles, F. B., Leather, G. R., Forrence, L. E., and Craker, L. E. 1971. Abscission: Regulation of senescence, protein synthesis and enzyme secretion by ethylene. *Hortic. Sci.* **6:**371–376.

Adams, R., Geissman, T. A., and Edwards, J. D. 1960. Gossypol, a pigment of cottonseed. *Chem. Rev.* **60:**555–574.

Addicott, F. T., and Lyon, J. L. 1969. Physiology of abscisic acid and related substances. *Annu. Rev. Plant Physiol.* **20:**139–164.

Aharoni, N., Marco, S., and Levy, D. 1977. Involvement of gibberellins and abscisic acid in the suppression of hypocotyl elongation in CMV-infected cucumbers. *Physiol. Plant Pathol.* **11:**189–194.

Aist, J. R. 1976a. Cytology of penetration and infection. In *Physiological Plant Pathology* (R. Heitefuss and P. M. Williams, eds.), Vol. 4, pp. 197–221. Springer-Verlag, New York.

Aist, J. R. 1976b. Papillae and related wound plugs of plants. *Annu. Rev. Phytopathol.* **14:**145–156.

Aist, J. R. 1977. Mechanically induced wall appositions of plant cells can prevent penetration by a parasitic fungus. *Science* **197:**568–571.

Aist, J. R., and Israel, H. W. 1977a. Timing and significance of papilla formation during host penetration by *Olpidium brassicae*. *Phytopathology* **67:**187–194.

Aist, J. R., and Israel, H. W. 1977b. Effects of heat–shock inhibition of papilla formation on compatible host penetration by two obligate parasites. *Physiol. Plant Pathol.* **10:**13–20.

Aist, J. R., and Williams, P. H. 1971. The cytology and kinetics of cabbage root hair penetration by *Plasmodiophora brassicae*. *Can. J. Bot.* **49:**2023–2034.

Aist, J. R., Kunoh, H., and Israel, H. W. 1979. Challenge appressoria of *Erysiphe graminis* fail to breach preformed papillae of a compatible barley cultivar. *Phytopathology* **69:**1245–1250.

Akai, S. 1959. Histology of defense in plants. In *Plant Pathology* (J. G. Horsfall and A. E. Diamond, eds.), Vol. 1, pp. 391–434. Academic Press, New York.

Akai, S., and Fukutomi, M. 1980. Preformed internal defenses. In *Plant Disease, An Advanced Treatise* (J. G. Horsfall and E. B. Cowling, eds.), Vol. 5, pp. 139–159. Academic Press, New York.

Akai, S., Kunoh, H., and Fukutomi, M. 1968. Histochemical changes of the epidermal cell walls of barley leaves infected by *Erysiphe graminis hordei*. *Mycopathol. Mycol. Appl.* **35:**175–180.

Akai, S., Horino, O., Fukutomi, M., Nakata, A., Kunoh, H., and Shiraishi, M. 1971. Cell wall reaction to infection and resulting change in cell organelles. In *Morphological and Biochemical Events in Plant Parasite Interaction* (S. Akai and S. Ouchi, eds.), pp. 329–347. Phytopathological Society of Japan, Tokyo.

Albersheim, P. 1965. Biogenesis of the cell wall. In *Plant Biochemistry* (J. Bonner and J. E. Varner, eds.), 2nd ed., pp. 298–321. Academic Press, New York.

Albersheim, P. 1969. Biochemistry of the cell wall in relation to infective processes. *Annu. Rev. Phytopathol.* **7:**171–194.

Albersheim, P. 1975. The walls of growing plant cells. *Sci. Am.* **232:**81–95.

Albersheim, P. 1976. The primary cell wall. In *Plant Biochemistry* (J. Bonner and J. E. Varner, eds.), 3d ed., pp. 225–274. Academic Press, New York.

Albersheim, P. 1977. An introduction to the research on elicitors of Phytoalexin accumulation. In *Cell Wall Biochemistry Related to Specificity in Host–Plant Pathogen Interactions* (B. Solheim and J. Raa, eds.), pp. 129–136. Universitetsforlaget, Oslo, Norway.

Albersheim, P. 1978. Concerning the structure and biosynthesis of the primary cell walls of plants. In *Biochemistry of Carbohydrates II* (D. J. Manners, ed.), pp. 127–150. University Park Press, Baltimore.

Albersheim, P., and Anderson, A. 1971. Host pathogen interactions. III. Proteins from plant cell walls inhibit polygalacturonase secreted by plant pathogens. *Proc. Natl. Acad. Sci. U.S.A.* **68:**1815–1819.

Albersheim, P., and Anderson–Prouty, A. J. 1975. Carbohydrates, proteins, cell surfaces, and the biochemistry of pathogens. *Annu. Rev. Plant Physiol.* **26:**31–52.

Albersheim, P., and Valent, B. S. 1978. Host pathogen interactions in plants. Plants, when exposed to oligosaccharides of fungal origin, defend themselves by accumulating antibiotics. *J. Cell. Biol.* **78:**627–643.

Albersheim, P., Neukom, H., and Deuel, H. 1960. Uber die Bildung von ungesattigten Abbauprodukten durch ein pektinabbauendes Enzym. *Helv. Chim. Acta.* **43:**1422–1426.

Albersheim, P., Ayres, A. R., Jr., Valent, B. S., Ebel, J., Hahn, M., Wolpert, J., and Carlson, R. 1977. Plants interact with microbial polysaccharides. *J. Supramol. Struct.* **6:**599–616.

Albersheim, P., McNeil, M., Darvill, A. G., Valent, B. S., Hahn, M. G., Robertsen, B. K., and Aman, P. 1981. Structure and function of complex carbohydrates active in regulating the interactions of plants and their pests. In *Recent Advances in Phytochemistry: The Phytochemistry of Cell Recognition and Cell Surface Interactions* (F. A. Loewus and C. A. Ryan, eds.), Vol. 15, pp. 37–58. Plenum Press, New York.

Allen, E. H., and Kuć, J. 1968. α-Solanine and α-Chaconine as fungitoxic compounds in extracts of Irish potato tubers. *Phytopathology* **58:**776–781.

Allen, P. J. 1942. Changes in the metabolism of wheat leaves induced by infection with powdery mildew. *Am. J. Bot.* **29:**425–435.

Allen, P. J. 1953. Toxins and tissue respiration. *Phytopathology* **43:**221–229.

Amasino, R. M., and Miller, C. O. 1982. Hormonal control of tobacco crown gall tumor morphology. *Plant Physiol.* **69:**389–392.

Ames, I. H., and Mistretta, P. W. 1975. Auxin: Its role in genetic tumor induction. *Plant Physiol. (Bethesda)* **56:**744–746.

Andebrahan, T., and Wood, R. K. S. 1980. The effect of ultraviolet radiation on the reaction of *Phaseolus vulgaris* to species of *Colletotrichum*. *Physiol. Plant Pathol.* **17:**105–110.

Anderson, A. J. 1978. Isolation from three species of *Colletotrichum* of glucan-containing polysaccharides that elicit browning and phytoalexin production in bean. *Phytopathology* **68:**189–194.

Anderson, W. P., ed. 1973. *Ion Transport in Plants*. Academic Press, New York.

Antonelli, E., and Daly, J. M. 1966. Decarboxylation of indoleacetic acid by near-isogenic lines of wheat resistant or susceptible to *Puccinia graminis tritici*. *Phytopathology* **56:**610–618.

Aranda, G., Berville, A., Cassini, R., Fetizon, M., and Poiret, B. 1978. Recherches sur les metabolites produits par les races T et O d' *Helminthosporium maydis*. *Annu. Rev. Phytopathol.* **10:**375–379.

Arneson, P. A., and Durbin, R. D. 1968. The sensitivity of fungi to α-tomatine. *Phytopathology* **58:**536–537.

Arntzen, C. J., Haugh, H. F., and Bobic, S. 1973a. Induction of stomatal closure by *Helminthosporium maydis* pathotoxin. *Plant Physiol. (Bethesda)* **52:**569–574.

Arntzen, C. J., Koeppe, D. E., Miller, R. J., and Peverly, J. H. 1973b. The effect of pathotoxin from *Helminthosporium maydis* (race T) on energy-linked processes of corn seedlings. *Physiol. Plant Pathol.* **3:**79–90.

Asada, Y., Ohguchi, T., and Matsumoto, I. 1979. Induction of lignification in response to fungal infection. In *Recognition and Specificity in Plant Host–Parasite Interactions* (J. M. Daly and I. Uritani, eds.), pp. 99–112. Japan Scientific Societies Press, Tokyo, and University Park Press, Baltimore.

Aspinall, G. O. 1970. *Polysaccharides*. Pergamon Press, New York.

Atkinson, D. E. 1968. The energy charge of the adenylate pool as a regulatory parameter. Interaction with feedback modifiers. Biochemistry **7:**4030–4034.

Atkinson, D. E. 1971. Adenine nucleotides as stoichiometric coupling agents in metabolism and as regulatory modifiers: the adenylate energy charge. In *Metabolic Pathways* (H. J. Vogel, ed.), Vol. 5, pp. 1–21. Academic Press, New York.

Aust, H. J., Domes, W., and Kranz, J. 1977. Influence of $CO_2$ uptake of barley leaves on inoculation period of powdery mildew under different light intensities. *Phytopathology* **67:**1469–1472.

Axelrood–McCarthy, P. E., and Linderman, R. G. 1981. Ethylene production by cultures of *Cylindrocladium floridanum* and *C. scoparium*. *Phytopathology* **71:**825–830.

Ayres, A. E., Valent, B., Ebel, J., and Albersheim, P. 1976. Host–pathogen interactions. XI. Composition and structure of wall-released elicitor fractions. *Plant Physiol. (Bethesda)* **57:**766–774.

Ayres, A. R., Ebel, J., Finelli, F., Berger, N., and Albersheim, P. 1976a. Host–pathogen interactions IX. Quantitative assays of elicitor activity and characterization of the elicitor present in the extracellular medium of cultures of *Phytophthora megasperma* var. *sojae*. *Plant Physiol. (Bethesda)* **57:**751–759.

Ayers, A. R., Ebel, J., Valent, B., and Albersheim, P. 1976b. Host–pathogen interactions. X. Fractionation and biological activity of an elicitor isolated from the mycelial walls of *Phytophthora megasperma* var. *sojae*. *Plant Physiol. (Bethesda)* **57:**760–765.

Ayers, A. R., Valent, B., Ebel, J., and Albersheim, P. 1976c. Host–pathogen interactions. XI. Composition and structure of wall-released elicitor fractions. *Plant Physiol. (Bethesda)* **57:**766–774.

Ayers, A. R., Ayers, S. B., and Goodman, R. N. 1979. Extracellular polysaccharide of *Erwinia amylovora:* a correlation with virulence. *Appl. Environ. Microbiol.* **38:**659–666.

Ayres, P. G. 1977. Effects of powdery mildew *Erysiphe pisi* and water stress upon the water relations of pea. *Physiol. Plant Pathol.* **10:**139–145.

Ayres, P. G. 1978. The water relations of diseased plants. In *Water Deficits and Plant Growth* (T. T. Kozlowski, ed.), Vol. 5, pp. 1–60. Academic Press, New York.

Ayres, P. G. 1981. Effects of disease on plant water relation. In *Effects of Disease on the Physiology of the Growing Plant* (P. G. Ayres, ed.), pp. 131–148. Cambridge University Press, Cambridge.

Ayres, P. G., and Woolacott, B. 1980. Effect of soil water level on the development of adult plant resistance to powdery mildew in barley. *Ann. of Appl. Biol.* **94:**255–263.

Ayers, S. B., Goodman, R. N., and Stoffl, P. 1977. A simple, highly reproducible and sensitive bioassay for the fire-blight toxin, amylovorin. *Proc. Am. Phytopathol. Soc.* **4:**107 (Abstr.).

Bachmann, O., and Blaich, R. 1979. Vorkommen und Eigenschaften kondensierter Tannine in Vitaceen. *Vitis* **18:**106–116.

Backman, P. A., and DeVay, J. E. 1971. Studies on the mode of action of and biogenesis of the phytotoxin syringomycin. *Physiol. Plant Pathol.* **1:**215–234.

Bailey, J. A. 1974. The relationship between symptom expression and phytoalexin concentration in hypocotyls of *Phaseolus vulgaris* infected with *Colletotrichum lindemuthianum*. *Physiol. Plant Pathol.* **4:**477–488.

Bailey, J. A., and Burden, R. S. 1973. Biochemical changes and phytoalexin accumulation in *Phaseolus vulgaris* following cellular browning caused by tobacco necrosis virus. *Physiol. Plant Pathol.* **3:**171–177.

Bailey, J. A., and Deverall, B. 1971. Formation and activity of phaseollin in the interaction between bean hypocotyls (*Phaseolus vulgaris*) and physiological races of *Colletotrichum lindemuthianum. Physiol. Plant Pathol.* **1:**435–449.

Bailey, J. A., and Ingham, J. L. 1971. Phaseollin accumulation in bean (*Phaseolus vulgaris*) in response to infection by tobacco necrosis virus and the rust *Uromyces appendiculatus. Physiol. Plant Pathol.* **1:**451–456.

Bailey, J. A., Vincent, G. G., and Burden, R. S. 1974. Diterpenes from *Nicotiana glutinosa* and their effect on fungal growth. *J. Gen. Microbiol.* **85:**57–64.

Bailey, J. A., Burden, R. S., Mynett, A., and Brown, C. 1977. Metabolism of phaseollin by *Septoria nodorum* and other non-pathogens of *Phaseolus vulgaris. Phytochemistry* **16:**1541–1544.

Bailiss, K. W. 1974. The relationship of gibberellin content to cucumber mosaic virus infection of cucumber. *Physiol. Plant Pathol.* **4:**73–80.

Bailiss, K. W., and Wilson, I. M. 1967. Growth hormones and the creeping thistle rust. *Ann. Bot. (Lond.)* **31:**195–211.

Baker, C. J., and Bateman, D.F. 1978. Cutin degradation by plant pathogenic fungi. *Phytopathology* **68:**1577–1584.

Baker, C. J., Whalen, C. H., and Bateman, D. F. 1977. Xylanase from *Trichoderma pseudokoningii:* Purification, characterization, and effects on isolated plant cell walls. *Phytopathology* **67:**1250–1258.

Baker, C. J., Whalen, C. H., Korman, R. Z., and Bateman, D. F. 1979. α-L-arabinofuranosidase from *Sclerotinia sclerotiorum:* Purification, characterization and effects on plant cell walls and tissue. *Phytopathology* **69:**789–793.

Balázs, E., Gáborjányi, R., Tóth, A., and Király, Z. 1969. Ethylene production in Xanthi tobacco after systemic and local virus infections. *Acta. Phytopathol. Acad. Sci. Hung.* **4:**355–358.

Balázs, E., Gáborjányi, R., and Király, Z. 1973. Leaf senescence and increased virus susceptibility in tobacco: The effect of abscisic acid. *Physiol. Plant Pathol.* **3:**341–346.

Balázs, E., Sziroki, I., and Király, Z. 1977. The role of cytokinins in the systemic acquired resistance of tobacco hypersensitive to tobacco mosaic virus. *Physiol. Plant Pathol.* **11:**29–37.

Ballio, A. 1978. Fusicoccin, the vivotoxin of *Fusicoccum amygdali*. Chemical properties and biological activity. *Ann. Phytopathol.* **10:**145–156.

Ballio, A., Brufani, M., Casinovi, C. G., Cerrini, S., Fedeli, W., Pellicciari, R., Santurbano, B., and Vaciago, A. 1968. The structure of Fusicoccin A. *Experientia* (*Basel*) **24:**631–635.

Ballio, A., Casinovi, C. G., Framondino, M., Grandolini, G., Randazzo, G., and Rossi, C. 1972. The structure of three isomers of monodeacetylfusicoccin. *Experientia (Basel)* **28:**1150–1151.

Barna, B., Érsek, T., and Mashaal, S. F. 1974. Hypersensitive reaction of rust-infected wheat in compatible host–parasite relationships. *Acta Phytopathol. Acad. Sci. Hung.* **9:**293–300.

Barrs, H. D. 1968. Determination of water deficits in plant tissues. In *Water Deficits and Plant Growth* (T. T. Kozlowski, ed.), Vol. 1, pp. 235–268. Academic Press, New York and London.

Barz, W., and Hoesel, W. 1979. Metabolism and degradation of phenolic compounds in plants. In *Recent Advances in Phytochemistry: Biochemistry of Plant Phenolics* (T. Swain, J. B. Harborne, and C. F. Van Sumere, eds.), Vol. 12, pp. 339–369. Plenum Press, New York.

Basham, H. G. 1974. The role of pectolytic enzymes in the death of plant cells. Ph.D. Thesis. Cornell University, Ithaca, New York.

Basham, H. G., and Bateman, D. F. 1975a. Relationship of cell death in plant tissue treated with a homogeneous endo-pectate lyase to cell wall degradation. *Physiol. Plant Pathol.* **5:**249–261.

Basham, H. G., and Bateman, D. F. 1975b. Killing of plant cells by pectic enzymes: The lack of direct injurious interaction between pectic enzymes or their soluble reaction products and plant cells. *Phytopathology* **65:**141–153.

Bashan, Y., Okon, Y., and Henis, Y. 1980. Ammonia causes necrosis in tomato leaves infected with *Pseudomonas tomato* (Okabe) Alstatt. *Physiol. Plant Pathol.* **17:**111–119.

Bateman, D. F. 1964a. Cellulase and the *Rhizoctonia* disease of bean. *Phytopathology* **54:**1372–1377.

Bateman, D. F. 1964b. An induced mechanism of tissue resistance to polygalacturonase in *Rhizoctonia*-infected hypocotyls of bean. *Phytopathology* **54:**438–445.

Bateman, D. F. 1966. Hydrolytic and *trans*-eliminative degradation of pectic substances by extracellular enzymes of *Fusarium solani* f. *phaseoli*. *Phytopathology* **56:**238–244.

Bateman, D. F. 1969. Some characteristics of the cellulase system produced by *Sclerotium rolfsii*. (Sacc.). *Phytopathology* **59:**37–42.

Bateman, D. F. 1976. Plant cell wall hydrolysis by pathogens. In *Biochemical Aspects of Plant–Parasite Relationships* (J. Friend and D. R. Threlfall, eds.), pp. 79–103. Academic Press, New York.

Bateman, D. F. 1978. The dynamic nature of disease. In *Plant Disease: An Advanced Treatise* (J. G. Horsfall and E. B. Cowling, eds.), Vol. 3, pp. 53–83. Academic Press, New York.

Bateman, D. F., and Basham, H. C. 1976. Degradation of plant cell walls and membranes by microbial enzymes. In *Physiological Plant Pathology* (R. Heitefuss and P. H. Williams, eds.), Vol. 4, pp. 316–355. Springer-Verlag, Berlin, Heidelberg, and New York.

Bateman, D. F., and Daly, J. M. 1967. The respiratory pattern of *Rhizoctonia*-infected bean hypocotyls in relation to lesion maturation. *Phytopathology* **57:**127–131.

Bateman, D. F., and Jones, T. M. 1976. Depletion of noncellulosic polysaccharides in cell walls of *Phaseolus vulgaris* L. during pathogenesis by *Sclerotium rolfsii*. *Proc. Am. Phytopathol. Soc.* **2:**131 (Abstr.).

Bateman, D. F., and Millar, R. L. 1966. Pectic enzymes in tissue degradation. *Annu. Rev. Phytopathol.* **4:**119–146.

Bateman, D. F., Van Etten, H. D., English, P. D., Nevins, D. J., and Albersheim, P. 1969. Susceptibility to enzymatic degradation of cell walls from bean plants resistant and susceptible to *Rhizoctonia solani* Kuhn. *Plant Physiol.* (*Bethesda*) **44:**641–648.

Bateman, D. F., Jones, T. M., and Yoder, O. C. 1973. Degradation of corn cell walls by extracellular enzymes produced by *Helminthosporium maydis* race T. *Phytopathology* **63:**1523–1529.

Bates, D. C., and Chant, S. R. 1970. Alterations in peroxidase activity and peroxidase isozymes in virus-infected plants. *Ann. Appl. Biol.* **65:**105–110.

Bauer, W. D., Talmadge, K. W., Keegstra, K., and Albersheim, P. 1973. The structure of plant cell walls. II. The hemicellulose of the walls of suspension-cultured sycamore cells. *Plant Physiol.* (*Bethesda*) **51:**174–187.

Bauer, W. D., Bateman, D. F., and Whalen, C. H. 1977. Purification of an endo-β-1,4 galactanase produced by *Sclerotinia sclerotiorum:* Effects on isolated plant cell walls and potato tissue. *Phytopathology* **67:**862–868.

Bauer, W. D., Bhuvaneswari, T. V., Bhagwat, A. A., and Kettering, C. F. 1980. Transient susceptibility of root cells in five common legumes to infection by Rhizobia. *Plant Physiol.* (*Bethesda*) **65** (Suppl. 1:136.)

Beardsley, R. E. 1972. The inception phase in the crown-gall disease. *Prog. Exp. Tumor Res.* **15:**1–75.

Beckman, C. H. 1964. Host responses to vascular infection. *Annu. Rev. Phytopathol.* **2:**231–252.

Beckman, C. H. 1969. The mechanics of gel formation by swelling of stimulated plant cell wall membranes and perforation plates of banana root vessels. *Phytopathology* **59:**837–843.

Beckman, C. H. 1971. The plasticizing of plant cell walls and tylose formation—A model. *Physiol. Plant Pathol.* **1:**1–10.

Beckman, C. H. 1980. Defenses triggered by the invader: Physical defenses. In *Plant Disease: An Advanced Treatise* (J. G. Horsfall and E. B. Cowling, eds.), Vol. 5, pp. 225–245. Academic Press, New York.

Beckman, C. H., and Halmos, S. 1962. Relation of vascular occluding reactions in banana roots to pathogenicity of root-invading fungi. *Phytopathology* **52:**893–897.

Beckman, C. H., Kuntz, J. E., Riker, A. J., and Berbee, J. G. 1953. Host responses associated with the development of oak wilt. *Phytopathology* **43:**448–454.

Beczner, L., and Lund, B. M. 1975. The production of lubimin by potato tubers inoculated with *Erwinia carotovora* var. *atroseptica*. *Acta Phytopathol. Acad. Sci. Hung.* **10:**269–274.

Bedbrook, J. R., and Matthews, R. E. F. 1972. Changes in the proportions of early products of photosynthetic carbon fixation induced by TYMV infection. *Virology* **48:**255–258.

Bednarski, M. A., Izawa, S., and Scheffer, R. P. 1977. Reversible effects of toxin from *Helminthosporium maydis* race T on oxidative phosphorylation by mitochondria from maize. *Plant Physiol. Bethesda* **59:**540–545.

Beer, S. V., and Aldwinckle, H. S. 1976. Lack of correlation between susceptibility to *Erwinia amylovora* and sensitivity to amylovorin in apple cultivars. *Proc. Am. Phytopathol. Soc.* **3:**300 (Abstr.).

Beer, S. V., and Woods, A. C. 1978. Distribution of *Erwinia amylovora* and amylovorin in apple (*Malus pumila*) shoots inoculated with the fire–blight pathogen. *Proc. 4th Int. Conf. Plant Path. Bact.*, Angers.

Beer, S. V., Baker, C. J., Woods, A. C., and Sjulin, T. M. 1977. Amylovorin production *in vitro* and partial characterization. *Proc. Am. Phytopathol. Soc.* **4:**182 (Abstr.).

Beffagna, N., Cocucci, S., and Marré, E. 1977. Stimulating effect of fusicoccin on $K^+$-activated ATPase in plasmalemma preparations from higher plant tissues. *Plant Sci. Lett.* **8:**91–98.

Begg, J. E., and Turner, N. Ç. 1970. Water potential gradients in field tobacco. *Plant Physiol. (Bethesda)* **46:**343–346.

Beiderbeck, R. 1973. Bakterienwand and Tumorinduktion durch *Agrobacterium tumefaciens*. *Z. Naturforsch. Sect. C Biosci.* **28**c:198–201.

Beier, H., Siler, D. J., Russell, M. L., and Bruening, G. 1977. Survey of susceptibility to cowpea mosaic virus among protoplasts of intact plants from *Vigua sinensis* lines. *Phytopathology* **67:**917–921.

Beier, H., Bruening, G., Russell, M. L., and Tucker, C. L. 1979. Replication of cowpea mosaic virus in protoplasts isolated from immune lines of cowpeas. *Virology* **95:**165–175.

Beijersbergen, J. C. M., and Lemmers, C. B. G. 1978. Influence of ethylene produced by *Fusarium oxysporum* on the postulated mechanism underlying the resistance of *Tulipa* sp. tissues to this fungus. Proceedings of the 3rd International Congress of Plant Pathology, Munich, 1978, p. 229.

Bell, A. A. 1964. Respiratory metabolism of *Phaseolus vulgaris* infected with alfalfa mosaic and southern bean mosaic virus. *Phytopathology* **54:**914–922.

Bell, A. A. 1967. Formation of gossypol in infected or chemically irritated tissues of *Gossypium* species. *Phytopathology* **57:**759–764.

Bell, A. A. 1980. The time sequence of defense. In *Plant Disease: An Advanced Treatise* (J. G. Horsfall and E. B. Cowling, eds.), Vol. 5, pp. 53–73. Academic Press, New York.

Bell, A. A. 1981. Biochemical mechanisms of disease resistance. *Annu. Rev. Plant Physiol.* **32:**21–81.

Bell, A. A., and Mace, M. E. 1980. Biochemistry and physiology of resistance. In *Fungal Wilt Diseases of Plants* (M. E. Mace, A. A. Bell, and C. H. Beckman, eds), pp. 431–486. Academic Press, New York.

Bell, A. A., and Presley, J. T. 1969a. Temperature effects upon resistance and phytoalexin synthesis in cotton inoculated with *Verticillium albo-atrum*. *Phytopathology* **59:**1141–1146.

Bell, A. A., and Presley, J. T. 1969b. Heat-inhibited or heat-killed conidia of *Verticillium albo-atrum* induce disease resistance and phytoalexin synthesis in cotton. *Phytopathology* **59:**1147–1151.

Bell, A. A., and Stipanovic, R. D. 1978. Biochemistry of disease and pest resistance in cotton. *Mycopathologia* **65:**91–106.

Ben–Tal, Y. and Marco, S. 1980. Qualitative changes in cucumber gibberellins following cucumber mosaic virus infection. *Physiol. Plant Pathol.* **16:**327–336.

*Bergey's Manual of Determinative Bacteriology*. 1974. (R. E. Buchanan and N. E. Gibbons, eds.). 8th ed. Williams and Wilkins, Baltimore.

Bergman, B. H. H. 1966. Presence of a substance in the white skin of young tulip bulbs which inhibits growth of *Fusarium oxysporum*. *Neth. J. Plant Pathol.* **72:**222–230.

Bergman, B. H. H., and Beijersbergen, J. C. M. 1968. A fungitoxic substance extracted from tulips and its possible role as a protectant against disease. *Neth. J. Plant Pathol.* **74** (1968 Suppl. 1):157–162.

Bhattacharya, P. K., and Shaw, M. 1967. The physiology of host–parasite relations, XVIII. Distribution of tritium-labelled cytidine, uridine, and leucine in wheat leaves infected with stem rust fungus. *Can. J. Bot.* **45:**555–563.

Bhattacharya, P. K., and Shaw, M. 1968. The effect of rust infection on DNA, RNA, and protein in nuclei of Khapli wheat leaves. *Can. J. Bot.* **46:**96–99.

Bhattacharya, P. K., Naylor, J. M., and Shaw, M. 1965. Nucleic acid and protein changes in wheat leaf nuclei during rust infection. *Science* **150:**1605–1607.

Bhuvaneswari, T. V., and Bauer, W. D. 1978. Role of lectins in plant–microorganism interactions. 3. Influence of rhizosphere–rhizoplane culture conditions on soybean lectin-binding properties of rhizobia. *Plant Physiol. (Bethesda)* **62:**71–74.

Bhuvaneswari, T. V., Pueppke, S. G., and Bauer, W. D. 1977. Role of lectins in plant–microorganism interactions. I. Binding of soybean lectin to *Rhizobia*. *Plant Physiol. (Bethesda)* **60:**486–491.

Billett, E. E., and Burnett, J. H. 1978. The host–parasite physiology of the maize smut fungus, *Ustilago maydis*. II. Translocation of $^{14}C$-labelled assimilates in smutted maize plants. *Physiol. Plant Pathol.* **12:**103–112.

Black, H. S., and Wheeler, H. 1966. Biochemical effects of victorin on oat tissues and mitochondria. *Am. J. Bot.* **53:**1108–1112.

Black, L. L., Gordon, D. T., and Williams, P. H. 1968. Carbon dioxide exchange by radish tissue infected with *Albugo candida* measured with an infrared $CO_2$ analyzer. *Phytopathology* **58:**173–178.

Blackhurst, F. M., and Wood, R. K. S. 1963. Resistance of tomato plants to *Verticillium albo-atrum*. *Trans. Br. Mycol. Soc.* **46:**385–392.

Blakeman, J. P. 1971. The chemical environment of the leaf surface in relation to growth of pathogenic fungi. In *Ecology of Leaf Surface Microorganisms* (T. F. Preece and C. H. Dickinson, eds.), pp. 255–268. Academic Press, New York.

Blakeman, J. P., and Atkinson, P. 1976. Evidence for a spore germination inhibitor co-extracted with wax from leaves. In *Microbiology of Aerial Plant Surfaces* (C. H. Dickinson and T. F. Preece, eds.), pp. 441–449. Academic Press, New York.

Blakeman, J. P., and Sztejnberg, A. 1973. Effect of surface wax on inhibition of germination of *Botrytis cinerea* spores on beet root leaves. *Physiol. Plant Pathol.* **3:**269–278.

Bloom, B. R. 1979. Games parasites play: How parasites evade immune surveillance. *Nature (Lond.)* **279:**21–26.

Bohlool, B. B., and Schmidt, E. L. 1974. Lectins: A possible basis for specificity in the *Rhizobium*–legume root nodule symbiosis. *Science* **185:**269–271.

Bomhoff, G., Klapwijk, P. M., Kester, H. C. M., Schilperoort, R. A., Hernalsteens, J. P., and Schell, J. 1976. Octopine and nopaline synthesis and breakdown genetically controlled by a plasmid of *Agrobacterium tumefaciens*. *Mol. Gen. Genet.* **145:**177–181.

Bonde, M. R., Millar, R. L., and Ingham, J. L. 1973. Induction and identification of sativan and vesitol as two phytoalexins from *Lotus cornicalatus*. *Phytochemistry* **12:**2957–2959.

Bottini, A. T., and Gilchrist, D. G. 1981. Phytotoxins. I. A l-aminodimethyl-heptadecapentol from *Alternaria alternata* f. sp. *lycopersici*. *Tetrahedron Lett.* **22:**2719–2722.

Bottini, A. T., Bowen, J. R., and Gilchrist, D. G. 1981. Phytotoxins. II. Characterization of a phytotoxic fraction from *Alternaria alternata* f. sp. *lycopersici*. *Tetrahedron Lett.* **22:**2723–2726.

Boyer, J. S. 1967. Leaf water potentials measured with a pressure chamber. *Plant Physiol. (Bethesda)* **42:**133–137.

Boyer, J. S. 1971. Resistance to water transport in soybean, bean, and sunflower. *Crop Sci.* **11:**403–407.

Bracker, C. E., and Littlefield, L. J. 1973. Structural concepts of host–pathogen interfaces. In *Fungal Pathogenicity and the Plant's Response* (R. J. W. Byrde and C. V. Cutting, eds.), pp. 159–318. Academic Press, New York and London.

Branton, D. 1969. Membrane structure. *Annu. Rev. Plant Physiol.* **20:**209–238.

Brants, D. H. 1965. Relation between ectodesmata and infection of leaves by $C^{14}$-labelled tobacco mosaic virus. *Virology* **26:**554–557.

Braun, A. C. 1969. Abnormal growth in plants. In *Plant Physiology* (F. C. Steward, ed.), Vol. 5B, pp. 379–420. Academic Press, New York and London.

Braun, A. C. 1972. The relevance of plant tumor systems to an understanding of the basic cellular mechanisms underlying tumorigenesis. *Prog. Exp. Tumor Res.* **15:**165–187.

Braun, A. C., and White, P. R. 1943. Bacteriological sterility of tissues derived from secondary crown-gall tumors. *Phytopathology* **33:**85–100.

Braun, A. C., and Wood, H. N. 1976. Suppression of neoplastic state with the acquisition of specialized functions of cells, tissues and organs of crown gall teratomas of tobacco. *Proc. Natl. Acad. Sci. U.S.A.* **73:**496–500.

Brenneman, I. A., and Black, L. L. 1979. Respiration and terminal oxidases in tomato leaves infected by *Phytophthora infestans*. *Physiol. Plant Pathol.* **14:**281–290.

Brian, P. W. 1976. The phenomenon of specificity in plant disease. In *Specificity in Plant Diseases* (R. K. S. Wood and A. Graniti, eds.), pp. 15–22. Plenum Press, New York.

Brian, P. W., Elson, G. W., Hemming, H. G., and Bradley, M. 1954. The plant growth promoting properties of gibberellic acid, a metabolic product of the fungus *Gibberella, Fujikuroi*. *J. Sci. Food. Agric.* **5:**601–612.

Bridge, M. A., and Klarman, W. L. 1973. Soybean phytoalexin hydroxyphaseollin, induced by ultraviolet irradiation. *Phytopathology* **63:**606–609.

Brookhouser, L. W., Hancock, J. G., and Weinhold, A. R. 1980. Characterization of endopolygalacturonase produced by *Rhizoctonia solani* in culture and during infection of cotton seedlings. *Phytopathology* **70:**1039–1042.

Brown, E. A., and Hendrix, F. F. 1981. Pathogenicity and histopathology of *Botryosphaeria dothidea* on apple stems. *Phytopathology* **71:**375–379.

Brown, J. F., Shipton, W. A., and White, N. N. 1966. The relationship between hypersensitive tissue and resistance in wheat seedlings infected with *Puccinia graminis tritici*. *Ann. Appl. Biol.* **58:**279–290.

Brown, N. A. 1923. Experiment with Paris daisy and rose to produce resistance to crown–gall. *Phytopathology* **13:**87–99.

Brown, R. W. 1970. Measurement of water potential with thermocouple psychrometers: Construction and applications. *U.S. Dept. Agr. Forest Serv. Res. Pap.* INT–80. 27 pp.

Brown, W. 1915. Studies in the physiology of parasitism. 1. The action of *Botrytis cinerea*. *Ann. Bot.* (Lond.) **29:**313–348.

Bruegger, B. B., and Keen, N. T. 1979. Specific elicitors of glyceollin accumulation in the *Pseudomonas glycinea*–soybean host–parasite system. *Physiol. Plant Pathol.* **15:**43–51.

Brueske, C. H. 1980. Phenylalanine ammonia lyase activity in tomato roots infected and resistant to the root–knot nematode, *Meloidogyne incognita*. *Physiol. Plant Pathol.* **16:**409–414.

Burden, R. S., and Bailey, J. A. 1975. Structure of the phytoalexin from soybean. *Phytochemistry* **14:**1389–1390.

Burden, R. S., Bailey, J. A., and Dawson, G. W. 1972. Structures of three new isoflavonoids from *Phaseolus vulgaris* infected with tobacco necrosis virus. *Tetrahedron Lett.* **41:** 4175–4178.

Bushnell, W. R. 1976. Reactions of cytoplasm and organelles in relation to host–parasite specificity. In *Specificity in Plant Diseases* (R. K. S. Wood and A. Graniti, eds.), pp. 131–150. Plenum Press, New York.

Bushnell, W. R. 1979. The nature of basic compatibility: Comparisons between pistil–pollen and host–parasite interaction. In *Recognition and Specificity in Plant–Host Parasite Interactions* (J. M. Daly and I. Uritani, eds.), pp. 211–227. Japan Scientific Societies Press, Tokyo and University Park Press, Baltimore.

Bushnell, W. R., and Bergquist, S. E. 1975. Aggregation of host cytoplasm and the formation of papillae and haustoria in powdery mildew of barley. *Phytopathology* **65:**310–318.

Bushnell, W. R., and Rowell, J. B. 1968. Premature death of adult rusted wheat plants in relation to carbon dioxide evolution by root systems. *Phytopathology* **58:**651–658.

Bushnell, W. R., and Rowell, J. B. 1981. Suppressors of defense reactions: A model for roles in specificity. *Phytopathology* **71:**1012–1014.

Byrde, R. F. J., and Fielding, A. H. 1968. Pectin methyl-trans-eliminase as the maceration factor of *Sclerotinia fructigena* and its significance in brown rot of apple. *J. Gen. Microbiol.* **52:**287–297.

Byrde, R. J. W., Fielding, A. H., Archer, S. A., and Davies, E. 1973. The role of extracellular enzymes in the rotting of fruit tissue by *Sclerotinia fructigena*. In *Fungal Pathogenicity and the Plant's Response* (R. J. W. Byrde and C. V. Cutting, eds.), pp. 39–54. Academic Press, New York.

Callow, J. A. 1975. Plant lectins. *Curr. Adv. Plant Sci.* **18:**181–193.

Callow, J. A. 1977. Recognition, resistance and role of plant lectins in host–parasite interactions. *Adv. Bot. Res.* **4:**1–49.

Callow, J. A., Long, D. E., and Lithgow, E. D. 1980. Multiple molecular forms of invertase in maize smut infections. *Physiol. Plant Pathol.* **16:**93–107.

Camm, E. L., and Towers, G. H. N. 1977. Phenylalanine ammonia lyase. In *Progress in Phytochemistry* (L. Reinhold, J. B. Harborne, and T. Swain, eds.), Vol. 4, pp. 169–188. Pergamon Press, Oxford.

Campbell, C. L., Huang, J., and Payne, G. A. 1980. Defense at the perimeter: The outer walls and the gates. In *Plant Disease: An Advanced Treatise* (J. G. Horsfall and E. B. Cowling, eds), Vol. 5, pp. 103–120. Academic Press, New York.

Campbell, G. K., and Deverall, B. J. 1980. The effects of light and a photosynthetic inhibitor on the expression of the Lr20 gene for resistance to leaf rust in wheat. *Physiol. Plant Pathol.* **16:**415–423.

Campbell, R. N., and Grogan, R. G. 1963. Big-vein virus of lettuce and its transmission by *Olipidium brassicae*. *Phytopathology* **53:**252–259.

Carlson, R. W., Sanders, R. E., Napoli, C., and Albersheim, P. 1978. Host–pathogen interactions. 13. Purification and partial characterization of *Rhizobium* lipopolysaccharides. *Plant Physiol. (Bethesda)* **62:**912–917.

Carlton, B. C., Peterson, C. E., and Tolbert, M. E. 1961. Effects of ethylene and oxygen production of a bitter compound by carrot roots. *Plant Physiol. (Bethesda)* **36:**550–552.

Carrasco, A., Boudet, A. M., and Marigo, C. 1978. Enhanced resistance of tomato plants to *Fusarium* by controlled stimulation of their natural phenolic production. *Physiol. Plant Pathol.* **12:**225–232.

Carroll, R. B., Lukezic, F. L., and Levine, R. G. 1972. Absence of a common antigen relationship between *Corynebacterium insidiosum* and *Medicago sativa* as a factor in disease development. *Phytopathology* **62:**1351–1360.

Carroll, T. W., and Kosuge, T. 1969. Changes in structure of chloroplasts accompanying necrosis of tobacco leaves systemically infected with tobacco mosaic virus. *Phytopathology* **59:**953–962.

Caruso, F. 1977. Protection of cucumber against *Colletotrichum lagenarium* by *Pseudomonas lachrymans*. *Proc. Am. Phytopathol. Soc.* **3:**159 (Abstr.).

Caruso, F. L., and Kuć, J. 1977a. Protection of watermelon and muskmelon against *Colletotrichum lagenarium* by *Colletotrichum lagenarium*. *Phytopathology* **67:**1285–1289.

Caruso, F. L., and Kuć, J. 1977b. Field protection of cucumber, watermelon, and muskmelon against *Colletotrichum lagenarium* by *Colletotrichum lagenarium*. *Phytopathology* **67:**1290–1292.

Cason, E. T., Richardson, P. E., Essenberg, M. K., Brinkerhoff, L. A., Johnson, W. M., and Venere, R. J. 1978. Ultrastructural cell wall alterations in immune cotton leaves inoculated with *Xanthomonas malvacearum*. *Phytopathology* **68:**1015–1021.

Castric, P. A., and Strobel, G. A. 1969. Cyanide metabolism of *Bacillus megaterium*. *J. Biol. Chem.* **244:**4089–4094.

Chakravorty, A. K., and Scott, K. J. 1979. Changes in barley leaf ribonucleases during early stages of infection by *Erysiphe graminis* f. sp. *hordei*. *Phytopathology* **69:**369–371.

Chakravorty, A. K., and Shaw, M. 1977. The role of RNA in host–parasite specificity. *Annu. Rev. Phytopathol.* **15:**135–151.

Chakravorty, A. K., Scott, K. J., and Simpson, R. S. 1978. Changes in the action of barley leaf ribonucleases on messenger and ribosomal RNA during powdery mildew infection. *Aust. Plant Pathol. Soc. Newslett.* **7:**44 (Abstr.).

Chakravorty, A. K., Shaw, M., and Scrubb, L. A. 1974a. Ribonuclease activity of wheat leaves and rust infection. *Nature (Lond.)* **247:**577–580.

Chakravorty, A. K., Shaw, M., and Scrubb, L. A. 1974b. Changes in ribonuclease activity during rust infection. I. Characterization of multiple molecular forms of ribonuclease from flax rust grown in host-free media. *Physiol. Plant Pathol.* **4:**313–334.

Chalova, L. I., Ozeretskovskaya, O. L., Yurganova, L. A., Baramidze, V. G., Protsenko, M. A., D'Yakov, Y. T., and Metlitskii, L. V. 1976. Metabolites of phytopathogenic fungi as elicitors of defense reactions in plants (illustrated by the interrelations of potatoes with *Phytophthora infestans*). *Dokl. Akad. Nauk. SSSR* **230:**722–725.

Chalutz, E. 1979. No role for ethylene in the pathogenicity of *Penicillium digitatum*. *Physiol. Plant Pathol.* **14:**259–262.

Chalutz, E., and DeVay, J. E. 1969. Production of ethylene *in vitro* and *in vivo* by *Ceratocystis fimbriata* in relation to disease development. *Phytopathology* **59:**750–755.

Chalutz, E., and Stahmann, M. A. 1969. Induction of pisatin by ethylene. *Phytopathology* **59:**1972–1973.

Chalutz, E., DeVay, J. E., and Maxie, E. C. 1969. Ethylene-induced isocoumarin formation in carrot root tissue. *Plant Physiol. (Bethesda)* **44:**235–241.

Chalutz, E., Lieberman, M., and Sisler, H. D. 1977. Methionine-induced ethylene production by *Penicillium digitatum*. *Plant Physiol. (Bethesda)* **60:**402–406.

Chamberlain, D. W., and Paxton, J. D. 1968. Protection of soybean plants by phytoalexin. *Phytopathology* **58:**1349–1350.

Charudattan, R., and DeVay, J. E. 1972. Common antigens among varieties of *Gossypium hirsutum* and isolates of *Fusarium* and *Verticillium* species. *Phytopathology* **62:**230–234.

Charudattan, R., and DeVay, J. E. 1981. Purification and partial characterization of an antigen from *Fusarium oxysporum* f. sp. *vasinfectum* that cross-reacts with antiserum to cotton (*Gossypium hirsutum*) root antigens. *Physiol. Plant Pathol.* **18:**289–295.

Chatterjee, A. K., and Starr, M. P. 1977. Donor strains of the soft-rot bacterium *Erwinia chrysanthemi* and conjugational transfer of the pectolytic activity. *J. Bacteriol.* **132:**862–869.

Chen, A. P., and Phillips, D. A. 1976. Attachment of *Rhizobium* to legume roots as the basis for specific interactions. *Physiol. Plant Pathol.* **38:**83–88.

Chet, I., Zilberstein, Y., and Henis, Y. 1973. Chemotaxis of *Pseudomonas lachrymans* to plant extracts and to water droplets from leaf surfaces of resistant and susceptible plants. *Physiol. Plant Pathol.* **3:**473–479.

Chi, C. C., and Sabo, F. E. 1978. Chemotaxis of zoospores of *Phytophthora megasperma* to primary roots of alfalfa seedlings. *Can. J. Bot.* **56:**795–800.

Chilton, M. D., Farrand, S. K., Eden, F., Currier, T., Bendich, A. J., Gordon, M. P., and Nester, E. W. 1975. Is there foreign DNA in crown gall tumor DNA. *2nd Ann. John Innes Symp. Modification of the Information Content of Plant Cells.* (R. Markham, D. R. Davies, D. A. Hopwood, and R. W. Horne, eds.), pp. 253–268. American Elsevier, New York.

Chilton, M. D., Farrand, S. K., Levin, R., and Nester, E. W. 1976. RP4 promotion of transfer of a large *Agrobacterium* plasmid which confers virulence. *Genetics* **83:**609–618.

Chilton, M. D., Drummond, M. H., Merlo, D. J., Sciaky, D., Montoya, A. 1., Gordon, M. P., and Nester, E. W. 1977. Stable incorporation of plasmid DNA into higher plant cells—the molecular basis of crown gall tumorigenesis. *Cell* **11:**263–271.

Chilton, M. D., Drummond, M. H., Merlo, D. J., and Sciaky, D. 1978. Highly conserved DNA of Ti-plasmids overlaps T-DNA maintained in plant tumors. *Nature (Lond.)* **275:**147–149.

Chilton, M. D., Saiki, R. K., Yadav, M., Gordon, M. P., and Quetier, F. 1980. T-DNA from *Agrobacterium* Ti-plasmid is in the nuclear DNA fraction of crown gall tumor cells. *Proc. Natl. Acad. Sci. U. S. A.* **77:**4060–4064.

Christman, R. F., and Oglesby, R. T. 1971. Microbiological degradation and the formation of humus. In *Lignins: Occurrence, Formation, Structure and Reactions* (K. V. Sarkamen and C. H. Ludwig, eds.), pp. 769–795. John Wiley and Sons, New York.

Chylinska, K., and Knypl, J. 1975. Decreased phenylalanine ammonia-lyase and ribonuclease activity in side roots of carrots infested with the northern root–knot nematode. *Nematologica* **21:**129–133.

Clancy, F. G., and Coffey, M. D. 1980. Patterns of translocation, changes in invertase activity, and polyol formation in susceptible and resistant flax infected with the rust fungus *Melampsora lini*. *Physiol. Plant Pathol.* **17:**41–52.

Clancy, M., Madill, K. A., and Wood, J. M. 1981. Genetic and biochemical requirements for chemotaxis to L-proline in *Escherichia coli*. *J. Bacteriol.* **146:**902–906.

Clarke, A. E., and Knox, R. B. 1978. Cell recognition in flowering plants. *Q. Rev. Biol.* **53:**3–28.

Cline, K., Wade, M., and Albersheim, P. 1978. Host–pathogen interactions. XV. Fungal glucans which elicit phytoalexin accumulation in soybean also elicit the accumulation of phytoalexins in other plants. *Plant Physiol. (Bethesda)* **62:**918–921.

Cocking, E. C. 1970. Virus uptake, cell wall regeneration, and virus multiplication in isolated plant protoplasts. *Int. Rev. Cytol.* **28:**89–124..

Cohen, J., and Loebenstein, G. 1975. An electron microscope study of starch lesions in cucumber cotyledons infected with tobacco mosaic virus. *Phytopathology* **65:**32–39.

Cole, A. I. J., and Bateman, D. F. 1969. Arabinase production by *Sclerotium rolfsii*. *Phytopathology* **59:**1750–1753.

Cole, A. I. J., and Sturdy, M. L. 1973. Hemicellulolytic enzymes associated with infection of potato tubers by *Fusarium caeruleum* and *Phytophthora erythroseptica*. (Abstr.) *2nd Int. Cong. Plant Pathol.*, Minneapolis, MN 964.

Collins, R. P., and Scheffer, R. P. 1958. Respiratory responses and systemic effects in *Fusarium*-infected tomato plants. *Phytopathology* **48:**349–355.

Comai, L., and Kosuge, T. 1980. Involvement of plasmid deoxyribonucleic acid in indoleacetic acid synthesis in *Pseudomonas savastanoi*. *J. Bacteriol.* **143:**950–957.

Comstock, J. C., and Scheffer, R. P. 1973. Role of host–selective toxin in colonization of corn leaves by *Helminthosporium carbonum*. *Phytopathology* **63:**24–29.

Comstock, J. C., Martinson, C. A., and Gengenback, B. G. 1973. Host specificity of a toxin from *Phyllosticta maydis* for Texas cytoplasmically male-sterile maize. *Phytopathology* **63:**1357–1361.

Cooper, R. M., and Wood, R. K. S. 1975. Regulation of synthesis of cell wall-degrading enzymes by *Verticillium albo-atrum* and *Fusarium oxysporum* F. sp. *lycopersici*. *Physiol. Plant Pathol.* **5:**135–156.

Cooper, R. M., and Wood, R. K. S. 1980. Cell wall degrading enzymes of vascular wilt fungi. III. Possible involvement of endo-pectin lyase in *Verticillium* wilt of tomato. *Physiol. Plant Pathol.* **16:**285–300.

Corsini, D. L., and Pavek, J. J. 1980. Phenylalanine ammonia lyase activity and fungitoxic metabolites produced by potato cultivars in response to *Fusarium* tuber rot. *Physiol. Plant Pathol.* **16:**63–72.

Costantino, P., Mauro, M. L., Micheli, G., Risuleo, G., Hooykaas, P. J. J., and Schilplroort, R. 1981. Fingerprinting and sequence homology of plasmids from different variant strains of *Agrobacterium rhizogenes*. *Plasmid* **5:**170–182.

Cowling, E. B., and Kirk, T. K. 1976. Properties of cellulose and lignocellulosic materials as substrates for enzymatic conversion processes. *Biotechnol. Bioeng. Symp.* **6:**95–123.

Coxon, D. T., Curtis, R. F., Price, K. R., and Howard, B. 1974. Phytuberin: A novel antifungal terpenoid from potato. *Tetrahedron Lett.* **27:**2363–2366.

Creasy, L. L., and Zucker, M. 1974. Phenylalanine ammonia–lyase and phenolic metabolism. *Recent Adv. Phytochem.* **8:**1–19.

Crosse, J. E. 1956. Bacterial canker of stone fruits II. Leaf scar infection of cherry. *J. Hort. Sci.* **31:**212–224.

Crosse, J. E., Goodman, R. N., and Shaffer, W. H., Jr. 1972. Leaf damage as a predisposing factor in the infection of apple shoots by *Erwinia amylovora*. *Phytopathology* **62:**176–182.

Crosthwaite, L. M., and Sheen, S. J. 1979. Inhibition of ribulose 1, 5-biphosphate carboxylase by a toxin isolated from *Pseudomonas tabaci*. *Phytopathology* **69:**376–379.

Cruickshank, I. A. M. 1962. Studies on phytoalexins IV. The antimicrobial spectrum of pisatin. *Aust. J. Biol. Sci.* **15:**147–159.

Cruickshank, I. A. M. 1963. Phytoalexins. *Annu. Rev. Phytopathol.* **1:**351–374.

Cruickshank, I. A. M. 1980. Defenses triggered by the invader: chemical defenses. In *Plant Disease: An Advanced Treatise* (J. G. Horsfall and E. B. Cowling, eds.), Vol. 5, pp. 247–267. Academic Press, New York.

Cruickshank, I. A. M., and Perrin, D. R. 1963. Studies on phytoalexins. VI. Pisatin: The effect of some factors on its formation in *Pisum sativum* L., and the significance of pisatin in disease resistance. *Aust. J. Biol. Sci.* **16:**111–128.

Cruickshank, I. A. M., and Perrin, D. R. 1964. Pathological function and phenolic compounds in plants. In *Biochemistry of Phenolic Compounds* (J. B. Harborne, ed.), pp. 511–544. Academic Press, New York.

Cruickshank, I. A. M., and Perrin, D. R. 1965. Studies on phytoalexins. VIII. The effect of some further factors on the formation, stability, and localization of pisatin *in vivo*. *Aust. J. Biol. Sci.* **18:**817–828.

Cruickshank, I. A. M., and Perrin, D. R. 1971. Studies on phytoalexins. XI. The induction of antimicrobial spectrum and chemical assay of phaseollin. *Phytopathol. Z.* **70:**209–229.

Cruickshank, I. A. M., Perrin, D. R., and Mandryk, M. 1977. Fungitoxicity of duvatriendiols associated with the cuticular wax of tobacco leaves. *Phytopathol. Z.* **90:**243–249.

Currier, W. W. 1974. Characterization of the induction and suppression of terpenoid accumulation in the potato–*Phytophthora* interaction. Ph.D. Thesis. Purdue University, Lafayette, Indiana.

Daly, J. M. 1967. Some metabolic consequences of infection by obligate parasites. In *The Dynamic Role of Molecular Constituents in Plant–Parasite Interaction* (C. J. Mirocha and I. Uritani, eds.), pp. 144–164. Bruce Publishing, St. Paul, MN.

Daly, J. M. 1972. The use of near-isogenic lines in biochemical studies of the resistance of wheat to stem rust. *Phytopathology* **62:**392–400.

Daly, J. M. 1976a. The carbon balance of diseased plants: Changes in respiration, photosynthesis, and translocation. In *Physiological Plant Pathology, Encyclopedia of Plant Physiology, New Series* (R. Heitefuss and P. H. Williams, eds.), Vol. 4, pp. 450–479. Springer-Verlag, Berlin.

Daly, J. M. 1976b. Specific interactions involving hormonal and other changes. In *Specificity in Plant Diseases* (R. K. S. Wood and A. Graniti, eds.), pp. 151–167. Plenum Press, New York.

Daly, J. M. 1976c. Some aspects of host–pathogen interactions. In *Physiological Plant Pathology: Encyclopedia of Plant Physiology, New Series* (R. Heitefuss and P. H. Williams, eds.), Vol. 4, pp. 27–50. Springer-Verlag, Berlin.

Daly, J. M. 1976d. Summary of points from discussions. In *Specificity in Plant Diseases* (R. K. S. Wood and A. Graniti, eds.), pp. 231–234. Plenum Press, New York.

Daly, J. M. 1976e. Induced susceptibility or induced resistance as the basis of host–parasite specificity. In *Biochemistry and Cytology of Plant–Parasite Interaction* (K. Tomiyama, J. M. Daly, I. Uritani, H. Oku, and S. Ouchi, eds.), pp. 144–156. Kodansha Ltd., Tokyo and Elsevier Scientific Publishing Co., Amsterdam.

Daly, J. M. 1981. Mechanisms of action. In *Toxins in Plant Disease* (R. D. Durbin, ed.), pp. 331–394. Academic Press, New York.

Daly, J. M., and Deverall, B. J. 1963. Metabolism of indoleacetic acid in rust diseases. I. Factors influencing rates of decarboxylation. *Plant Physiol. (Bethesda)* **38:**741–750.

Daly, J. M., and Inman, R. E. 1958. Changes in auxin levels in safflower hypocotyls infected with *Puccinia carthami*. *Phytopathology* **48:**91–97.

Daly, J. M., and Knoche, H. W. 1976. Hormonal involvement in metabolism of host–parasite interactions. In *Biochemical Aspects of Plant–Parasite Relationships* (J. Friend and D. R. Threlfall, eds.), pp. 117–133. Academic Press, London.

Daly, J. M., and Sayre, R. M. 1957. Relations between growth and respiratory metabolism in safflower infected by *Puccinia carthami*. *Phytopathology* **47:**163–168.

Daly, J. M., Seevers, P. M., and Ludden, P. 1970. Studies on wheat stem rust resistance controlled at the Sr6 locus. III. Ethylene and disease reaction. *Phytopathology* **60:**1648–1652.

Daly, J. M., Ludden, P., and Seevers, P. M. 1971. Biochemical comparisons of resistance to wheat stem rust disease controlled by the Sr6 or Sr11 alleles. *Physiol. Plant Pathol.* **1:**397–407.

Daniels, D. L., and Hadwiger, L. A. 1976. Pisatin-inducing components in filtrates of virulent and avirulent *Fusarium solani* cultures. *Physiol. Plant Pathol.* **8:**9–19.

Daub, M. E., and Hagedorn, D. J. 1979. Resistance of *Phaseolus* line WBR 133 to *Pseudomonas syringae*. *Phytopathology* **69:**946–951.

Daub, M. E., and Hagedorn, D. J. 1980. Growth kinetics of interactions of *Pseudomonas syringae* with susceptible and resistant bean tissues. *Phytopathology* **70:**429–436.

Davies, W. J. 1981. Transpiration and the water balance of plants. In *Plant Physiology* (J. F. Sutcliffe and F. C. Steward, eds.), Vol. 7, pp. 201–213. Academic Press, London.

Day, P. R. 1974. *Genetics of Host–Parasite Interaction*. W. H. Freeman and Company, San Francisco.

Day, P. R. 1976. Gene functions in host–parasite systems. In *Specificity in Plant Diseases* (R. K. S. Wood and A. Graniti, eds.), pp. 65–73. Plenum Press, New York.

Day, P. R. 1979. Modes of gene expression in disease reaction. In *Recognition and Specificity in Plant Host–Parasite Interactions* (J. M. Daly and I. Uritani, eds.), pp. 19–31. Japan Scientific Societies Press, Tokyo, and University Park Press, Baltimore.

Dazzo, F. B. 1980. Adsorption of microorganisms to roots and other plant surfaces. In *Adsorption of Microorganisms to Surfaces* (G. Bitton and K. C. Marshall, eds.), pp. 253–316. John Wiley and Sons, New York.

Dazzo, F. B., and Hubbell, D. H. 1975. Cross-reactive antigens and lectin as determinants of symbiotic specificity in the *Rhizobium*–clover association. *Appl. Microbiol.* **30:**1017–1033.

Dazzo, F. B., Yanke, W. E., and Brill, W. J. 1978. Trifoliin: A *Rhizobium* recognition protein from white clover. *Biochim. Biophys. Acta.* **539:**276–286.

Dazzo, F. B., Urbano, M. R., and Brill, W. J. 1979. Transient appearance of lectin receptors on *Rhizobium trifolii. Curr. Microbiol.* **2:**15–20.

DeBary, A. 1886. Über einige Sclerotinien und Sclerotienkrankheiten. *Z. Bot.* **44:**433–441.

Dehority, B. A., Johnson, R. R., and Conrad, H. R. 1962. Digestibility of forage hemicellulose and pectin by rumen bacteria *in vitro* and the effect of lignification thereon. *J. Dairy Sci.* **45:**508–512.

Dekhuijzen, H. M. 1976. Endogenous cytokinins in healthy and diseased plants. In *Physiological Plant Pathology* (R. Heitefuss and P. H. Williams, eds.), pp. 527–559. Springer-Verlag, Berlin.

Dekhuijzen, H. M., and Overeem, J. C. 1971. The role of cytokinins in clubfoot formation. *Physiol. Plant Pathol.* **1:**151–162.

Dekhuijzen, H. M., and Staples, R. C. 1968. Mobilization factors in uredospores and bean leaves infected with bean rust fungus. *Contrib. Boyce Thompson Inst.* **24:**39–52.

DeLaat, A. M. M., and Van Loon, L. C. 1982. Regulation of ethylene biosynthesis in virus-infected tobacco leaves. *Plant Physiol.* **69:**240–245.

Delon, R. 1974. Localization of polyphenoloxidase and peroxidase activity in root-cells of tomato plants after infection with *Pyrenochaeta lycopersici. Phytopathol. Z.* **80:**199–208.

Depicker, A., Van Montagu, M., and Schell, J. 1978. Homologous DNA sequences in different Ti-plasmids are essential for oncogenicity. *Nature (Lond.)* **275:**150–153.

DeVay, J. E. 1976. Protein specificity in plant disease development: Protein sharing between host and parasite. In *Specificity in Plant Diseases* (R. K. S. Wood and A. Graniti, eds.), pp. 199–212. Plenum Press, New York.

DeVay, J. E., and Adler, H. E. 1976. Antigens common to hosts and parasites. *Annu. Rev. Microbiol.* **30:**147–168.

DeVay, J. E., Schnathorst, W. C., and Foda, M. S. 1967. Common antigens and host–parasite interactions. In *The Dynamic Role of Molecular Constituents in Plant–Parasite Interaction* (C. J. Mirocha and I. Uritani, eds.), pp. 313–328. Bruce Publishing, Minneapolis, MN.

DeVay, J. E., Lukezic, F. L., Sinden, S. L., English, H., and Coplin, D. L. 1968. A biocide produced by pathogenic isolates of *Pseudomonas syringae* and its possible role in the bacterial canker disease of peach trees. *Phytopathology* **58:**95–101.

DeVay, J. E., Charudattan, R., and Wimalajeewa, D. L. S. 1972. Common antigenic determinants as a possible regulator of host–pathogen compatibility. *Am. Nat.* **106:**185–194.

DeVay, J. E., Wakeman, R. J., Kavanagh, J. A., and Charudattan, R. 1981. The tissue and cellular location of a major cross-reactive antigen shared by cotton and soil-borne-fungal parasites. *Physiol. Plant Pathol.* **18:**59–66.

Deverall, B. J. 1976. Current perspectives in research on phytoalexins. In *Biochemical Aspects of Plant–Parasite Relationships* (J. Friend and D. R. Threlfall, eds.), pp. 207–223. Academic Press, New York.

Deverall, B. J. 1977. Defense mechanisms of plants. *Cambridge Monographs in Experimental Biology, No. 19.* Cambridge University Press, Cambridge.

Deverall, B. J., and Wood, R. K. S. 1961. Chocolate spot of beans (*Vicia faba* L.): Interactions between phenolase of host and pectic enzymes of the pathogen. *Ann. Appl. Biol.* **49**:473–487.

Deverall, B. J., Smith, I. M., and Makris, S. 1968. Disease resistance in *Vicia faba* and *Phaseolus vulgaris*. *Neth. J. Plant Pathol.* **74** (Suppl. 1): 137–148.

De Witt, P. J. G. M., and Roseboom, P. H. M. 1980. Isolation, partial characterization, and specificity of glycoprotein elicitors from culture filtrates, mycelium and cell walls of *Cladosporium fulvum* (syn. *Fulvia fulva*). *Physiol. Plant Pathol.* **16**:391–408.

DeWit-Elshove, A., and Fuchs, A. 1971. The influence of the carbohydrate source on pisatin breakdown by fungi pathogenic to pea *(Pisum sativum)*. *Physiol. Plant Pathol.* **1**:17–24.

DeZoeten, G. A. 1976. Cytology of virus infection and virus transport. In *Physiological Plant Pathology* (R. Heitefuss and P. M. Williams, eds.), Vol. 4, pp. 129–149. Springer–Verlag, Berlin.

DeZoeten, G. A., and Gaard, G. 1969. Possibilities of inter- and intra-cellular translocation of some isocahedral plant viruses. *J. Cell Biol.* **40**:814–823.

Diener, T. D. 1961. Virus infection and other factors affecting ribonuclease activity of plant leaves. *Virology* **14**:177–189.

Dimond, A. E. 1955. Pathogenesis in the wilt disease. *Annu. Rev. Plant Physiol.* **6**:329–350.

Dimond, A. E. 1966. Pressure and flow relations in vascular bundles of the tomato plant. *Plant Physiol. (Bethesda)* **41**:119–131.

Dimond, A. E. 1970. Biophysics and biochemistry of the vascular wilt syndrome. *Annu. Rev. Phytopathol.* **8**:301–322.

Dimond, A. E., and Turner, N. C. 1972. The water status of leaves of healthy and *Fusarium*-infected tomato plants. *Phytopathology* **62**:494 (Abstr.).

Dimond, A. E., and Waggoner, P. E. 1953a. On the nature and role of vivotoxins in plant disease. *Phytopathology* **43**:229–235.

Dimond, A. E., and Waggoner, P. E. 1953b. The cause of epinastic symptoms in *Fusarium* wilt of tomatoes. *Phytopathology* **43**:663–669.

Dixon, G. R., and Pegg, G. F. 1969. Hyphal lysis and tylose formation in tomato cultivars infected by *Verticillium albo-atrum*. *Trans. Br. Mycol. Soc.* **53**:109–118.

Dodman, R. L. 1979. How the defenses are breached. In *Plant Disease: An Advanced Treatise* (J. G. Horsfall and E. B. Cowling, eds.), pp. 135–153. Academic Press, New York.

Doke, N. 1972. Incorporation of $^{14}CO_2$ into free and bound amino acids in tobacco leaves infected with tobacco mosaic virus. *Phytopathol. Z.* **73**: 215–226.

Doke, N. 1975. Preventions of the hypersensitive reaction of potato cells to infection with an incompatible race of *Phytophthora infestans* by constituents of the zoospores. *Physiol. Plant Pathol.* **7**:1–7.

Doke, N., and Hirai, T. 1969. Starch metabolism in tobacco leaves infected with tobacco mosaic virus. *Phytopathol. Z.* **65**:307–317.

Doke, N., and Tomiyama, K. 1977. Effect of high-molecular-weight substances released from zoospores of *Phytophthora infestans* on hypersensitive response of potato tubers. *Phytopathol. Z.* **90**:236–242.

Doke, N., and Tomiyama, K. 1980a. Effect of hyphal wall components from *Phytophthora infestans* on protoplasts of potato tuber tissues. *Physiol. Plant Pathol.* **16**:169–176.

Doke, N., and Tomiyama, K. 1980b. Suppression of hypersensitive response of potato tuber protoplasts to hyphal wall components by water soluble glucans isolated from *Phytophthora infestans*. *Physiol. Plant. Pathol.* **16**:177–186.

Doke, N., Garas, N. A., and Kuć, J. 1979. Partial characterization and aspects of the mode of action of a hypersensitivity-inhibiting factor (HIF) isolated from *Phytophthora infestans*. *Physiol. Plant Pathol.* **15**:127–140.

Doke, N., Garas, N. A., and Kuć, J. 1980. Effect on host hypersensitivity of supressors released during the germination of *Phytophthora infestans* cytospores. *Phytopathology* **70:**35–39.

Doodson, J. K., Manners, J. G., and Myers, A. 1965. Some effects of yellow rust (*Puccinia striiformis*) on[14]carbon assimilation and translocation in wheat. *J. Exp. Bot.* **16:**304–317.

Dorn, F., and Arigoni, D. 1972. The structure of victoxinine. *J. Chem. Soc. Chem. Commun.* **1972:**1342–1343.

Doubly, J. A., Flor, H. H., and Clagett, C. O. 1960. Relation of antigens of *Melampsora lini* and *Linum usitatissimum* to resistance and susceptibility. *Science* **131:**229.

Drlica, K. A., and Kado, C. I. 1974. Quantitative estimation of *Agrobacterium tumefaciens* DNA in crown gall tumor cells. *Proc. Natl. Acad. Sci. U. S. A.* **71:**3677–3681.

Dropkin, V. H., Helgeson, J. P., and Upper, C. D. 1969. The hypersensitivity reaction of tomatoes resistant to *Meloidogyne incognita:* Reversal by cytokinins. *J. Nematol.* **1:**55–61.

Drummond, M. 1979. Crown gall disease. *Nature (Lond.)* **281:**343–347.

Drummond, M. H., Gordon, M. P., Nester, E. W., and Chilton, M. D. 1977. Foreign DNA of bacterial plasmid origin is transcribed in crown gall tumors. *Nature (Lond.)* **269:**535–536.

Drysdale, R. B., and Langcake, P. 1973. Response of tomato to infection by *Fusarium oxysporum* f. *lycopersici*. In *Fungal Pathogenicity and the Plant's Response* (R. J. W. Byrde and C. V. Cutting, eds.), pp. 423–436. Academic Press, London and New York.

Duczek, L. J., and Higgins, V. J. 1976. The role of medicarpin and maackiain in the response of red clover leaves to *Helminthosporium carbonum, Stemphylium botryosum,* and *S. sarcinaeforme*. *Can. J. Bot.* **54:**2609–2619.

Duniway, J. M. 1971a. Resistance to water movement in tomato plants infected with *Fusarium*. *Nature (Lond.)* **230:**252–253.

Duniway, J. M. 1971b. Water relations of *Fusarium* wilt in tomato. *Physiol. Plant Pathol.* **1:**537–546.

Duniway, J. M. 1975. Water relations in safflower during wilting induced by *Phytophthora* root rot. *Phytopathology* **65:**886–891.

Duniway, J. M. 1976. Water status and imbalance. In *Encyclopedia of Plant Physiology: Physiological Plant Pathology* (R. Heitefuss and P. H. Williams, eds.), Vol. 4, pp. 430–449. Springer-Verlag, Berlin.

Duniway, J. M. 1977. Changes in resistance to water transport in safflower during the development of *Phytophthora* root rot. *Phytopathology* **67:**331–337.

Duniway, J. M. 1979. Water relations of water molds. *Annu. Rev. Phytopathol.* **17:**431–460.

Duniway, J. M., and Durbin, R. D. 1971a. Some effects of *Uromyces phaseoli* on the transpiration rate and stomatal response of bean leaves. *Phytopathology* **61:**114–119.

Duniway, J. M., and Durbin, R. D. 1971b. Detrimental effect of rust infection on water relations of bean. *Plant Physiol (Bethesda),***48:**69–72.

Duniway, J. M., and Slatyer, R. O. 1971. Gas exchange studies on the transpiration and photosynthesis of tomato leaves infected by *Fusarium oxysporum* f. sp. *lycopersici*. *Phytopathology* **61:**1377–1381.

Dunkle, L. D., and Wolpert, T. J. 1981. Independence of milo disease symptoms and electrolyte leakage induced by the host-specific toxin from *Periconia circinata*. *Physiol. Plant Pathol.* **18:**315–323.

Durbin, R. D. 1967. Obligate parasites: Effect on the movement of solutes and water. In *The Dynamic Role of Molecular Constitutents in Plant–Parasite Interactions* (C. J. Mirocha and I. Uritani, eds.). Bruce Publishing, St. Paul, MN.

Durbin, R. D. 1971. Chlorosis-inducing pseudomonas toxins: Their mechanism of action and structure. In *Morphology and Biochemical Events in Plant–Parasite Interaction* (S. Akai and S. Ouchi, eds.), pp. 369–386. Mochizuki, Tokyo.

Durbin, R. D. 1972. Bacterial phytotoxins: A survey of occurrence, mode of action and composition. In *Phytotoxins in Plant Diseases* (R. K. S. Wood, A. Ballio and A. Graniti, eds.), pp. 19–33. Academic Press, London and New York.

Durbin, R. D., ed. 1981. *Toxins in Plant Disease*. Academic Press, New York.

Durbin, R. D., and Sinden, S. L. 1967. The effect of light on the symptomatology of oat halo blight. *Phytopathology* **57:**1000–1001.

Durbin, R. D., and Steele, J. A. 1979. What art thou, O specificity? In *Recognition and Specificity in Plant–Host–Parasite Interactions* (J. M. Daly and I. Uritani, eds.), pp. 115–131. Japan Scientific Society Press, Tokyo, and University Park Press, Baltimore.

Durbin, R. D., Uchytil, T. F., Steele, J. A., and Ribeiro, R. de L. D. 1978. Tabtoxinine-β-lactam from *Pseudomonas tabaci*. *Phytochemistry* **17:**147.

Dwurazna, M. M., and Weintraub, M. 1969. The respiratory pathways of tobacco leaves infected with potato virus X. *Can. J. Bot.* **47:**731–736.

Dye, M. H., Clark, G., and Wain, R. L. 1962. Investigations on the auxins in tomato crown-gall tissue. *Proc. R. Soc. Lond., Biol. Sci.* **155:**478–492.

Easton, C. Z., and Hanchey, P. 1972. Localization of crystals in diseased oats treated with uranyl acetate. *Plant Physiol (Bethesda)* **50:**706–712.

Edwards, H. H. 1970a. A basic staining material associated with the penetration process in resistant and susceptible powdery mildewed barley. *New Phytol.* **69:**299–301.

Edwards, H. H. 1970b. Biphasic inhibition of photosynthesis in powdery mildewed barley. *Plant Physiol.* **47:**324–328.

Edwards, H. H. 1971. Translocation of carbon in powdery mildewed barley. *Plant Physiol.* **47:**324–328.

Edwards, J. D., and Cashaw, J. L. 1957. Studies in the haphthalene series III. Synthesis of apogossypol hexamethyl ether. *J. Am. Chem. Soc.* **79:**2283–2285.

Ehrlich, M. A., and Ehrlich, H. G. 1971. Fine structure of the host–parasite interfaces in mycoparasitism. *Annu. Rev. Phytopathol.* **9:**155–184.

El Khalifa, M. D., and Lippincott, J. A. 1968. The influence of plant-growth factors on the initiation and growth of crown gall tumors on primary pinto bean leaves. *J. Exp. Bot.* **19:**749–759.

Ellingboe, A. H. 1976. Genetics of host–parasite interactions. In *Physiological Plant Pathology, Encyclopedia of Plant Physiology* (R. Heitefuss and P. H. Williams, eds.), pp. 761–778. Springer–Verlag, Berlin.

Ellingboe, A. H. 1981. Changing concepts in host–pathogen genetics. *Annu. Rev. Phytopathol.* **19:**125–143.

Ellis, M. A., Ferree, D. C., and Spring, D. E. 1981. Photosynthesis, transpiration, and carbohydrate content of apple leaves infected by *Podosphaera leucotricha*. *Phytopathology* **71:**392–395.

Endo, B. Y. 1975. Pathogenesis of nematode-infected plants. *Annu. Rev. Phytopathol.* **13:**213–238.

Engler, G., Holsters, M., Van Montagu, M., Schell, J., Hernalsteens, J. P., and Schilperoort, R. 1975. Agrocin 84 sensitivity: A plasmid determined property in *Agrobacterium tumefaciens*. *Mol. Gen. Genet.* **138:**345–349.

English, P. D., Jurale, J. B., and Albersheim, P. 1971. Host–pathogen interactions. II. Parameters affecting polysaccharide-degrading enzyme secretion by *Colletotrichum lindemuthianum* grown in culture. *Plant Physiol. (Bethesda),***47:**1–6.

Esau, K. 1968. *Viruses in Plant Hosts, Distribution and Pathological Effects*. University of Wisconsin Press, Madison.

Esau, K., and Hoefert, L. L. 1971. Cytology of beet yellows virus infection in *Tetrangonia*. I. Parenchyma cells in infected leaf. *Protoplasma* **72:**255–273.

Eyjolfsson, R. 1970. Recent advances in the chemistry of cyanogenic glycosides. *Fortschr. Chem. Org. Naturst.* **28:**74–108.

Farkas, G. L., and Király, Z. 1958. Enzymological aspects of plant diseases. I. Oxidative enzymes. *Phytopathol. Z.* **31:**251–272.

Farkas, G. L., and Király, Z. 1962. Role of phenolic compounds in the physiology of plant diseases and disease resistance. *Phytopathol. Z.* **44:**8–150.

Farkas, G. L., and Szirmai, J. 1969. Increase in phenylalanine ammonia-lyase activity in bean leaves infected with tobacco mosaic virus. *Neth. J. Plant Pathol.* 75: 82–85.

Fawcett, C. H, Firn, R. D., and Spencer, D. M. 1971. Wyerone increase in leaves of broad bean (*Vicia faba* L.) after infection by *Botrytis fabae*. *Physiol. Plant Pathol.* **1:**163–166.

Ferguson, A. R., and Johnston, J. S. 1980. Phaseolotoxin: Chlorosis, ornithine accumulation and inhibition of ornithine carbamoyltransferase in different plants. *Physiol. Plant Pathol.* **16:**269–275.

Fernow, K. H. 1967. Tomato as a test plant for detecting mild strains of potato spindle tuber virus. *Phytopathology* **57:**1347–1352.

Feung, C. S., Hamilton, R. H., and Mumma, R. O. 1976. Metabolism of indole-3-acetic acid. III. Identification of metabolites isolated from crown gall callus tissue. *Plant Physiol. (Bethesda)* **58:**666–669.

Fisher, R. W.. and Corke, A. T. K. 1971. Infection of Yarlington Mill Fruit by an apple scab fungus. *Can. J. Plant Sci.* **51:**535–542.

Fletcher, R. A. 1969. Retardation of leaf senescence by benzyladenine in intact bean plants. *Planta (Berl.).* **89:**1–7.

Fletcher, R. A., Teo, C., and Ali, A. 1973. Stimulation of chlorophyll synthesis in cucumber cotyledons by benzyladenine. *Can. J. Bot.* **51:**937–939.

Fliege, R., Flugge, U. I., Werdan, K., and Heldt, H. W. 1978. Specific transport of inorganic phosphate, 3-phosphoglycerate and triosephosphates across the inner membrane of the envelope in spinach chloroplasts. *Biochim. Biophys. Acta.* **502:**232–247.

Flor, H. H. 1955. Host–parasite interaction in flax rust—its genetics and other implications. *Phytopathology* **45:**680–685.

Flor, H. H. 1971. Current status of the gene-for-gene concept. *Annu. Rev. Phytopathol.* **9:**275–296.

Floss, H. G. 1979. The shikimate pathway. In *Recent Advances in Phytochemistry: Biochemistry of Plant Phenolics* (T. Swain, J. B. Harborne, and C. F. Van Sumere, eds.), Vol. 12, pp. 59–89. Plenum Press, New York.

Flugge, U. I., Freisl, M., and Heldt, H. W. 1980. Balance between metabolite accumulation and transport relation to photosynthesis by isolated spinach chloroplasts. *Plant Physiol. (Bethesda)* **65:**574–577.

Flynn, J. G., Chakravorty, A. K., and Scott, K. J. 1976. Changes in the transcription pattern of wheat leaves during the early stages of infection by *Puccinia graminis tritici*. *Proc. Aust. Biochem. Soc.* **9:** 44 (Abstr.).

Foster, J. W. 1958. An evaluation of the role of molds in the comparative biochemistry of carbohydrate oxidation. *Tex. Rep. Biol. Med.* **16:**79–100.

Fox, R. t. V., Manners, J. G., and Myers, A. 1972. Ultrastructure of tissue disintegration and host reactions in potato tubers infected by *Erwinia carotovora* var. *atroseptica*. *Potato Res.* **15:**130–145.

Fraser, L., and Matthews, R. E. F. 1979. Strain-specific pathways of cytological changes in individual chinese cabbage protoplasts infected with turnip yellow mosaic virus. *J. Gen. Virol.* **45:**623–630.

Frazier, W., and Glazer, L. 1979. Surface components and cell recognition. *Annu. Rev. Biochem.* **48:**491–523.

Freudenberg, K. 1968. The constitution and biosynthesis of lignin. In *Molecular Biology, Biochemistry and Biophysics* (A. Kleinzeller, G. F. Springer, and H. C. Witmann, eds.), pp. 45–125. Springer-Verlag, New York.

Frič, F. 1969. Phenolische Stoffe und Oxydasen vom Standpunkt der Resistenz der Gerste gegen Mehltau (*Erysiphe graminis* f. sp. *hordei* Marchal). *Biológia (Bratisl.)* **24:**54–67.

Frič, F. 1976. Oxidative enzymes. In *Physiological Plant Pathology, Encyclopedia of Plant Pathology* (R. Heitefuss and P. H. Williams, eds.), Vol. 4, pp. 617–631. Springer-Verlag, Berlin.

Frič, F., and Fuchs, W. H. 1970. Veränderungen der Aktivität einiger Enzyme im Weizenblatt in Abhängigkeit von der temperaturlabilen Verträglichkeit für *Puccinia graminis tritici*. *Phytopathol. Z.* **67:**161–174.

Frič, F., and Heitefuss, R. 1970. Immunochemische und elektrophoretische untersuchungen der peroteine von Weizenblättern nach infektion mit *Puccinia graminis tritici*. *Phytopathol. Z.* **69:**236–246.

Friend, J. 1973. Resistance of potato to *Phytophthora*. In *Fungal Pathogenicity and the Plant's Response* (R. J. W. Byrde and C. V. Cuttings, eds.), pp. 383–396. Academic Press, London and New York.

Friend, J. 1976. Lignification in infected tissue. In *Biochemical Aspects of Plant–Parasite Relationships* (J. Friend and D. R. Threlfall, eds.), pp. 291–303. Academic Press, London and New York.

Friend, J. 1977. Lignification. In *Cell Wall Biochemistry Related to Specificity in Host–Plant Pathogen Interactions* (B. Solheim and J. Raa, eds.), pp. 277–279. Universitetsforlaget, Oslo, Norway.

Friend, J. 1979. Phenolic substances and plant disease. In *Recent Advances in Phytochemistry: Biochemistry of Plant Phenolics* (T. Swain, J. B. Harborne, and C. F. Van Sumere, eds.), Vol. 12, pp. 557–588. Plenum Press, New York.

Friend, J., Reynolds, S. B., and Aveyard, M. A. 1973. Phenylalanine ammonia-lyase, chlorogenic acid and lignin in potato tuber tissue inoculated with *Phytophthora infestans*. *Physiol. Plant Pathol.* **3:**495–507.

Fritig, B., Gosse, J., Legrand, M., and Hirth, L. 1973. Changes in phenylalanine ammonia-lyase during the hypersensitive reaction of tobacco to TMV. *Virology* **55:**371–379.

Fry, W. E., and Evans, P. H. 1977. Association of formamide hydro-lyase with fungal pathogenicity to cyanogenic plants. *Phytopathology* **67:**1001–1006.

Fry, W. E., and Millar, R. L. 1972. Cyanide degradation by an enzyme from *Stemphylium loti*. *Arch. Biochem. Biophys.* **151:**468–474.

Fry, W. E., and Munch, D. C. 1975. Hydrogen cyanide detoxification by *Gloeocercospora sorghi*. *Physiol. Plant Pathol.* **7:**23–33.

Fuchs, W. H. 1976. History of physiological plant pathology. In *Physiological Plant Pathology* (R. Heitefuss and P. H. Williams, eds.), pp. 1–26. Springer-Verlag, Heidelberg.

Fuchs, A. Jorsen, J. A., and Wouts, W. M. 1965. Arabanases in phytopathogenic fungi. *Nature (Lond.)* **206:**714–715.

Furuichi, N., Tomiyama, K., and Doke, N. 1980. The role of potato lectin in the binding of germ tubes of *Phytophthora infestans* to potato cell membrane. *Physiol. Plant Pathol.* **16:**249–256.

Garas, N. A., Doke, N., and Kuć, J. 1979. Suppression of the hypersensitive reaction in potato tubers by mycelial components of *Phytophthora infestans*. *Physiol. Plant Pathol.* **15:**117–126.

Gardner, J. M., and Kado, C. I. 1977. Studies on *Agrobacterium tumefaciens* VI. α-DNA polymerases of crown-gall tumor and normal cells, and of the bacterium. *Physiol. Plant Pathol.* **11:**79–86.

Gardner, J. M., and Scheffer, R. P. 1973. Effects of cycloheximide and sulfhydryl-binding compounds on sensitivity of oat tissues to *Helminthosporium victoriae* toxin. *Physiol. Plant Pathol.* **3:**147–157.

Gardner, J. M., Mansour, I. S., and Scheffer, R. P. 1972. Effects of the host-specific toxin of *Periconia circinata* on some properties of sorghum plasma membranes. *Physiol. Plant Pathol.* **2:**197–206.

Gardner, J. M., Scheffer, R. P., and Higinbotham, N. 1974. Effects of host-specific toxins on electropotentials of plant cells. *Plant Physiol. (Bethesda).* **54:**246–249.

Gardner, W. R., and Ehlig, C. F. 1965. Physical aspects of the internal water relations of plant leaves. *Plant Physiol. (Bethesda).* **40:**706–710.

Garfinkel, D. J., and Nester, E. W. 1980. Mutants of *Agrobacterium tumefaciens* affected in crown gall tumorigenesis and octopine catabolism. *J. Bacteriol.* **144:**732–743.

Garibaldi, A., and Bateman, D. F. 1970. Association of pectolytic and cellulytic enzymes with slow bacterial wilt of carnation caused by *Erwinia chrysanthemi* Burkh. McFad. et Dim. *Phytopathol. Mediterr.* **9:**136–144.

Garibaldi, A., and Bateman, D. F. 1971. Pectic enzymes produced by *Erwinia chrysanthemi* and their effects on plant tissue. *Physiol. Plant Pathol.* **1:**25–40.

Gäumann, E. 1958. The mechanism of fusaric acid injury. *Phytopathology* **48:**670–686.

Geigert, J., Stermitz, F. R., Johnson, G., Magg, D. D., and Johnson, D. K. 1973. Two phytoalexins from sugarbeets (*Beta vulgaris*) leaves. *Tetrahedron* **29:**2703–2706.

Gelvin, S. B., Gordon, M. P., Nester, E. W., and Aronson, A. I. 1981. Transcription of the *Agrobacterium* Ti Plasmid in the bacterium and in crown gall tumors. *Plasmid* **6:**17–29.

Genetello, C., Van Larebeke, N., Holsters, M., DePicker, A., Van Montagu, M., and Schell, J. 1977. Ti plasmids of *Agrobacterium* as conjugative plasmids. *Nature (Lond.)* **265:**561–563.

Gengenbach, B. G., Miller, R. J., Koeppe, D. E., and Arntzen, C. J. 1973. The effect of toxin from *Helminthosporium maydis* (race T) on isolated corn mitochondria: swelling. *Can. J. Bot.* **51:**2119–2125.

Gengenbach, B. G., Green, C. E., and Donovan, C. M. 1977. Inheritance of selected pathotoxin resistance in maize plants regenerated from cell cultures. *Proc. Natl. Acad. Sci. U. S. A.* **74:**5113–5117.

Gentile, I. A., and Matta, A. 1975. Production of and some effects of ethylene in relation to *Fusarium* wilt of tomato. *Physiol. Plant Pathol.* **5:**27–35.

Gerlach, W. W. P., Hoitink, H. A. J., and Schmitthener, A. F. 1976. *Phytophthora citrophthora* on *Pyrus japonica:* Infection, sporulation, and dissemination. *Phytopathology* **66:**302–308.

Gerola, F. M., Bassi, M., and Betto, E. 1969. A submicroscopical study of leaves of alfalfa, basil and tobacco experimentally infected with lucerne mosaic virus. *Protoplasma* **67:**307–318.

Ghabrial, S. A., and Pirone, T. P. 1967. Physiology of tobacco etch virus-induced wilt of tobasco peppers. *Virology* **31:**154–162.

Giebel, J. 1973. Phenylalanine and tyrosine ammonia-lyase activities in potato roots and their significance in potato resistance to *Heterodera rostochiensis*. *Nematologica* **19:**1–6.

Gil, F., and Gay, J. L. 1977. Ultrastructural and physiological properties of the host interfacial components of haustoria of *Erysiphe pisi in vivo* and *in vitro*. *Physiol. Plant Pathol.* **10:**1–12.

Gilchrist, D. G., and Grogan, R. G. 1976. Production and nature of a host-specific toxin from *Alternaria alternata* f. sp. *lycopersici*. *Phytopathology* **66:**165–171.

Gitaitis, R. D., Samuelson, D. A., and Strandberg, J. O. 1981. Scanning electron microscopy of the ingress and establishment of *Pseudomonas alboprecipitans* in sweet corn leaves. *Phytopathology* **71:**171–175.

Glazener, J. A. 1982. Accumulation of phenolic compounds in cells and formation of lignin-like polymers in cell walls of young tomato fruits after inoculation with *Botrytis cinerea*. *Physiol. Plant Pathol.* **20:**11–25.

Glazener, J. A., and Van Etten, H. D. 1978. Phytotoxicity of phaseollin to, and alteration of phaseollin by, cell suspension cultures of *Phaseolus vulgaris*. *Phytopathology* **68:** 111–117.

Glogowski, W., and Galsky, A. G. 1978. *Agrobacterium tumefaciens* site attachment as a necessary prerequisite for crown gall tumor formation on potato discs. *Plant Physiol. (Bethesda)* **61:**1031–1033.

Gnanamanickam, S. S., and Patil, S. S. 1975. Induction of phytoalexins in bean by *Pseudomonas phaseolicola*. *Proc. Am. Phytopathol. Soc.* **2:**64 (Abstr.).

Gnanamanickam, S. S., and Patil, S. S. 1976. Bacterial growth, toxin production, and levels of ornithine carbamoyltransferase in resistant and susceptible varieties of beans inoculated with *Pseudomonas phaseolicola*. *Phytopathology* **66:**290–294.

Gnanamanickam, S. S., and Patil, S. S. 1977a. Accumulation of antibacterial isoflavonoids in hypersensitivity responding bean leaf tissues inoculated with *Pseudomonas phaseolicola*. *Physiol. Plant Pathol.* **10:**159–168.

Gnanamanickam, S. S., and Patil, S. S. 1977b. Phaseotoxin suppresses bacterially induced hypersensitive reaction and phytoalexin synthesis in bean cultivars. *Physiol. Plant Pathol.* **10:**169–179.

Gnanamanickam, S. S., and Smith, D. A. 1980. Selective toxicity of isoflavonoid phytoalexins to Gram-positive bacteria. *Phytopathology* **70:**894–896.

Goffeau, A., and Bove, J. M. 1965. Virus infection and photosynthesis. *Virology* **27:**243–252.

Gonzalez, C. F., DeVay, J. E., and Wakeman, R. J. 1981. Syringotoxin: A phytotoxin unique to citrus isolates of *Psueodmonas syringae*. *Physiol. Plant Pathol.* **18:**41–50.

Goodenough, P. W., Kempton, R. J., and Maw, G. A. 1976. Studies on the root rotting fungus *Pyrenochaeta lycopersici:* Extracellular enzyme secretion by the fungus grown on cell-wall material from susceptible and tolerant tomato plants. *Physiol. Plant Pathol.* **8:**243–251.

Goodman, R. N. 1967. Protection of apple stem tissue against *Erwinia amylovora* infection by avirulent strains and three other bacterial species. *Phytopathology* **57:**22–24.

Goodman, R. N. 1968. The hypersensitivity reaction in tobacco: A reflection of changes in host cell permeability. *Phytopathology* **58:**872–873.

Goodman, R. N. 1972. Phytotoxin-induced ultrastructural modifications of plant cells. In *Phytotoxins in Plant Diseases* (R. K. S. Wood, A. Ballio, and A. Graniti, eds.), pp. 311–329. Academic Press, London and New York.

Goodman, R. N. 1976. Physiological and cytological aspects of the bacterial infection process. In *Physiological Plant Pathology* (R. Heitefuss and P. H. Williams, eds.), pp. 172–196. Springer-Verlag, Berlin and New York.

Goodman, R. N. 1980. Defenses triggered by previous invaders: Bacteria. In *Plant Disease: An Advanced Treatise* (J. G. Horsfall and E. B. Cowling, eds.), Vol. 5, pp. 305–317. Academic Press, New York.

Goodman, R. N., Király, Z., and Zaitlin, M. 1967. *The Biochemistry and Physiology of Infectious Plant Disease*. D. Van Nostrand, Princeton, New Jersey.

Goodman, R. N., Huang, J. S., and Huang, P. 1974. Host specific phytotoxic polysaccharide from apple tissue infected by *Erwinia amylovora*. *Science* **183:**1081–1082.

Goodman, R. N., Huang, P. Y., Huang, J. S., and Thiapanich, V. 1976a. Induced resistance to bacterial infection. In *Biochemistry and Cytology of Plant–Parasite Interaction* (K. Tomiyama, J. M. Daly, I. Uritani, H. Oku, and S. Ouchi, eds.), pp. 35–42. Kodansha, Tokyo.

Goodman, R. N., Huang, P. Y., and White, J. A. 1976b. Ultrastructural evidence for immobilization of an incompatible bacterium, *Pseudomonas pisi,* in tobacco leaf tissue. *Phytopathology* **66:**754–764.

Goodman, R. N., Politis, D. M., and White, J. A. 1977. Ultrastructural evidence of an "active" immobilization process of incompatible bacteria in tobacco tissue: A resistance reaction. In *Cell Wall Biochemistry Related to Specificity in Host–Plant Pathogen Interactions* (B. Solheim and J. Raa, eds.), pp. 423–429. Universitetsforlaget, Oslo, Norway.

Goodman, R. N., Stoffl, P. R., and Ayers, S. M. 1978. The utility of the fireblight toxin, amylovorin, for the detection of resistance of apple, pear and quince to *Erwinia amylovora*. *Acta. Hortic. (The Hague)* **86:**51–56.

Gordon, M., Stoessl, A., and Stothers, J. B. 1973. Post-infectional inhibitors from plants. IV. The structure of capsidiol, an antifungal sesquiterpene from sweet peppers. *Can. J. Chem.* **51:**748–752.

Gordon, T. R., and Duniway, J. M. 1982. Effect of powdery mildew infection on the efficiency of $CO_2$ fixation and light utilization by sugarbeet leaves. *Plant Physiol.* **69:**139–142.

Goto, M., Yaguchi, Y., and Hyodo, H. 1980. Ethylene production in citrus leaves infected with *Xanthomonas citri* and its relation to defoliation. *Physiol. Plant Pathol.* **16:**343–350.

Gracen, V. E., Grogan, C. O., and Forster, M. J. 1972. Permeability changes induced by *Helminthosporium maydis,* race T, toxin. *Can. J. Bot.* **50:**2167–2170.

Graham, T. L., Sequeira, L., and Huang, T. R. 1977. Bacterial lipopolysaccharides as inducers of disease resistance in tobacco. *Appl. and Env. Microbiol.* **34:**424–432.

Graniti, A. 1962. Azione fitotossica di *Fusicoccum amygdali* Del. su Mandorla (*Prunus amygdali* St.). *Phytopathol. Mediterr.* **1:**182–185.

Graniti, A. 1972. The evolution of the toxin concept in plant pathology. In *Phytotoxins in Plant Diseases* (R. K. S. Wood, A. Ballio, and A. Graniti, eds.), pp. 1–18. Academic Press, London and New York.

Grassmann, W., Zechmeister, L., Bender, R., and Toth, G. 1934. Über die Chitin-spaltung durch Emulsin-präparate. (III. Mitteil über enzymatische Spaltung von Polysacchariden). *Bericht* **67:**1–5.

Gregory, G. F. 1971. Correlation of isolability of the oak wilt pathogen with leaf wilt and vascular water flow resistance. *Phytopathology* **61:**1003–1005.

Gregory, P., Matthews, D. E., York, D. W., Earle, E. D., and Gracen, V. E. 1978. Southern corn leaf blight disease: Studies on mitochondrial biochemistry and ultrastructure. *Mycopathologia* **66:**105–112.

Gregory, P., Earle, E. D., and Gracen, V. E. 1980. Effects of purified *Helminthosporium maydis* race T toxin on the structure and function of corn mitochondria and protoplasts. *Plant Physiol. (Bethesda)* **66:**477–481.

Grisebach, H. 1977. Biochemistry of lignification. *Naturwissenschaften* **64:**619–625.

Gross, D. C., and DeVay, J. E. 1977a. Role of syringomycin in holcus spot of maize and systemic necrosis of cowpea caused by *Pseudomonas syringae. Physiol. Plant Pathol.* **11:**1–11.

Gross, D. C., and DeVay, J. E. 1977b. Population dynamics and pathogenesis of *Pseudomonas syringae* in maize and cowpea in relation to the *in vitro* production of syringomycin. *Phytopathology* **67:**475–483.

Gross, D. C., DeVay, J. E., and Stadtman, F. H. 1977. Chemical properties of syringomycin and syringotoxin: Toxigenic peptides produced by *Pseudomonas syringae. J. Appl. Bacteriol.* **43:**453–463.

Gross, G. G. 1979. Recent advances in the chemistry and biochemistry of lignin. In *Recent Advances in Phytochemistry: Biochemistry of Plant Phenolics* (T. Swain, J. B. Harborne, and C. F. Van Sumere, eds.), Vol. 12, pp. 177–220. Plenum Press, New York.

Gruen, H. 1959. Auxin and fungi. *Annu. Rev. Plant Physiol.* **10:**405–440.

Grzelinska, A. 1969. Changes in protein level and activities of several enzymes in susceptible and resistant tomato plants after infection by *Fusarium oxysporum* f. *lycopersici* (Sacc.). Snyder et Hansen. *Phytopathol. Z.* **66:**374–380.

Gulyás, A., Barna, B., Klement, Z., and Farkas, G. L. 1979. Effect of plasmolytica on hypersensitive reaction induced by bacteria in tobacco: A comparison with the virus-induced hypersensitive reaction. *Phytopathology* **69:**121–124.

Gurley, W. B., Kemp, J. D., Albert, M. J., Sutton, D. W., and Callis, J. 1979. Transcription of Ti-plasmid derived sequences in three octopine-type crown gall tumor lines. *Proc. Natl. Acad. Sci. U. S. A.* **76:**2828–2832.

Gustine, D. L., Sherwood, R. T., and Vance, C. P. 1978. Regulation of phytoalexin synthesis in Jackbean callus cultures. *Plant Physiol. (Bethesda)* **61:**226–230.

Haard, N. F. 1978. Isolation and partial characterization of auxin protectors from *Synchytrium endobioticum* incited tumors in potato. *Physiol. Plant Pathol.* **13:**223–232.

Hack, E., and Kemp, J. D. 1977. Comparison of octopine, histopine, lysopine, and octopinic acid synthesizing activities in sunflower crown gall tissues. *Biochem. Biophys. Res. Commun.* **78:**785–791.

Hadwiger, L. A. 1979. Chitosan formation in *Fusarium solani* macrocondia on tissue. *Plant Physiol.* **63** (Suppl.):133.

Hadwiger, L. A., and Adams, M. J. 1978. Nuclear changes associated with the host–parasite interaction between *Fusarium solani* and peas. *Physiol. Plant Pathol.* **12:**63–72.

Hadwiger, L. A., and Beckman, J. M. 1980. Chitosan as a component of Pea–*Fusarium solani* interactions. *Plant Physiol. (Bethesda)* **66:**205–211.

Hadwiger, L. A., and Martin, A. R. 1971. Induced formation of phenylalanine ammonia lyase and pisatin by chlorpromazine and other phenothiazine derivatives. *Biochem. Pharmacol.* **20:**3255–3261.

Hadwiger, L. A., and Schwochau, M. E. 1969. Host resistance responses—an induction hypothesis. *Phytopathology* **59:**223–227.

Hadwiger, L. A., and Schwochau, M. E. 1971. Ultraviolet light-induced formation of pisatin and phenylalanine ammonia lyase. *Plant Physiol. (Bethesda)* **47:**588–590.

Hadwiger, L. A., Hess, S. L., and Von Broembsen, S. V. 1970. Stimulation of phenylalanine ammonia-lyase activity and phytoalexin production. *Phytopathology* **60:**332–336.

Hadwiger, L. A., Loschke, D. C., and Teasdale, J. R. 1977. An evaluation of pea histones as disease resistance factors. *Phytopathology* **67:**755–758.

Hahlbrock, K., and Grisebach, H. 1979. Enzymic controls in the biosynthesis of lignin and flavonoids. *Annu. Rev. Plant Physiol.* **30:**105–130.

Hahlbrock, K., Lamb, C. J., Purwin, C., Ebel, J., Fautz, E., and Schäfer, E. 1981. Rapid responses of suspension-cultured parsley cells to the elicitor from *Phytophthora megasperma* var. *sojae*. *Plant Physiol. (Bethesda)* **67:**768–773.

Hahn, M. G., and Albersheim, P. 1978. Host–pathogen interactions XIV. Isolation and partial characterization of an elicitor from yeast extract. *Plant Physiol. (Bethesda)* **62:**107–111.

Hahn, M. G., Darvill, A. G., and Albersheim, P. 1981. Host–pathogen interactions XIX: The endogenous elicitor, a fragment of a plant cell wall polysaccharide that elicits phytoalexin accumulation in soybeans. Submitted for publication.

Hale, M. G., Moore, L. D., and Griffin, G. J. 1978. Root exudates and exudation. In *Interaction Between Non-pathogenic Soil Microorganisms and Plants* (Y. R. Dommergues and S. V. Krupa, eds.), pp. 163–197. American Elsevier, New York.

Hall, A. E., and Loomis, R. S. 1972. An explanation for the difference in photosynthetic capabilities of healthy and beet yellows virus-infected sugar beets (*Beta vulgaris* L.). *Plant Physiol. (Bethesda)* **50:**576–580.

Hall, J. A., and Wood, R. K. S. 1970. Plant cells killed by soft rot parasites. *Nature (Lond.)* **227:**1266–1267.

Hall, J. A., and Wood, R. K. S. 1973. The killing of plant cells by pectolytic enzymes. In *Fungal Pathogenicity and the Plant's Response* (R. J. W. Byrde and C. V. Cutting, eds.), pp. 19–38. Academic Press, New York.

Hall, R., and Busch, L. V. 1971. *Verticillium* wilt of chrysanthemum: Colonization of leaves in relation to symptom development. *Can. J. Bot.* **49:**181–185.

Halliwell, G., and Mohammed, R. 1971. Interactions between components of the cellulase complex of *Trichoderma koningii* on native substrates. *Arch. Mikrobiol.* **78:**295–309.

Halloin, J. M., Comstock, J. C., Martinson, C. A., and Tipton, C. L. 1973. Leakage from corn tissues induced by *Helminthosporium maydis* race T toxin. *Phytopathology* **63:**640–642.

Hamblin, J., and Kent, S. P. 1973. Possible role of phytohaemagglutinin in *Phaseolus vuglaris* L. *Nature (Lond.).* **245:**28–30.

Hamilton, R. I. 1980. Defenses triggered by previous invaders: Viruses. In *Plant Disease: An Advanced Treatise* (J. G. Horsfall and E. B. Cowling, eds.), Vol. 5, pp. 279–303. Academic Press, New York.

Hammerschmidt, R., and Kuć, J. 1976. Protection of cucumber against *Colletotrichum lagenarium* and *Cladosporium cucumerinum*. *Phytopathology* **66:**790–793.

Hampton, R. E., Hopkins, D. L., and Nye, T. G. 1966. Biochemical effects of tobacco leaf tissue. I. Protein synthesis by isolated chloroplasts. *Phytochemistry* **5:**1181–1185.

Hanchey, P. 1980. Histochemical changes in oat cell walls after victorin treatment. *Phytopathology* **70**:377–381.

Hanchey, P., and Wheeler, H. 1979. The role of host cell membranes. In *Recognition and Specificity in Plant Host–Parasite Interactions* (J. M. Daly and I. Uritani, eds.), pp. 193–210. Japan Scientific Societies Press, Tokyo, and University Park Press, Baltimore.

Hanchey, P., Wheeler, H., and Luke, H. H. 1968. Pathological changes in ultrastructure: effects of victorin on oat roots. *Am. J. Bot.* **55**:53–61.

Hancock, J. G. 1966. Pectate lyase production by *Colletotrichum trifolii* in relation to changes in pH. *Phytopathology* **56**:1112–1113.

Hancock, J. G. 1967. Hemicellulose degradation in sunflower hypocotyls infected with *Sclerotinia sclerotiorum*. *Phytopathology* **57**:203–206.

Hancock, J. G. 1968. Effect of infection by *Hypomyces solani* f. sp. *Cucurbitae* on apparent free space, cell membrane permeability, and respiration of squash hypocotyls. *Plant Physiol. (Bethesda)* **43**:1666–1672.

Hancock, J. G. 1969. Uptake of 3-0-methylglucose by healthy and hypomyces-infected squash hypocotyls. *Plant Physiol.* **44**:1267–1272.

Hancock, J. G. 1972. Changes in cell membrane permeability in sunflower hypocotyls infected with *Sclerotinia sclerotiorum*. *Plant Physiol.* **49**:358–364.

Hancock, J. G. 1976. Multiple forms of endo-pectate lyase formed in culture and in infected squash hypocotyls by *Hypomyces solani* f. sp. *cucurbitae*. *Phytopathology* **66**:40–45.

Hancock, J. G. 1977. Evidence for the role of a toxin in the wilt syndrome associated with *Sclerotinia* infection of a sunflower. *Proc. Am. Phytopathol. Soc.* **4**:206.

Hancock, J. G., Millar, R. L., and Lorbeer, J. W. 1964. Pectolytic and cellulolytic enzymes produced by *Botrytis allii, B. cinerea,* and *B. squamosa in vitro* and *in vivo*. *Phytopathology* **54**:928–931.

Harborne, J. B., Ingham, J. L., King, L., and Payne, M. 1976. The isopentenyl isoflavone luteone as a pre-infectional antifungal agent in the genus *Lupinus*. *Phytochemistry* **15**:1485–1487.

Hargreaves, J. A., and Bailey, J. A. 1978. Phytoalexin production by hypocotyls of *Phaseolus vulgaris* in response to constitutive metabolites released by damaged bean cells. *Physiol. Plant Pathol.* **13**:89–100.

Harrison, B. D. 1955. Studies on virus multiplication in inoculated leaves. Ph.D. Thesis. London University, London, England.

Harrison, J. A. C. 1971. Transpiration in potato plants infected with *Verticillium* spp. *Ann. Appl. Biol.* **68**:159–168.

Harvey, A. E., Chakrovorty, A. K., Shaw, M., and Scrubb, L. A. 1974. Changes in ribonuclease activity in *Ribes* leaves and pine tissue culture infected with blister rust, *Cronartium ribicola*. *Physiol. Plant Pathol.* **4**:359–371.

Hazelbauer, G. L., and Parkinson, J. S. 1977. Bacterial chemotaxis. In *Microbial Interactions (Receptors and Recognition)* (J. Reissig, ed.), pp. 59–98. Chapman and Hall, London.

Heale, J. B., and Gupta, D. P. 1972. Mechanism of vascular wilting induced by *Verticillium alboatrum*. *Trans. Br. Mycol. Soc.* **58**:19–28.

Heale, J. B., and Sharman, S. 1977. Induced resistance to *Botrytis cinerea* in root slices and tissue cultures of carrot (*Daucus carota* L.). *Physiol. Plant Pathol.* **10**:51–61.

Heath, M. C. 1974. Light and electron microscope studies of the interactions of host and nonhost plants with cowpea rust–*Uromyces phaseoli* var. *vignae*. *Physiol. Plant Pathol.* **4**:403–414.

Heath, M. C. 1979. Effects of heat shock, actinomycin D, cycloheximide and blasticidin S on nonhost interactions with rust fungi. *Physiol. Plant Pathol.* **15**:211–218.

Heath, M. C., 1980a. Effects of infection by compatible species or injection of tissue extracts on the susceptibility of non-host plants to rust fungi. *Phytopathology* **70**:356–360.

Heath, M. C. 1980b. Reactions of nonsuscepts to fungal pathogens. *Annu. Rev. Phytopathol.* **18:**211–236.

Heath, M. C. 1981a. A generalized concept of host–parasite specificity. *Phytopathology* **71:**1121–1123.

Heath, M. C. 1981b. Nonhost resistance. In *Plant Disease Control: Resistance and Susceptibility* (R. C. Staples and G. H. Toenniessen, eds.), pp. 201–217. John Wiley and Sons, New York.

Heath, M. C. 1981c. The absence of active defense mechanisms in compatible host–pathogen interactions. In *Active Defense Mechanisms in Plants* (R. K. S. Wood, ed.), pp. 143–156. Plenum Press, London and New York.

Heath, M. C., and Higgins, V. J. 1973. *In vitro* and *in vivo* conversion of phaseollin and pisatin by an alfalfa pathogen *Stemphylium botryosum*. *Physiol. Plant Pathol.* **3:**107–120.

Heber, U. 1974. Metabolic exchange between chloroplasts and cytoplasm. *Annu. Rev. Plant Physiol.* **25:**393–421.

Hecht, E. I., and Bateman, D. F. 1964. Non-specific acquired resistance to pathogens resulting from localized infections by *Thielaviopsis basicola* or viruses in tobacco leaves. *Phytopathology* **54:**523–530.

Heichel, G. H., and Turner, N. C. 1972. Carbon dioxide and water vapour exchange of bean leaves responding to *Fusicoccin*. *Physiol. Plant Pathol.* **2:**375–382.

Heitefuss, R., and Wolf, G. 1976. Nucleic acid in host–parasite interactions. In *Encyclopedia of Plant Physiology, Physiological Plant Pathology* (R. Heitefuss and P. H. Williams, eds.), Vol. 4, pp. 480–508. Springer-Verlag, Berlin.

Heldt, H. W., Chon, C. J., Maronde, D., Herold, A., Stankovic, Z. S., Walker, D. A., Kraminer, A., Kirk, M. R., and Heber, U. 1977. Role of orthophosphate and other factors in the regulation of starch formation in leaves and isolated chloroplasts. *Plant Physiol. (Bethesda)* **59:**1146–1155.

Helgeson, J. P. 1978. Alteration of growth by disease. In *Plant Disease: An Advanced Treatise* (J. G. Horsfall and E. B. Cowling, eds.), Vol. 3, pp. 183–200. Academic Press, New York.

Helgeson, J. P., and Leonard, N. J. 1966. Cytokinins: Identification of compounds isolated from *Corynebacterium fascians*. *Proc. Natl. Acad. Sci. U. S. A.* **56:**60–63.

Helgeson, J. P., Krueger, S. M., and Upper, C. D. 1969. Control of logarithmic growth rates of tobacco callus tissue by cytokinins. *Plant Physiol. (Bethesda)* **44:**193–198.

Henfling, J. W. D. M., Lisker, N., and Kuć, J. 1978. The effect of ethylene on phytuberin and phytuberol accumulation in potato tuber slices. *Phytopathology* **68:**857–862.

Henfling, J. W. D. M., Bostock, R. M., and Kuć, J. 1980a. Cell walls of *Phytophthora infestans* contain an elicitor of terpene accumulation in potato tubers. *Phytopathology* **70:**772–776.

Henfling, J. W. D. M., Bostock, R., and Kuć, J. 1980b. Effect of abscisic acid on rishitin and Lubimin accumulation and resistance to *Phytophthora infestans* and *Cladosporium cucumerinum* in potato tuber tissue slices. *Phytopathology* **70:**1074–1078.

Henry, M. E., and Nyns, E. J. 1975. Cyanide insensitive respiration. An alternative mitochondrial pathway. *Sub-Cell. Biochem.* **4:**1–65.

Herridge, E. A., and Schlegel, D. E. 1962. Autoradiographic studies of tobacco mosaic virus inoculations on host and non-host species. *Virology* **18:**517–523.

Hewitt, W. B., Raski, D. J., and Goheen, A. C. 1958. Nematode vector of soil-borne fanleaf virus of grapevines. *Phytopathology* **48:**586–595.

Hickman, C. J. 1970. Biology of *Phytophthora* zoospores. *Phytopathology* **60:**1128–1135.

Higgins, V. J. 1978. The effect of some pterocarpanoid phytoalexins on germ tube elongation of *Stemphylium botryosum*. *Phytopathology* **68:**339–345.

Higgins, V. J., and Ingham, J. L. 1981. Demethylmedicarpin, a product formed from medicarpin by *Colletotrichum coccodes*. *Phytopathology* **71:**800–803.

Higgins, V. J., and Millar, R. L. 1969a. Degradation of alfalfa phytoalexin by *Stemphylium botryosum*. *Phytopathology* **59:**1500–1506.

Higgins, V. J., and Millar, R. L. 1969b. Comparative ability of *Stemphylium botryosum* and *Helminthosporium turcicum* to induce and to degrade a phytoalexin from alfalfa. *Phytopathology* **59:**1493–1499.

Higinbotham, N. 1973. Electropotentials of plant cells. *Annu. Rev. Plant Physiol.* **24:**25–46.

Hildebrand, D. C., and Schroth, M. N. 1964. Arbutin–hydroquinone complex in pear as a factor in fire blight development. *Phytopathology* **54:**640–645.

Hildebrand, D. C., Aloci, M. C., and Schroth, M. N. 1980. Physical entrapment of pseudomonads in bean leaves by films formed at air–water interfaces. *Phytopathology* **70:**98–109.

Hill, G. 1977. Fruhphase der Pathogenese von *Botrytis cinerea* auf unterscheidliche Entwicklungsstadien vegetativer und generativer Organe von *Vitis vinifera*. Ph.D. Thesis. University of Giessen, Giessen, West Germany.

Hill, R., and Van Heyningen, R. 1951. Ranunculin: The precursor of the vesicant substance of the buttercup. *Biochem. J.* **49:**332–335.

Hinchi, J. M., and Clarke, A. E. 1980. Adhesion of fungal zoospores to root surfaces is mediated by carbohydrate determinants of the root slime. *Physiol. Plant Pathol.* **16:**303–307.

Hirai, A., and Wildman, S. G. 1969. Effect of TMV multiplication on RNA and protein synthesis in tobacco chloroplasts. *Virology* **38:**73–82.

Hirata, K. 1967. Notes on haustoria, hyphae and conidia of the powdery mildew fungus of barley, *Erysiphe graminis* f. sp. *hordei*. *Mem. Fac. Agric. Niigata Univ.* **6:**207–259.

Hiruki, C. 1977. Cell wall alterations in localized plant virus infections. In *Cell Wall Biochemistry Related to Specificity in Host–Plant Pathogen Interactions* (B. Solheim and J. Raa, eds.), pp. 267–270. Universitetsforlaget, Oslo, Norway.

Hislop, E. C., Hoad, G. V., and Archer, S. A. 1973. The involvement of ethylene in plant diseases. In *Fungal Pathogenicity and the Plant's Response* (R. J. W. Byrde and C. V. Cutting, eds.), pp. 87–117. Academic Press, New York.

Hislop, E. C., Keon, J. P. R., and Fielding, A. H. 1978. Effects of pectin lyase from *Monilinia fructigena* on viability, ultrastructure and localization of acid phosphatase of cultured apple cells. *Physiol. Plant Pathol.* **14:**371–381.

Ho, H. H., and Zentmyer, G. A. 1977. Infection of avocado and other species of *Persea* by *Phytophthora cinnamomi*. *Phytopathology* **67:**1085–1089.

Holsters, M., Dewaele, D., Depicker, A., Messens, E., Van Montagu, M., and Schell, J. 1978. Transfection and transformation of *Agrobacterium tumefaciens*. *Mol. Gen. Genet.* **163:**181–189.

Holsters, M., Silva, B., Van Vliet, F., Genetello, C., DeBlock, M., Dhaese, P., Villarroel, R., Van Montagu, M., and Schell, J. 1980. The functional organization of the nopaline *A. tumefaciens* plasmid pTiC58. *Plasmid* **3:**212–230.

Holter, H. 1960. Pinocytosis. *Int. Rev. Cytol.* **8:**481–504.

Hooker, A. I., Smith, D. R., Lim, S. M., and Beckett, J. B. 1970. Reaction of corn seedlings with male-sterile cytoplasm to *Helminthosporium maydis*. *Plant Dis. Rep.* **54:**708–712.

Hooykaas, P. J. J., Klapwijk, P. M., Nuti, M. P., Schilperoort, R. A., and Rorsch, R. 1977. Transfer of the *Agrobacterium tumefaciens* Ti-plasmid to avirulent agrobacteria and *Rhizobium ex planta*. *J. Gen. Microbiol.* **98:**477–484.

Hoppe, H. H. 1973. Untersuchungen zur Regulation des Kohlenhydratstoffwechsels in Weizenkeimpflanzen nach Infektion mit *Puccinia graminis tritici*. Ph.D. Thesis, University of Göttingen. Göttingen, West Germany.

Hoppe, H. H., and Heitefuss, R. 1975a. Permeability and membrane lipid metabolism of *Phaseolus vulgaris* infected with *Uromyces phaseoli*. IV. Phospholipids and phospholipid fatty acids in healthy and rust-infected bean leaves resistant and susceptible to *Uromyces phaseoli*. *Physiol. Plant Pathol.* **5:**263–271.

Hoppe, H. H., and Heitefuss, R. 1975b. Permeability and membrane lipid metabolism of *Phaseolus vulgaris* infected with *Uromyces phaseoli*. V. Sterols in healthy and rust-infected bean leaves resistant and susceptible to *Uromyces phaseoli*. *Physiol. Plant Pathol.* **5:**273–281.

Horikawa, T., Tomiyama, K., and Doke, N. 1976. Accumulation and transformation of rishitin and lubimin in potato tuber tissue infected by an incompatible race of *Phytophthora infestans*. *Phytopathology* **66:**1186–1191.

Horsfall, J. G., and Dimond, A. E. 1959. Prologue—the diseased plant. In *Plant Pathology: An Advanced Treatise* (J. G. Horsfall and A. E. Dimond, eds.), Vol. 1, pp. 1–17. Academic Press, New York.

Howell, C. R. 1976. Use of enzyme-deficient mutants of *Verticillium dahliae* to assess the importance of pectolytic enzymes in symptom expression of *Verticillium* wilt of cotton. *Physiol. Plant Pathol.* **9:**279–283.

Hsu, S. T., and Goodman, R. N. 1978. Agglutinating activity in apple cell suspension cultures inoculated with a virulent strain of *Erwinia amylovora*. *Phytopathology* **68:**355–360.

Huang, H. C., and Tinline, R. D. 1976. Histology of *Cochliobolus sativus* infection in subcrown internodes of wheat and barley. *Can. J. Bot.* **54:**1344–1354.

Huang, J. S., and Goodman, R. N. 1975. Further characterization and mode of action of amylovorin. *Proc. Am. Phytopathol. Soc.* **1:**117 (Abstr.).

Huang, P. Y. 1974. Ultrastructural modifications by and pathogenicity of *Erwinia amylovora* in apple tissues. Ph.D. Thesis. Department of Plant Pathology. University of Missouri–Columbia.

Huang, P. Y., and Goodman, R. N. 1976. Ultrastructural modifications in apple stems induced by *Erwinia amylovora* and the fire blight toxin. *Phytopathology* **66:**269–276.

Huang, P. Y., and Huang, J. S. 1975. Resistance mechanisms of apple shoots to an avirulent strain of *Erwinia amylovora*. *Physiol. Plant Pathol.* **6:**283–287.

Hubbard, J. P., Williams, J. D., Niles, R. M., and Mount, M. S. 1978. The relationship between glucose repression of endo-polygalacturonate trans-eliminase and adenasine 3, 5′ -cyclic monophosphate levels in *Erwinia carotovora*. *Phytopathology* **68:**95–99.

Husain, A., and Kelman, A. 1958. Relation of slime production to mechanism of wilting and pathogenicity of *Pseudomonas solanacearum*. *Phytopathology* **48:**155–165.

Hutzinger, O., and Kosuge, T. 1967. Microbial synthesis and degradation of indole-3-acetic acid. II. The source of oxygen in the conversion of L-tryptophan to indole-3-acetamide. *Biochim. Biophys. Acta* **136:**389–391.

Ichihara, A., Shiraishi, K., Sato, H., Sakamura, S., Nishiyama, K., Sakai, R., Frusaki, A., and Matsumuro, T. 1977a. The structure of coronatine. *J. Am. Chem. Soc.* **99:**636–637.

Ichihara, A., Shiraishi, K., Sakamura, S., Nishiyama, K., and Sakai, R. 1977b. Partial synthesis and stereochemistry of coronatine. *Tetrahedron Lett.* **3:**269–272.

Ilag, L., and Curtis, R. W. 1968. Production of ethylene by fungi. *Science* **159:**1357–1358.

Imolehin, E. D., and Grogan, R. G. 1980. Effect of oxygen, carbon dioxide, and ethylene on growth, sclerotial production, germination, and infection by *Sclerotinia minor*. *Phytopathology* **70:**1158–1161.

Ingham, J. L., and Millar, R. L. 1973. Sativin: An induced isoflavan from the leaves of *Medicago sativa* L. *Nature (Lond.)* **242:**125–126.

Ingham, J. L., Keen, N. T., and Hymowitz, T. 1977. A new isoflavone phytoalexin from fungus-inoculated stems of *Glycine wightii*. *Phytochemistry* **16:**1943–1946.

Ingram, D. S. 1969. Growth of *Plasmodiophora brassicae* in host callus. *J. Gen. Microbiol.* **55:**9–18.

Ingram, D. S. 1978. Cell death and resistance to biotrophs. *Ann. Appl. Biol.* **89:**291–295.

Ingram, D. S., Sargent, J. A., and Tommerup, I. S. 1976. Structural aspects of infection by biotrophic fungi. In *Biochemical Aspects of Plant–Parasite Relationships* (J. Friend and D. R. Threlfall, eds.), pp. 43–78. Academic Press, London.

Inman, R. E. 1962. Relationships between disease intensity and stage of disease development on carbohydrate levels of rust-affected bean leaves. *Phytopathology* **52:**1207–1211.

Ioannou, N., Schneider, R. W., and Grogan, R. G. 1977a. Effect of oxygen, carbon dioxide, and ethylene on growth, sporulation, and production of microsclerotia by *Verticillium dahliae*. *Phytopathology* **67:**645–650.

Ioannou, N., Schneider, R. W., and Grogan, R. G. 1977b. Effect of flooding on the soil gas composition and the production of microsclerotia by *Verticillium dahliae* in the field. *Phytopathology* **67:**651–656.

Irwin, J. A. G. 1976. Observations on the mode of infection of lucerne roots by *Phytophthora megasperma*. *Aust. J. Bot.* **24:**447–451.

Ishiguri, Y., Tomiyama, K., Doke, N., Murai, A., Katsui, N., Yagihashi, F., and Masamune, T. 1978. Induction of rishitin-metabolizing activity in potato tuber tissue disks by wounding and identification of rishitin metabolites. *Phytopathology* **68:**720–725.

Israel, H. W., and Ross, A. F. 1967. The fine structure of local lesions by tobacco virus in tobacco. *Virology* **33:**272–286.

Israel, H. W., Wilson, R. G., Aist, J. R., and Kunoh, H. 1980. Cell wall appositions and plant disease resistance. Acoustic microscopy of papillae that block fungal ingress. *Proc. Natl. Acad. Sci. U. S. A.* **77:**2046–2049.

Jackson, W. A., and Volk, R. J. 1970. Photorespiration. *Annu. Rev. Plant Physiol.* **21:**385–452.

Jenns, A. E., and Kuć, J. 1977. Localized infection with tobacco necrosis virus protects cucumber against *Colletotrichum lagenarium*. *Physiol. Plant Pathol.* **11:**207–212.

Jenns, A. E., and Kuć, J. 1980. Characteristics of anthracnose resistance induced by localized infection of cucumber with tobacco necrosis virus. *Physiol. Plant Pathol.* **17:**81–91.

Jensen, R. G., and Bahr, J. T. 1977. Ribulose 1, 5-bisphosphate carboxylase-oxygenase. *Annu. Rev. Plant Physiol.* **28:**379–400.

Johnson, G., Maag, D. D., Johnson, D. K., and Thomas, R. D. 1976. The possible role of phytoalexins in the resistance of sugarbeet (*Beta vulgaris*) to *Cercospora beticola*. *Physiol. Plant Pathol.* **8:**225–230.

Johnson, H. W., and Sproston, T., Jr. 1965. The inhibition of fungus infection pegs in *Ginkgo biloba*. *Phytopathology* **55:**225–227.

Johnson, L. B., and Cunningham, B. A. 1972. Peroxidase activity in healthy and leaf-rust-infected wheat leaves. *Phytochemistry* (*Oxf.*) **11:**547–551.

Johnson, L. B., and Lee, R. F. 1978. Peroxidase changes in wheat isolines with compatible and incompatible leaf rust infections. *Physiol. Plant Pathol.* **13:**173–181.

Johnston, C. O., and Huffman, M. D. 1958. Evidence of local antagonism between two cereal rust fungi. *Phytopathology* **48:**69–70.

Jones, D. R., and Deverall, B. J. 1978. The use of leaf transplants to study the cause of hypersensitivity to leaf rust, *Puccinia recondita,* in wheat carrying the *Lr20* gene. *Physiol. Plant Pathol.* **12:**311–319.

Jones, H. G. 1973. Estimation of plant water status with the beta-gauge. *Agric. Meteorol.* **11:**345–355.

Jones, L. R. 1909. The bacterial soft rots of certain vegetables. II. Pectinase, the cytolytic enzyme produced by *Bacillus carotovorus* and certain other soft-rot organisms. *Tech. Bull. Vt. Agric. Exp. Sta.* **147:**283–360.

Jones, R. A., Rupert, E. A., and Barnett, D. W. 1981. Virus infection of *Trifolium* species in cell suspension cultures. *Phytopathology* **71:**116–119.

Jones, T. M., and Albersheim, P. 1972. A gas chromatographic method for determination of aldose and uronic acid constituents of plant cell wall polysaccharides. *Plant Physiol.* **49:**926–936.

Kado, C. I. 1976. The tumor-inducing substance of *Agrobacterium tumefaciens*. *Annu. Rev. Phytopathol.* **14:**265–308.

Kado, C. I., and Lurquin, P. F. 1975. Studies on *Agrobacterium tumefaciens*. IV. Nonreplication of the bacterial DNA in mung bean (*Phaseolus aureus*). *Biochem. Biophys. Res. Commun.* **64:**175–183.

Kaiss-Chapman, R. W., and Morris, R. O. 1977. Trans-zeatin in culture filtrates of *Agrobacterium tumefaciens*. *Biochem. Biophys. Res. Comm.* **76:**453–459.

Kaplan, D. T., Keen, N. T., and Thompson, I. J. 1980a. Association of glyceollin with the incompatible response of soybean roots to *Meloidogyne incognita*. *Physiol. Plant Pathol.* **16:**309–318.

Kaplan, D. T., Keen, N. T., and Thompson, I. J. 1980b. Studies on the mode of action of glyceollin in soybean incompatibility to the root knot nematode, *Meloidogyne incognita*. *Physiol. Plant Pathol.* **16:**319–325.

Karr, A. I., Karr, D. B., and Strobel, G. A. 1974. Isolation and partial purification of four host-specific toxins of *Helminthosporium maydis* (race T). *Plant Physiol.* (*Bethesda*) **53:**250–257.

Kato, S., and Misawa, T. 1971. Studies on the infection and multiplication of plant viruses. *Ann. Phytopathol. Soc. Jpn.* **37:**272–282.

Katsui, N., Murai, A., Takasugi, M., Imaizumi, K., Masamune, T., and Tomiyama, K. 1968. The structure of rishitin, a new antifungal compound from diseased potato tubers. *Chem. Commun.* (*J. Chem. Soc. Sect. D*) **1968:**43–44.

Katsura, K., Egawa, H., Masuko, M., and Ueyama, A. 1969. On the plant hormones in healthy and *Plasmodiophora* infected roots of *Brassica rapa* var. Neosuguki Kitam II. *Proc. Kansai Plant Prot. Soc.* **11:**23–27.

Kazmaier, H. E. 1960. Some pathophysiological aspects of premature defoliation associated with rose blackspot. *Diss. Abstr.* **21:**21.

Keck, R. W., and Hodges, T. K. 1973. Membrane permeability in plants: Changes induced by host-specific pathotoxins. *Phytopathology* **63:**226–230.

Keegstra, K., Talmadge, K. W., Bauer, W. D., and Albersheim, P. 1973. The structure of plant cell walls. III. A model of the wall of suspension-cultured sycamore cells based on interconnections of the macromolecular components. *Plant Physiol.* (*Bethesda*) **51:**188–197.

Keen, N. T. 1971. Hydroxyphaseollin production by soybeans resistant and susceptible to *Phytophthora megasperma* var. *sojae*. *Physiol. Plant Pathol.* **1:**265–275.

Keen, N. T. 1975. Specific elicitors of phytoalexin production: Determinants of race specificity in pathogens? *Science* **187:**74–75.

Keen, N. T. 1978. Surface glycoproteins of *Phytophthora megasperma* var. *sojae* function as race specific glyceollin elicitors in soybeans. *Phytopathol. News* **12:**221 (Abstr).

Keen, N. T. 1981. Evaluation of the role of phytoalexins. In *Plant Disease Control* (R. C. Staples, ed.), pp. 155–177. John Wiley and Sons, New York.

Keen, N. T., and Bruegger, B. 1977. Phytoalexins and chemicals that elicit their production in plants. In *Host Plant Resistance to Pests* (P. A. Hedin, ed.), pp. 1–26. American Chemical Society, Washington, D. C.

Keen, N. T., and Horton, J. C. 1966. Induction and repression of endo-polygalacturonase synthesis by *Pyrenochaeta terrestris*. *Can. J. Microbiol.* **12:**443–453.

Keen, N. T., and Kennedy, B. W. 1974. Hydroxyphaseollin and related isoflavonoids in the hypersensitive resistant reaction of soybean against *Pseudomonas glycinea*. *Physiol. Plant Pathol.* **4:**173–185.

Keen, N. T., and Littlefield, L. J. 1979. The possible association of phytoalexins with resistance gene expression in flax to *Melampsora lini*. *Physiol. Plant Pathol.* **14:**265–280.

Keen, N. T., and Williams, P. H. 1971. Chemical and biological properties of a lipomucopolysaccharide from *Pseudomonas lachrymans*. *Physiol. Plant Pathol.* **1:**247–264.

Keen, N. T., Wang, M. C., Bartnicki–Garcia, S., and Zentmyer, G. A. 1975. Phytoxicity of mycolaminarans-β-1,3-glucans from *Phytophthora* spp. *Physiol. Plant Pathol.* **7**:91–97.

Keen, N. T. Ersek, T., Long, M., Bruegger, B., and Holliday, M. 1981. Inhibition of hypersensitive reaction of soybean leaves to incompatible *Pseudomonas* spp. by blasticidin S, streptomycin or elevated temperature. *Physiol. Plant Pathol.* **18**:325–337.

Kehr, A. E., and Smith, H. H. 1954. Genetic tumors in *Nicotiana hybrids*. *Brookhaven Symp. Biol.* **6**:55–76.

Kelman, A., and Cowling, E. B. 1965. Cellulase of *Pseudomonas solanacearum* in relation to pathogenesis. *Phytopathology* **55**:148–155.

Kelman, A., and Hruschka, J. 1973. The role of motility and aerotaxis in the selective increase of avirulent bacteria in still broth cultures of *Pseudomonas solanacearum*. *J. Gen. Microbiol.* **76**:177–188.

Kemp, J. D. 1976. Octopine as a marker for the induction of tumorous growth by *Agrobacterium tumefaciens* strain $B^6$. *Biochem. Biophys. Res. Commun.* **69**:816–822.

Kemp, J. D. 1977. A new amino acid derivative present in crown gall tumor tissue. *Biochem. Biophys. Res. Commun.* **74**:862–868.

Kemp, J. D. 1978. *In vivo* synthesis of crown gall-specific *Agrobacterium tumefaciens*-directed derivatives of basic amino acids. *Plant Physiol.* (*Bethesda*) **62**:26–30.

Kende, H. 1971. The cytokinins. *Int. Rev. Cytol.* **31**:301–338.

Kenfield, D. S., and Strobel, G. A. 1977. Biochemical aspects of plant disease resistance and susceptibility. In *Host Plant Resistance to Pests* (P. A. Hedin, ed.), pp. 35–46. ACS Symposium Series 62. American Chemical Society, Washington, D. C.

Kenfield, D. S., and Strobel, G. A. 1981. α -Galactoside binding proteins from plant membranes: Isolation, characterization, and relation to helminthosporoside binding proteins of sugarcane. *Plant Physiol.* (*Bethesda*) **67**:1174–1180.

Kenning, L. A., and Hanchey, P. 1980. Ultrastructure of lesion formation in *Rhizoctonia*-infected bean hypocotyls. *Phytopathology* **70**:998–1004.

Kern, H. 1972. Phytotoxins produced by *Fusaria*. In *Phytotoxins in Plant Diseases* (R. K. S. Wood, A. Ballio, and A. Graniti, eds.), pp. 35–48. Academic Press, London and New York.

Kern, H., and Naef–Roth, S. 1971. Phytolysin, ein durch pflanzenpathogene Pilze gebildeter mazerierender Faktor. *Phytopath. Z.* **71**:231–246.

Kerr, A., and Htay, K. 1974. Biological control of crown gall through bacteriocin production. *Physiol. Plant Pathol.* **4**:37–44.

Kerr, A., and Roberts, W. P. 1976. *Agrobacterium:* Correlations between and transfer of pathogenicity, octopine and nopaline metabolism and bacteriocin 84 sensitivity. *Physiol. Plant Pathol.* **9**:205–211.

Kerr, A., Manigault, H. P., and Tempe, J. 1977. Transfer of virulence *in vivo* and *Agrobacterium*. *Nature* (*Lond.*) **265**:560–561.

Keskin, B., and Fuchs, W. H. 1969. Der Infektionsvorgang bei *Polymyxa betae*. *Arch. Mikrobiol.* **68**:218–226.

Khew, K. L., and Zentmyer, G. A. 1974. Electrotactic response of zoospores of seven species of *Phytophthora*. *Phytopathology* **64**:500–507.

Kimmins, W. C. 1977. Wound induced resistance to plant virus infections. In *Cell Wall Biochemistry Related to Specificity in Host–Plant Pathogen Interactions* (B. Solheim and J. Raa, eds.), pp. 271–276. Universitetsforlaget, Oslo, Norway.

Kimmins, W. C., and Wuddah, D. 1977. Hypersensitive resistance: Determination of lignin in leaves with a localized virus infection. *Phytopathology* **67**:1012–1016.

King, K. W., and Vessal, M. I. 1969. Enzymes of the cellulase complex. In *Cellulases and Their Applications. Advances in Chemistry* (R. F. Gould, ed.), Series No. 95, pp. 7–25. American Chemical Society, Washington, D. C.

King, N. J., and Fuller, D. B. 1968. The xylanase system of *Coniophora cerebella*. *Biochem. J.* **108:**571–576.

Király, Z. 1980. Defenses triggered by the invader: Hypersensitivity. In *Plant Disease: An Advanced Treatise* (J. G. Horsfall and E. B. Cowling, eds.), Vol. 5, pp. 201–224. Academic Press, New York.

Király, Z., Barna, B., and Érsek, T. 1972. Hypersensitivity as a consequence, not a cause, of plant resistance to infection. *Nature (Lond.)* **239:**456–457.

Király, Z., Hevesi, M., and Klement, Z. 1977. Inhibition of bacterial multiplication in incompatible host–parasite relationships in the absence of hypersensitive necrosis. *Acta Phytopathol. Acad. Sci. Hung.* **12:**247–256.

Kirk, T. K. 1971. Effects of microorganisms on lignin. *Annu. Rev. Phytopath.* **9:**185–210.

Kitazawa, K., and Tomiyama, K. 1969. Microscopic observations of infection of potato cells by compatible and incompatible races of *Phytophthora infestans*. *Phytopathol. Z.* **66:**317–324.

Kitazawa, K., Inagaki, H., and Tomiyama, K. 1973. Cinephotomicrographic observations on the dynamic responses of protoplasm of potato plant cell to infection by *Phytophthora infestans*. *Phytopathol. Z.* **76:**80–86.

Klämbt, D., Thies, G., and Skoog, F. 1966. Isolation of cytokinins from *Corynebacterium fascians*. *Proc. Natl. Acad. Sci. U. S. A.* **56:**52–59.

Klapwijk, P. M., Hooykaas, P. J. J., Kerster, H. C. M., Schilperoort, R. A., and Rorsch, A. 1976. Isolation and characterization of *Agrobacterium tumefaciens* mutants affected in the utilization of octopine, octopinic acid and lysopine. *J. Gen. Microbiol.* **96:**155–163.

Klapwijk, P. M., Van Bruekeler, J., Korevaar, K., Ooms, G., and Schilperoort, R. A. 1980. Transposition of Tn904 encoding streptomycin resistance into the octopine Ti plasmid of *Agrobacterium tumefaciens*. *J. Bacteriol.* **141:**129–136.

Klarman, W. L., and Gerdemann, J. W. 1963. Induced susceptibility in soybean plants genetically resistant to *Phytophthora sojae*. *Phytopathology* **53:**863–864.

Klein, R. M. 1965. The physiology of bacterial tumors in plants and of habitation. In *Handbuch der Pflanzenphysiologie* (W. Ruhland, ed.), Vol. 15, Part 2, pp. 209–235. Springer-Verlag, Berlin and New York.

Klement, Z. 1963. Method for the rapid detection of the pathogenicity of phytopathogenic pseudomonads. *Nature (Lond.)* **199:**299–300.

Klement, Z. 1971. The hypersensitive reaction of plants to bacterial infections. In *Biochemical and Ecological Aspects of Plant–Parasite Relations* (Z. Király and L. Szalay-Marzsó, eds.), pp. 115–118. Akadémiai Kiadó, Budapest.

Klement, Z., and Goodman, R. N. 1967. The role of the living bacterial cell and induction time in the hypersensitive reaction of tobacco plant. *Phytopathology* **57:**322–323.

Klement, Z., Király, Z., and Pozsár, B. I. 1966. Supression of virus multiplication and local lesion production in tobacco following inoculation with a saprophytic bacterium. *Acta Phytopahtol. Acad. Sci. Hung.* **1:**11–18.

Kloepper, J. W., Leong, J., Teintze, M., and Schroth, M. N. 1980. Enhanced plant growth by siderophores produced by plant growth-promoting rhizobacteria. *Nature (Lond.)* **286:**885–886.

Knee, M., and Friend, I. 1968. Extracellular "galactanase" activity from *Phytophthora infestans* (Mont.) de Bary. *Phytochemistry* **7:**1289–1291.

Kobayashi, K., and Ui, T. 1979. Phytotoxicity and antimicrobial activity of Graminin A, produced by *Cephalosporium gramineum,* the causal agent of *Cephalosporium* stripe disease of wheat. *Physiol. Plant Pathol.* **14:**129–133.

Koekman, B. P., Ooms, G., Klapwijk, P. M., and Schilperoort, R. A. 1979. Genetic map of an octopine Ti-plasmid. *Plasmid* **2:**347–357.

Kohmoto, K., Khan, I. D., Renbutsu, Y., Taniguchi, T., and Nishimura, S. 1976. Multiple host-specific toxins of *Alternaria mali* and their effects on the permeability of host cells. *Physiol. Plant Pathol.* **8:**141–153.

Kohmoto, K., Scheffer, R. P., and Whiteside, J. O. 1979. Host-selective toxins from *Alternaria citri*. *Phytopathology* **69**:667–671.

Kojima, M., and Uritani, I. 1974. The possible involvement of a spore agglutinating factor(s) in various plants in establishing host specificity by various strains of black rot fungus, *Ceratocystis fimbriata*. *Plant Cell Physiol.* **15**:733–737.

Kojima, M., Kawakita, K., and Uritani, I. 1982. Studies on a factor in sweet potato roots which agglutinates spores of *Ceratocystis fimbriata*, black rot fungus. *Plant Physiol.* **69**:474–478.

Köller, W., Allan, C. R., and Kolattukudy, P. E. 1982. Role of cutinase and cell wall degrading enzymes in infection of *Pisum sativum* by *Fusarium solani* f. sp. *pisi*. *Physiol. Plant Pathol.* **20**:47–60.

Kono, Y., and Daly, J. M. 1979. Characterization of the host-specific pathotoxin produced by *Helminthosporium madis*, race T, affecting corn with Texas male-sterile cytoplasm. *Bioorg. Chem.* **8**:391–397.

Korn, E. D. 1969. Cell membranes: Structure and synthesis. *Annu. Rev. Biochem.* **38**:263–288.

Kosuge, T. 1969. The role of phenolics in host response to infection. *Annu. Rev. Phytopathol.* **7**:195–222.

Kosuge, T. 1978. The capture and use of energy by diseased plants. In *Plant Disease: An Advanced Treatise* (J. G. Horsfall and E. B. Cowling, eds.), Vol. 3, pp. 85–116. Academic Press, New York.

Kosuge, T., and Gilchrist, D. G. 1976. Metabolic regulation in host–parasite interactions. In *Physiological Plant Pathology* (R. Heitefuss and P. H. Williams, eds.), Vol. 4, pp. 679–702. Springer-Verlag, Berlin and New York.

Kosuge, T., and Kimpel, J. A. 1981. Energy use and metabolic regulators in plant-pathogen interactions. In *Effects of Disease on the Physiology of the Growing Plant* (P. G. Ayres, ed.), pp. 29–45. Cambridge University Press, Cambridge.

Kozlowski, T. T. 1978. How healthy plants grow. In *Plant Disease: An Advanced Treatise* (J. G. Horsfall and E. B. Cowling, eds.), Vol. 3, pp. 19–51. Academic Press, New York.

Kraft, J. M., Endo, R. M., and Erwin, D. C. 1967. Infection of primary roots of bentgrass by zoospores of *Pythium aphanidermatum*. *Phytopathology* **57**:86–90.

Krenzer, E. G., Jr., Moss, D. N., and Crookston, R. K. 1975. Carbon dioxide-compensation points of flowering plants. *Plant Physiol. (Bethesda)* **56**:194–206.

Krikon, J., Chorin, M., and Vaadia, Y. 1971. Hormonal status of tomato plants infected by *Verticillium dahliae*. In *International Verticillium Symposium* (G. F. Pegg, ed.), pp. 22. University of London, London.

Krupinsky, J. M., Scharen, A. L., and Schillinger, J. A. 1973. Pathogenic variation in *Septoria nodorum* (Berk.) in relation to organ specificity, apparent photosynthetic rate and yield of wheat. *Physiol. Plant Pathol.* **3**:187–194.

Kuć, J. 1964. Phenolic compounds and disease resistance in plants. In *Phenolics in Normal and Diseased Fruits and Vegetables* (V. C. Runeckles, ed.), pp. 63–81. Proc. Plant Phenolics Group of North America Symp. Imperial Tobacco Co., Montreal, Quebec.

Kuć, J. 1967. Shifts in oxidative metabolism during pathogenesis. In *The Dynamic Role of Molecular Constituents in Plant–Parasite Interaction* (C. J. Mirocha and I. Uritani, eds.), pp. 183–202. Bruce Publishing, St. Paul, Mn.

Kuć, J. 1972. Phytoalexins. *Annu. Rev. of Phytopathol.* **10**:207–232.

Kuć, J. 1976a. Phytoalexins. In *Encyclopedia of Plant Physiology, New Series, Physiological Plant Pathology* (R. H. Heitefuss and P. H. Williams, eds.), Vol. 4, pp. 632–652. Springer-Verlag, Berlin, Heidelberg, and New York.

Kuć, J. 1976b. Phytoalexins and the specificty of plant–parasite interaction. In *Specificity in Plant Diseases* (R. K. S. Wood and A. Graniti, eds.), pp. 253–268. Plenum Press, New York.

Kuć, J. 1978. Changes in intermediary metabolism caused by disease. In *Plant Disease: An Advanced Treatise* (J. G. Horsfall and E. B. Cowling, eds.), Vol. 3, pp. 349–374. Academic Press, New York.

Kuć, J. 1981. Multiple mechanisms, reaction rates, and induced resistance in plants. In *Plant Disease Control: Resistance and Susceptibility* (R. C. Staples and G. H. Toenniessen, eds.), pp. 259–272. John Wiley and Sons, New York.

Kuć, J., and Caruso, F. L. 1977. Activated coordinated chemical defense against disease in plants. In *Host-Plant Resistance to Pests* (P. A. Hedin, ed.), pp. 78–89. American Chemical Society, Washington, D. C.

Kuć, J., Richmond, S. 1977. Aspects of the protection of cucumber against *Colletotrichum lagenarium* by *Colletotrichum lagenarium*. *Phytopathology* **67**:533–536.

Kuć, J., Shockley, G., and Kearney, K. 1975. Protection of cucumber against *Colletotrichum lagenarium* by *Colletotrichum lagenarium*. *Physiol. Plant Pathol.* **7**:195–199.

Kuć, J., Currier, W. W., Elliston, J., and McIntyre, J. L. 1976a. Determinant of plant disease resistance and susceptibility: A perspective based on three plant–parasite interactions. In *Biochemistry and Cytology of Plant–Parasite Interactions* (K. Tomiyama, M. Daly, I. Uritani, H. Oku, and S. Ouchi, eds.), pp. 168–180. Kadansha Ltd., Tokyo, and Elsevier Scientific Publishing Co., Amsterdam, The Netherlands.

Kuć, J., Currier, W., and Shih, M. J. 1976b. Terpenoid phytoalexins. In *Biochemical Aspects of Plant–Parasite Relationships* (J. Friend and D. R. Threlfall, eds.), pp. 225–237. Academic Press, New York.

Kühn, J. 1858. *Die Krankheiten der Kulturgewachse, ihre Ursachen und ihre Verhutung*. G. Bosselmann, Berlin.

Kuhn, P. J., and Smith, D. A. 1979. Isolation from *Fusarium solani* f. sp. *Phaseoli* of an enzymic system responsible for kievitone and phaseollidin detoxification. *Physiol. Plant Pathol.* **14**:179–190.

Kunoh, H., and Akai, S. 1969. Histochemical observation of the halo on the epidermal cell wall of barley leaves attacked by *Erysiphe graminis hordei*. *Mycopathol. Mycol. Appl.* **37**:113–118.

Kunoh, H., and Ishizaki, H. 1976. Accumulation of chemical elements around the penetration sites of *Erysiphe graminis hordei* on barley leaf epidermis: (III) Micromanipulation and X-ray microanalysis of silicon. *Physiol. Plant Pathol.* **8**:91–96.

Kuo, M. S., and Scheffer, R. P. 1964. Evaluation of fusaric acid as a factor in development of *Fusarium* wilt. *Phytopathology* **54**:1041–1044.

Kuo, M. S., and Scheffer, R. P. 1970. Comparative effects of host-specific toxins and *Helminthosporium* infections on respiration and carboxylation by host tissue. *Phytopathology* **60**:1391–1394.

Kuo, M. S., Yoder, O. C., and Scheffer, R. P. 1970. Comparative specificity of the toxins of *Helminthosporium carbonum* and *Helminthosporium victoriae*. *Phytopathology* **60**: 365–368.

Kuo, T., and Kosuge, T. 1969. Factors influencing the production and further metabolism of indole-3-acetic acid by *Pseudomonas savastanoi*. *J. Gen. Appl. Microbiol.* **15**:51–63.

Kuo T., Chien, M., and Li, H. 1969. Ethyl acetate produced by *Ceratocystis paradoxa* and *C. adiposum* and its role in the inhibition of the germination of sugarcane buds. *Can. J. Bot.* **47**:1459–1463.

Lai, M. T., Weinhold, A. R., and Hancock, J. G. 1968. Permeability changes in *Phaseolus aureus* associated with infection by *Rhizoctonia solani*. *Phytopathology* **58**:240–245.

Lamport, D. T. A. 1970. Cell wall metabolism. *Annu. Rev. Plant Physiol.* **21**:235–270.

Lamport, D. T. A., Katona, L., and Roerig, S. 1973. Galactosylserine in extensin. *Biochem. J.* **133**:125–132.

Lappe, U., and Barz, W. 1978. Degradation of pisatin by fungi of the genus *Fusarium*. *Z. Naturforsch. Sect. C. Biosci.* **33**:301–302.

Larkin, P. J., and Scrowcroft, W. R. 1981. Eyespot disease of sugarcane, induction of host-specific toxin and its interaction with leaf cells. *Plant Physiol. (Bethesda)* **67**:408–414.

Law, T. J., and Strijdom, B. W. 1977. Some observations on plant lectins and *Rhizobium* specificity. *Soil Biol. Biochem.* **9**:79–84.

Layne, R. E. C. 1967. Foliar trichomes and their importance as infection sites for *Corynebacterium michiganense* on tomato. *Phytopathology* **57**:981–985.

Leatham, G. F., King, V., and Stahmann, M. A. 1980. *In vitro* protein polymerization by quinones or free radicals generated by plant or fungal oxidative enzymes. *Phytopathology* **70**:1134–1140.

LeBeau, J. B., and Atkinson, T. G. 1967. Borax inhibition of cyanogenesis in snow mold of alfalfa. *Phytopathology* **57**:863–865.

Ledeboer, A. M. 1978. Large plasmids in Rhizobiaceae. I. Studies on the transcriptions of the tumor-inducing plasmid from *Agrobacterium tumefaciens* in sterile crown gall tumor cells. Ph.D. Thesis. University of Leiden, The Netherlands.

Lee, S. C., and West, C. A. 1981a. Polygalacturonase from *Rhizopus stolonifer* is an elicitor of casbene synthetase activity in castor bean (*Ricinus communis* L.) seedlings. *Plant Physiol. (Bethesda)* **67**:633–639.

Lee, S. C., and West, C. A. 1981b. Properties of *Rhizopus stolonifer* polygalacturonase, an elicitor of casbene synthetase activity in castor bean (*Ricinus communis* L.) seedlings. *Plant Physiol. (Bethesda)*, **67**:633–639.

Legrand, M., Fritig, B., and Hirth, L. 1976. Enzymes of the phenylpropanoid pathway and the necrotic reaction of hypersensitive tobacco to tobacco mosaic virus. *Phytochemistry* **15**:1353–1359.

Lehrer, R. I. 1969. Antifungal effects of peroxidase systems. *J. Bacteriol.* **99**:361–365.

Lemmers, M., De Beuckeleer, M., Holsters, M., Zambryski, P., Depicker, A., Hernalsteens, J. P., Van Montagu, M., and Schell, J. 1980. Internal organization, boundaries, and integration of Ti-plasmid DNA in nopaline crown gall tumors. *J. Mol. Biol.* **144**:353–376.

Leonard, K. J., and Czochor, R. J. 1980. Theory of genetic interactions among populations of plants and their pathogens. *Annu. Rev. Phytopathol.* **18**:237–258.

Letcher, R. M., Widdowson, D. A., Deverall, B. J., and Mansfield, J. W. 1970. Identification and activity of Wyerone acid as a phytoalexin in broad bean (*Vicia faba*) after infection by *Botrytis*. *Phytochemistry* **9**:249–252.

Lieberman, M. 1979. Biosynthesis and action of ethylene. *Annu. Rev. Plant Physiol.* **30**:533–591.

Liener, I. E. 1976. Phytohemagglutinins (Phytolectins). *Annu. Rev. Plant Physiol.* **27**:291–319.

Lin, B. C., and Kado, C. I. 1977. Studies on *Agrobacterium tumefaciens*. VII. Avirulence induced by temperature and ethidium bromide. *Can. J. Microbiol.* **23**:1554–1561.

Lin, T. S., and Kolattukudy, P. E. 1980. Isolation and characterization of a cuticular polyester (cutin) hydrolyzing enzyme from phytopathogenic fungi. *Physiol. Plant Pathol.* **17**:1–15.

Linderman, R. G. 1974. Production of ethylene by *Cylindrocladium*-infected azalea tissue. *Proc. Am. Phytopathol. Soc.* **1**:38–39 (Abstr.).

Ling, G. N. 1969. A new model for the living cell: A summary of the theory and recent experimental evidence in its support. *Int. Rev. Cytol.* **26**:1–61.

Ling, G. N., and Ochsenfeld, M. M. 1973. Mobility of potassium ion in frog muscle cells, both living and dead. *Science* **181**:78–81.

Lippincott, B. B., and Lippincott J. A. 1969. Bacterial attachment to a specific wound site as an essential stage in tumor initiation by *Agrobacterium tumefaciens*. *J. Bacteriol.* **97**:620–628.

Lippincott, B. B., Whatley, M. H., and Lippincott, J. A. 1977. Tumor induction by *Agrobacterium* involves attachment of the bacterium to a site on the host plant cell wall. *Plant Physiol.* **59**:388–390.

Lippincott, J. A., and Lippincott, B. B. 1965. Timing of events in crown gall tumor development on pinto bean leaves. *Dev. Biol.* **12**:309–327.

Lippincott, J. A., and Lippincott, B. B. 1975. The genus *Agrobacterium* and plant tumorogenesis. *Annu. Rev. Microbiol.* **29**:377–405.

Lippincott, J. A., and Lippincott, B. B. 1976. Morphogenic determinants as exemplified by crown gall disease. In *Physiological Plant Pathology* (R. Heitefuss and P. M. Williams, eds.), Vol. 4, pp. 357–388. Springer-Verlag, Berlin.

Lippincott, J. A., and Lippincott, B. B. 1977. Nature and specificity of the bacterium–host attachment in *Agrobacterium* infection. In *Cell Wall Biochemistry Related to Specificity in Host–Plant Pathogen Interactions* (B. Solheim and J. Raa, eds.), pp. 439–451. Universitetsforlaget, Oslo, Norway.

Lippincott, J. A., Beiderbeck, R., and Lippincott, B. B. 1973. Utilization of octopine and nopaline by *Agrobacterium. J. Bacteriol.* **116:**378–383.

Lisker, N., and Kuć, J. 1977. Elicitor of terpenoid accumulation in potato tuber slices. *Phytopathology* **67:**1356–1359.

Lisker, N., and Kuć, J. 1978. Terpenoid accumulation and browning in potato sprouts inoculated with *Phytophthora infestans. Phytopathology* **68:**1284–1287.

Littlefield, L. J. 1969. Flax rust resistance induced by prior inoculation with an avirulent race of *Melampsora lini. Phytopathology* **59:**1323–1328.

Littlefield, L. J., and Heath, M. C. 1979. *Ultrastructure of Rust Fungi.* Academic Press, New York.

Liu, S. T., and Kado, C. I. 1979. Indoleacetic acid production: A plasmid function of *Agrobacterium tumefaciens* C58. *Biochem. Biophys. Res. Commun.* **90:**171–178.

Livingston, R. S., and Scheffer, R. P. 1981. Isolation and characterization of host-selective toxin from *Helminthosporium sacchari. J. Biol. Chem.* **256:**1705–1710.

Livne, A. 1964. Photosynthesis in healthy and rust-affected plants. *Plant Physiol.* **39:**614–621.

Livne, A., and Daly, J. M. 1966. Translocation in healthy and rust-infected beans. *Phytopathology* **56:**170–175.

Lockhart, B. E. L., and Semanick, J. S. 1970. Growth inhibition, peroxidase and 3-indoleacetic acid oxidase activity, and ethylene production in cowpea mosaic virus-infected cowpea seedlings. *Phytopathology* **60:**553–554.

Loebenstein, G. 1962. Inducing partial protection in the host plant with native virus protein. *Virology* **17:**574–581.

Loebenstein, G. 1972. Localization and induced resistance in virus-infected plants. *Annu. Rev. Phytopathol.* **10:**177–206.

Loebenstein, G., and Lovrekovich, L. 1966. Interference with tobacco mosaic virus local lesion formation in tobacco by injecting heat-killed cells of *Pseudomonas syringae. Virology* **30:**587–591.

Loebenstein, G., Rabina, S., and Van Praagh, T. 1966. Induced interference phenomena in virus infections. In *Viruses of Plants* (A. B. R. Beemster and J. Djikstra, eds.), pp. 151–157. North Holland, Amsterdam.

Loegering, W. Q. 1978. Current Concepts in interorganismal genetics. *Annu. Rev. Phytopathol.* **16:**309–320.

Long, D. E., Fung, A. K., McGee, E. E. M., Cooke, R. C., and Lewis, D. H. 1975. The activity of invertase and its relevance to the accumulation of storage polysaccharides in leaves infected by biotrophic fungi. *New Phytol.* **74:**173–182.

Loper, J. E., and Kado, C. 1979. Host range conferred by the virulence-specifying plasmid of *Agrobacterium tumefaciens. J. Bacteriol.* **139:**591–596.

Lovrekovich, L., and Farkas, G. L. 1965. Induced protection against wildfire disease in tobacco leaves treated with heat-killed bacteria. *Nature (Lond.)* **205:**823–824.

Lovrekovich, L., Lovrekovich, H., and Stahmann, M. A. 1968. The importance of peroxidase in the wildfire disease. *Phytopathology* **58:**193–198.

Ludwig. A. R. 1960. Toxins. In *Plant Pathology: An Advanced Treatise* (J. G. Horsfall and A. E. Dimond, eds.), Vol. 2, pp. 315–357. Academic Press, New York and London.

Luke, H. H., and Freeman, T. E. 1965. Effects of victorin on Krebs cycle intermediates of a susceptible oat variety. *Phytopathology* **55:**967–969.

Luke, H. H., and Gracen, V. E. Jr. 1972a. Phytopathogenic toxins. In *Microbial Toxins* (S. Kadis, A. Ciegler, and S. J. Ajl, eds.), Vol. 8, pp. 131–137. Academic Press, New York and London.

Luke, H. H., and Gracen, V. E. Jr. 1972b. *Helminthosporium* toxins. In *Microbial Toxins* (S. Kadis, A. Ciegler, and S. J. Ajl, eds.), Vol. 8, pp. 139–168. Academic Press, New York and London.

Luke, H. H., and Wheeler, H. E. 1955. Toxin production by *Helminthosporium victoriae*. *Phytopathology* **45**:453–458.

Luke, H. H., and Wheeler, H. 1964. An intermediate reaction to victorin. *Phytopathology* **54**:1492–1493.

Luke, H. H., Warmke, H. E., and Hanchey, P. 1966. Effects of the pathotoxin victorin on ultrastructure of root and leaf tissue of *Avena* species. *Phytopathology* **56**:1178–1183.

Luke, H. H., Freeman, T. E., Garrard, L. A., and Humphreys, T. E. 1969. Leakage of phosphorylated sugars from oat tissue treated with victorin. *Phytopathology* **59**:1002–1004.

Lumsden, R. D. 1969. *Sclerotinia sclerotiorum* infection of beans and the production of cellulase. *Phytopathology* **59**:653–657.

Lumsden, R. D., and Bateman, D. F. 1968. Phosphatide-degrading enzymes associated with pathogenesis in *Phaseolus vulgaris* infected with *Thielaviopsis basicola*. *Phytopathology* **58**:219–227.

Lumsden, R. D., and Dow, R. L. 1973. Histopathology of *Sclerotinia sclerotiorum* infection of beans. *Phytopathology* **63**:708–715.

Lund, B. M. 1973. The effect of certain bacteria on ethylene production by plant tissue. In *Fungal Pathogenicity and the Plant's Response* (R. J. W. Byrde and C. B. Cutting, eds.), pp. 69–86. Academic Press, London and New York.

Lupton, F. G. H. 1956. Resistance mechanisms of species of *Triticum* and *Aegilops* and of amphidiploids between them to *Erysiphe graminis* DC. *Trans. Br. Mycol. Soc.* **39**:51–59.

Lupu, R., Grossman, S., and Cohen, Y. 1980. The involvement of lipoxygenase and antioxidants in pathogenesis of powdery mildew on tobacco plants. *Physiol. Plant Pathol.* **16**:241–248.

Lurie, S., and Hendrix, D. L. 1979. Differential ion stimulation of plasmalemma adenosine triphosphatase from leaf epidermis and mesophyll of *Nicotiana rustica* L. *Plant Physiol. (Bethesda)* **63**:936–939.

Luttrell, E. S. 1974. Parasitism of fungi on vascular plants. *Mycologia* **66**:1–15.

Lyon, F. M., and Wood, R. K. S. 1975. Production of phaseollin, coumestrol and related compounds in bean leaves inoculated with *Pseudomonas* spp. *Physiol. Plant Pathol.* **6**:117–124.

Lyon, F. M., and Wood, R. K. S. 1976. The hypersensitive reaction and other responses of bean leaves to bacteria. *Ann. Bot. (Lond.)* **40**:479–491.

Lyon, G. D. 1976. Metabolism of the phytoalexin rishitin by *Botrytis* spp. *J. Gen. Microbiol.* **96**:225–226.

Lyon, G. D., and Bayliss, C. E. 1975. The effect of rishitin on *Erwinia carotovora* var. *atroseptica* and other bacteria. *Physiol. Plant Pathol.* **6**:177–186.

Lyon, G. D., Lund, B. M., Bayliss, C. E., and Wyatt, G. M. 1975. Resistance of potato tubers to *Erwinia carotovora* and formation of rishitin and phytuberin in infected tissue. *Physiol. Plant Pathol.* **6**:43–50.

McClure, M. A., Misaghi, I., and Nigh, E. L. 1973. Shared antigens of parasitic nematodes and host plants. *Nature (Lond.)* **244**:306–307.

McDonald, P. W., and Strobel, G. A. 1970. Adenosine diphosphate-glucose pyrophosphorylase control of starch accumulation in rust-infected wheat leaves. *Plant Physiol. (Bethesda)* **46**:126–135.

McDonnell, K. 1958. Absence of pectolytic enzymes in a pathogenc strain of *Fusarium oxysporum* f. sp. *lycopersici*. *Nature (Lond.)* **182**:1025–1026.

Mace, M. E. 1964. Phenols and their involvement in Fusarium wilt pathogenesis. In *Phenolics in Normal and Diseased Fruits and Vegetables* (V. C. Runeckles, ed.), pp. 13–19. Imperial Tobacco Co., Montreal.

Mace, M. E. 1978. Contribution of tyloses and terpenoid aldehyde phytoalexins to *Verticillium* wilt resistance in cotton. *Physiol. Plant Pathol.* **12**:1–11.

McIntyre, J. L., and Dodds, J. A. 1979. Induction of localized and systemic protection against *Phytophthora parasitica* var. *nicotiana* by tobacco mosaic virus infection of tobacco hypersensitive to the virus. *Physiol. Plant Pathol.* **15:**321–330.

McIntyre, J. L., Kuć, J., and Williams, E. B. 1975. Protection of Barlett Pear against fireblight with deoxyribonucleic acid from virulent and avirulent *Erwinia amylovora*. *Physiol. Plant Pathol.* **7:**153–170.

McIntyre, J. L., Dodds, J. A., and Hare, J. D. 1981. Effects of localized infection of *Nicotiana tabacum* by tobacco mosaic virus on systemic resistance against diverse pathogens and an insect. *Phytopathology* **71:**297–301.

McKeen, W. E. 1974. Mode of penetration of epidermal cell walls of *Vicia faba* by *Botrytis cinerea*. *Phytopathology* **64:**461–467.

McKeen, W. E., and Rimmer, S. R. 1973. Initial penetration process in powdery mildew infection of susceptible barley leaves. *Phytopathology* **63:** 1049–1053.

McKeen, W. E., Smith, R., and Bhattacharya, P. K. 1969. Alterations of the host wall surrounding the infection peg of powdery mildew fungi. *Can. J. Bot.* **47:**701–706.

Maclean, D. J., and Tommerup, I. C. 1979. Histology and physiology of compatibility and incompatibility between lettuce and the downy mildew fungus, *Bremia lactucae* Regal. *Physiol. Plant Pathol.* **14:**291–312.

McLean, D. M. 1967. Interaction of race I and race II of *Colletotrichum orbiculare* on watermelon. *Plant Dis. Rep.* **51:**885–887.

Magie, A. R. 1963. Physiological factors involved in tumor production by the oleander knot pathogen, *Pseudomonas savastanoi*. Ph.D. Thesis, University of California, Davis, California.

Magie, A. R., Wilson, E. E., and Kosuge, T. 1963. Indoleacetamide as an intermediate in the synthesis of indoleacetic acid in *Pseudomonas savastanoi*. *Science* **141:**1281–1282.

Magyarosy, A. C. and Malkin, R. 1978. Effect of powdery mildew infection of sugar beet on the content of electron carriers in chloroplasts. *Physiol. Plant Pathol.* **13:** 183–188.

Magyarosy, A. C., Buchanan, B. B., and Schurmann, P. 1973. Effect of a systemic virus infection on chloroplast function and structure. *Virology* **55:**426–438.

Magyarosy, A. C., Schurmann, P., and Buchanan, B. B. 1976. Effect of powdery mildew infection on photosynthesis by leaves and chloroplasts of sugarbeets. *Plant Physiol. (Bethesda)* **57:**486–489.

Maheshwari, R., Allen, P. J., and Hildebrandt, A. C. 1967. Physical and chemical factors controlling the development of infection structures from uredospores germ tubes of rust fungi. *Phytopathology* **57:**855–862.

Maiti, I. B., and Kolattukudy, P. E. 1979. Prevention of fungal infection of plants by specific inhibitors of cutinase. *Science* **205:**507–508.

Majernik, O. 1971. A physiological study of the effects of $SO_2$ pollution, phenylmercuric acetate sprays, and parasitic infection on stomatal behavior and ageing in barley. *Phytopathol. Z.* **72:**255–268.

Malajczuk, N., and McComb, A. J. 1977. Root exudates from *Eucalyptus calophylla* R. Br. and *Eucalyptus marginata* Donn. ex Sm. seedlings and their effect on *Phytophthora cinnamomi* Rands. *Aust. J. Bot.* **25:**501–504.

Mandryk, M. 1963. Acquired systemic resistance to tobacco mosaic virus in *Nicotiana tabacum* evoked by stem injection with *Peronostora tabacina* Adam. *Aust. J. Agric. Res.* **14:**315–318.

Mankarios, A. T., and Friend, J. 1980. Polysaccharide-degrading enzymes of *Botrytis allii* and *Sclerotium cepivorum*. Enzyme production in culture and the effect of the enzymes on isolated onion cell walls. *Physiol. Plant Pathol.* **17:**93–104.

Maramorosch, K. 1957. Reversal of virus-caused stunting in plants by giberellic acid. *Science* **126:**651.

Marco, S., Levy, D., and Aharoni, N. 1976. Involvement of ethylene in the suppression of hypocotyl elongation in CMV-infected cucumbers. *Physiol. Plant Pathol.* **8:**1.

Marré, E. 1977. Effects of fusicoccin and hormones on plant cell membrane activities: Observations and hypotheses. In *Regulation of Cell Membrane Activities in Plants* (E. Marré and O. Ciferri, eds.), pp. 185–202. North-Holland, New York.

Marré, E. 1979. Fusicoccin: A tool in plant physiology. *Annu. Rev. Plant Physiol.* **30:**273–288.

Marré, E., Lado, P., Rasi–Caldogno, F., and Colombo, R. 1973a. Correlation between cell enlargement in pea internode segments and decrease in the pH of the medium of incubation. I. Effects of fusicoccin, natural and synthetic auxins and mannitol. *Plant Sci. Lett.* **1:**179–184.

Marré, E., Lado, P., Rasi–Caldogno, F., and Colombo, R. 1973b. Correlation between cell enlargement in pea internode segments and decrease in the pH of the medium of incubation. II. Effects of inhibitors of respiration, oxidative phosphorylation and protein syynthesis. *Plant Sci. Lett.* **1:**185–192.

Marré, E., Lado, P., and Rasi–Caldogno, F. 1974a. Correlation between proton extrusion and stimulation of cell enlargement. Effects of fusicoccin and of cytokinins on leaf fragments and isolated cotyledons. *Plant Sci. Lett.* **2:**139–150.

Marré, E. Lado, P., Rasi–Caldogno, F., Colombo, R., and DeMichelis, M. I. 1974b. Evidence for the coupling of proton extrusion to $K^+$ uptake in pea internode segments treated with fusicoccin or auxin. *Plant Sci. Lett.* **3:**365–379.

Marton, L., Wullems, G. J., Molendijk, L., and Schilperoort, R. A. 1979. *In vitro* transformation of cultured cells from *Nicotiana tabacum* by *Agrobacterium tumefaciens*. *Nature* (*Lond.*) **277:**129–131.

Mason, H. S., and Peterson, E. W. 1965. Melanoproteins. I. Reactions between enzyme-generated quinones and amino acids. *Biochim. Biophys. Acta.* **111:**134–146.

Massala, R., Legrand, M., and Fritig, B. 1980. Effect of α-aminooxyacetate, a competitive inhibitor of phenylalanine ammonia-lyase, on the hypersensitive resistance of tobacco mosaic virus. *Physiol. Plant Pathol.* **16:**213–226.

Matsubara, S., and Nakahira, R. 1967. Cytokinin activity in an extract from the gall of *Plasmodiophora*-infected root of *Brassica rapa*. *Bot. Mag. Tokyo* **80:**373–374.

Matta, A. 1971. Microbial penetration and immunization of uncongenial host plants. *Annu. Rev. Phytopathol.* **9:**387–410.

Matta, A. 1980. Defenses triggered by previous diverse invaders. In *Plant Disease: An Advanced Treatise* (J. G. Horsfall and E. B. Cowling, eds.), Vol. 5, pp. 345–361. Academic Press, New York.

Matta, A., and Gentile, I. 1964. Variation in auxin content induced in tomato by *Fusarium oxysporum* f. sp. *lycopersici*. *Rev. Appl. Mycol.* **44:**233.

Matthee, F. N., and Daines, R. H. 1969. The influence of nutrition on susceptibility of peach foliage to water congestion and infection. *Phytopathology* **59:**285–287.

Matthews, D. E., Gregory, P., and Gracen, V. E. 1979. *Helminthosporium maydis* race T toxin induces leakage of $NAD^+$ from T cytoplasm corn mitochondria. *Plant Physiol.* (*Bethesda*) **63:**1149–1153.

Matthews, R. E. F. 1970. *Plant Virology*. Academic Press, London and New York.

Matthews, R. E. F. 1980. Host plant responses to virus infection. In *Comprehensive Virology* (H. Fraenkel-Conrat and R. R. Wagner, eds.)., Vol. 16, pp. 297–359. Plenum Press, New York.

Matthysse, A. G., and Stump, A. J. 1976. The presence of *Agrobacterium tumefaciens* plasmid DNA in crown gall tumor cells. *J. Gen. Microbiol.* **95:**9–16.

Matthysse, A. G., Holmes, K. V., and Gurlitz, R. H. G. 1982. Binding of *Agrobacterium tumefaciens* to carrot protoplasts. *Physiol. Plant Pathol.* **20:**27–33.

Maugh, T. H. 1978. Acoustic microscopy: A new window to the world of small. *Science* **201:**1110–1114.

Maxwell, D. P., and Lumsden, R. D. 1970. Oxalic acid production by *Sclerotinia sclerotiorum* in infected bean and in culture. *Phytopathology* **60:**1395–1398.

Mayama, S., and Shishiyama, J. 1976a. Histological observation of cellular responses of barley leaves to powdery mildew infection by UV-fluorescence microscopy. *Ann. Phytopathol. Soc. Jpn.* **42:**591–596.

Mayama, S., and Shishiyama, J. 1976b. Detection of cellular collapse in albino barley leaves inoculated with *Erysiphe graminis hordei* by UV-fluorescence microscopy. *Ann. Phytopathol. Soc. Jpn.* **42:**618–620.

Mayama, S., and Shishiyama, J. 1978. Localized accumulation of fluorescent and UV-absorbing compounds at penetration sites in barley leaves infected with *Erysiphe graminis hordei. Physiol. Plant Pathol.* **13:**347–354.

Mazzucchi, U., and Puppillo, P. 1976. Prevention of confluent hypersensitive necrosis in tobacco leaves by a bacterial protein–lipopolysaccharide complex. *Physiol. Plant Pathol.* **9:**101–112.

Mazzucchi, U., Bazzi, C., and Pupillo, P. 1979. The inhibition of susceptible and hypersensitive reactions by protein–lipopolysaccharide complexes from phytopathogenic pseudomonads: Relationship to polysaccharide antigenic determinants. *Physiol. Plant Pathol.* **14:**19–30.

Meehan, F., and Murphy, H. C. 1947. Differential phytotoxicity of metabolic by-products of *Helminthosporium victoriae. Science* **106:**270–271.

Menage, A., and Morel, M. G. 1964. Sur la presence d'octopine dans les tissue de crown-gall. *C. R. Acad. Sci. Agric. Bulg.* **259:**4795–4796.

Mendgen, K. 1978. Attachment of bean rust cell wall material to host and nonhost plant tissue. *Arch. Microbiol.* **119:**113–117.

Merlo, D. J. 1978. Crown gall—A unique disease. In *Plant Disease: An Advanced Treatise* (J. G. Horsfall and E. B. Cowling, eds.), Vol. 3, pp. 261–313. Academic Press, New York.

Merlo, D. J. 1982. Crown gall—A multipotential disease. In *Advances in Plant Pathology* (P. H. Williams and D. S. Ingram, eds.), Vol. 1, (in press), Academic Press, New York.

Merlo, D. J., Nutter, R. C., Montoya, A. I., Garfinkel, D. J., Drummond, M. H., Chilton, M. D., Gordon, M. P., and Nester, E. W. 1980. The boundaries and copy numbers of Ti-plasmid T-DNA vary in crown gall tumors. *Mol. Gen. Genet.* **177:**637–643.

Mesibov, R., and Adler, J. 1972. Chemotaxis toward amino acids in *Escherichia coli. J. Bacteriol.* **112:**315–326.

Mesibov, R., Ordal, G. W., and Adler, J. 1973. The range of attractant concentrations for bacterial chemotaxis and the threshold and size of response over this range. *J. Gen. Physiol.* **62:**203–223.

Metlitskii, L. V. 1976. Phyto-immunity molecular mechanisms. Transl. from 31st Annu. Bakh. Symp., "Nauka," Moscow, pp. 1–50.

Metlitskii, L. V., and Ozeretskovskaya, O. L. 1968. *Plant Immunity: Biochemical Aspects of Plant Resistance to Parasitic Fungi.* Plenum Press, New York.

Meyer, W. L., Kuyper, L. F., Lewis, R. B., Templeton, G. E., and Woodhead, S. H. 1974. The amino acid sequence and configuration of tentoxin. *Biochem. Biophys. Res. Commun.* **56:**234–240.

Mignucci, J. S., and Boyer, J. S. 1979. Inhibition of photosynthesis and transpiration in soybean infected by *Microsphaera diffusa. Phytopathology* **69:**227–230.

Miles, A. A. 1955. The meaning of pathogenicity. In *Mechanism of Microbial Pathogenicity* (J. W. Howie and A. J. O'Hea, eds.). *Symp. Soc. Gen. Microbiol.* **5:**1–16.

Millar, R. L., and Hemphill, R. 1978. β-glucosidase associated with cyanogenesis in *Stemphylium* leafspot of birdsfoot trefoil. *Physiol. Plant Pathol.* **13:**259–270.

Millar, R. L., and Higgins, V. J. 1970. Association of cyanide with infection of birdsfoot trefoil by *Stemphylium loti. Phytopathology* **60:**104–110.

Miller, R. J., and Koeppe, D. E. 1971. Southern corn leaf blight: susceptible and resistant mitochondria. *Science* **173:**67–69.

Mills, L. J., and Van Staden, J. 1978. Extraction of cytokinins from maize, smut tumors of maize and *Ustilago maydis* cultures. *Physiol. Plant Pathol.* **13:**73–80.

Milne, R. G. 1966. Electron microscopy of tobacco mosaic virus in leaves of *Nicotiana glutinosa*. *Virology* **28:**527–532.

Minamikawa, T., and Uritani, I. 1965. Phenylalanine ammonia-lyase in sliced sweet potato root. Effect of antibiotics on the enzyme formation and its relation to the polyphenol biosynthesis. *Agric. Biol. Chem.* **29:**1021–1026.

Mirocha, C. J. 1972. Phytotoxins and metabolism. In *Phytotoxins in Plant Diseases* (R. K. S. Wood, A. Ballio, and A. Graniti, eds.), pp. 191–209. Academic Press, London and New York.

Misaghi, I. J., and McClure, M. A. 1974. Antigenic relationship of *Meloidogyne incognita, M. javanica* and *M. arenaria*. *Phytopathology* **64:**698–701.

Misaghi, I., DeVay, J. E., and Kosuge, T. 1972. Changes in cytokinin activity associated with the development of *Verticillium* wilt and water stress in cotton plants. *Physiol. Plant Pathol.* **2:**187–196.

Misaghi, I., McClure, M. A., and Kruk, T. H. 1975. Concentration of adenylates and energy charge values in cotton roots infected with *Meloidogyne incognita*. *Physiol. Plant Pathol.* **6:**85–91.

Misaghi, I. J., DeVay, J. E., and Duniway, J. M. 1978. Relationship between occlusion of xylem elements and disease symptoms in leaves of cotton plants infected with *Verticillium dahliae*. *Can. J. Bot.* **56:**339–342.

Misaghi, I. J., Grogan, R. G., Spearman, L. C., and Stowell, L. J. 1980. Antifungal activity of a fluorescent pigment produced by fluorescent pseudomonads. Am. Assoc. Adv. Sci., Pacific Division, Sixty-first Annual Meeting, p. 12 (Abstr.).

Misaghi, I. J., Stowell, L. J., Grogan, R. G., and Spearman, L. C. 1982. Fungistatic activity of water-soluble fluorescent pigments of fluorescent pseudomonads. *Phytopathology* **82:** 33–36.

Mitchell, R. E. 1976a. Isolation and structure of a chlorosis-inducing toxin of *Pseudomonas phaseolicola*. *Phytochemistry (Oxf.)* **15:**1941–1947.

Mitchell, R. E. 1976b. Bean halo-blight toxin. *Nature* (*Lond.*) **260:**75–76.

Mitchell, R. E. 1978. Halo blight of beans: Toxin production by several *Pseudomonas phaseolicola* isolates. *Physiol. Plant Pathol.* **13:**37–49.

Mitchell, R. E. 1981. Structure: Bacterial. In *Toxins in Plant Disease* (R. D. Durbin, ed.), pp. 259–291. Academic Press, New York.

Mitchell, R. E. 1982. Coronatine production by some phytopathogenc pseudomonads. *Physiol. Plant Pathol.* **20:**83–89.

Mitchell, R. E., and Bieleski, R. L. 1977. Involvement of phaseolotoxin in halo blight of beans: Transport and conversion to a functional toxin. *Plant Physiol.* (*Bethesda*) **60:**723–729.

Mitchell, R. E., and Parsons, E. A. 1977. A naturally occurring analogue of phaseolotoxin (bean halo blight toxin). *Phytochemistry (Oxf.)* **16:**280–281.

Mitchell, R. E., and Young, H. 1978. Identification of a chlorosis-inducing toxin of *Pseudomonas glycinea* as coronatine. *Phytochemistry* **17:**2028–2029.

Montalbini, P., and Buchanan, B. B. 1974. Effect of a rust infection on photophosphorylatin by isolated chloroplasts. *Physiol. Plant Pathol.* **4:**191–196.

Montalbini, P., Buchanan, B. B., and Hutcheson, S. W. 1981. Effect of rust infection on rates of photochemical polyphenol oxidation and latent polyphenol oxidase activity of *Vicia faba* chloroplast membranes. *Physiol. Plant Pathol.* **18:**51–57.

Montalbini, P., Koch, F., Burba, M., and Elstner, E. F. 1978. Increase in lipid-dependent carotene destruction as compared to ethylene formation and chlorophyllase activity following mixed infection of sugarbeet (*Beta vulgaris* L.) with beet yellows virus and beet mild yellowing virus. *Physiol. Plant Pathol.* **12:**211–223.

Montoya, A. I., Chilton, M. D., Gordon, M. P., Sciaky, D., and Nester, E. W. 1977. Octopine and nopaline metabolism in *Agrobacterium tumefaciens* and crown gall tumor cells: Role of plasmid genes. *J. Bacteriol.* **129:**101–107.

Moran, F., and Starr, M. P. 1969. Metabolic regulation of polygalacturonic acid *trans*-eliminase in *Erwinia*. *Eur. J. Biochem.* **II:** 291–295.

Morré, D. J. 1975. Membrane biogenesis. *Annu. Rev. Plant Physiol.* **26:**441–481.

Mount, M. S. 1978. Tissue is disintegrated. In *Plant Disease: An Advanced Treatise* (J. G. Horsfall and E. B. Cowling, eds.), Vol. 3, pp. 279–297. Academic Press, New York.

Mount, M. S., Bateman, D. F., and Bashman, H. G. 1970. Induction of electrolyte loss, tissue maceration, and cellular death of potato tissue by an endopolygalacturonate *trans*-eliminase. *Phytopathology* **60:**924–931.

Mount, M. S., Berman, P. M., Mortlock, R. P., and Hubbard, J. P. 1979. Regulation of endopolyglacturonate *trans*-eliminase in an adenosine 3′, 5′-cyclic monophosphate-deficient mutant of *Erwinia carotovora*. *Phytopathology* **69:**117–120.

Mullen, J. M. 1974. Enzymatic degradation of potato cell walls by *Fusarium roseum* (L. K.) Snyd. and Hans. Avenaceum. Ph.D. Thesis. Cornell University, Ithaca, New York.

Mullen, J. M., and Bateman, D. F. 1975. Polysaccharide degrading enzymes produced by *Fusarium roseum* 'Avenaceum' in culture and during pathogenesis. *Physiol. Plant Pathol.* **6:**233–246.

Müller, K. 1958. The formation and immunological significance of phytoalexin produced by *Phaseolis vulgaris* in response to infections with *Sclerotinia fructicola* and *Phytophthora infestans*. *Aust. J. Biol. Sci.* **11:**257–300.

Müller, K. O., and Börger, H. 1940. Experimentelle Untersuchungen über die *Phytophthora*-Resistenz der Kartoffel; zugleich ein Beitrag zum Problem der "erworbenen Resistenz" im Pflanzenreich. *Arb. Biol. Reichsanst. Landw. Forstw. Berlin-Dahlem* **23:**189–231.

Mulrean, E. N. 1980. Chemotaxis of *Pseudomonas syringae* pv. *phaseolicola*. Ph.D. Thesis. University of California, Berkeley.

Mulrean, E. N., and Schroth, M. N. 1979. *In vitro* and *in vivo* chemotaxis by *Pseudomonas phaseolicola*. *Phytopathology* **69:**1039 (Abstr.).

Murch, R. S., and Paxton, J. D. 1977. Glyceollin concentrations in Phytophthora resistant soybean: Light influences. *Proc. Am. Phytopathol. Soc.* **4:**135–136.

Mussell, H. W. 1973. Endopolygalacturonase: Evidence for involvement in *Verticillium* wilt of cotton. *Phytopathology* **63:**62–70.

Mussell, H. W. 1980. Tolerance to disease. In *Plant Disease: An Advanced Treatise* (J. G. Horsfall and E. B. Cowling, eds.), Vol. 5, pp. 39–52. Academic Press, New York.

Mussell, H. W., and Strand, L. L. 1977. Pectic enzymes: Involvement in pathogenesis and possible relevance to tolerance and specificity. In *Cell Wall Biochemistry Related to Specificity in Host-Plant Pathogen Interactions* (B. Solheim and J. Raa, eds.), pp. 31–77. Universitetsforlagert, Oslo, Norway.

Myers, D. F., and Fry, W. E. 1978. Enzymatic release and metabolism of hydrogen cyanide in sorghum infected by *Gloeocercospora sorghi*. *Phytopathology* **68:**1717–1722.

Nadolny, L., and Sequeira, L. 1980. Increases in peroxidase activities are not directly involved in induced resistance in tobacco. *Physiol. Plant Pathol.* **16:**1–8.

Naef–Roth, S., Gäumann, E., and Albersheim, P. 1961. Zur Bildung eines mazerierenden Fermentes durch *Dothidea ribesia* Fr. *Phytopathol. Z.* **40:**283–302.

Nakagaki, Y., Hirai, T., and Stahmann, M. A. 1970. Ethylene production by detached leaves infected with tobacco mosaic virus. *Virology* **40:**1–9.

Nakajima, T., Tomiyama, K., and Kinukawa, M. 1975. Distribution of rishitin and lubimin in potato tuber tissue infected by an incompatible race of *Phytophthora infestans* and the site where rishitin is synthesized. *Ann. Phytopathol. Soc. Jpn.* **41:**49–55.

Neish, A. C. 1964. Major pathways of biosynthesis of phenols. In *Biochemistry of Phenolic Compunds* (J. B. Harborne, ed.), pp. 295–359. Academic Press, London and New York.

Nelson, R. R. 1973. The meaning of disease resistance in plants. In *Breeding Plants for Disease Resistance: Concepts and Applications* (R. R. Nelson, ed.), pp. 13–25. University Park Press, London, and Pennsylvania State University Press.

Nelson, R. R., MacKenzie, D. R., and Scheifele, G. L. 1970. Interaction of genes for pathogenicity and virulence in *Trichometasphaeria turicica* with different numbers of genes for vertical resistance in *Zea mays*. *Phytopathology* **60:**1250–1254.

Nemeth, J., Klement, Z., and Farkas, G. L. 1969. An enzymological study of the hypersensitive reaction induced by *Pseudomonas syringae* in tobacco leaf tissues. *Phytopathol. Z.* **65:**265–276.

Nester, E. W. 1979. Molecular studies on crown gall tumors. In *Recognition and Specificity in Plant Host–Parasite Interactions* (J. M. Daly and I. Uritani, eds.), pp. 289–298. Japan Scientific Societies Press, Tokyo, and University Park Press, Baltimore.

Nester, E. W., and Kosuge, T. 1981. Plasmids specifying plant hyperplasias. *Annu. Rev. Microbiol.* **35:**531–565.

Nester, E. W., Chilton, M. D., Drummond, M., Merlo, D., Montoya, A., Sciaky, D., and Gordon, M. P. 1977. Search for bacterial DNA in crown gall tumors. In *Recombinant Molecules: Impact on Science and Society* (R. F. Beers and E. G. Basset, eds.), pp. 179–188. Raven Press, New York.

Niblett, C. L., Dickson, E., Fernow, K. H., Horst, R. K., and Zaitlin, M. 1979. Cross protection among four viroids. *Virology* **91:**198–203.

Nichols, E. J., Beckman, J. M., and Hadwiger, L. A. 1980. Glycosidic enzyme activity in pea tissue and pea-*Fusarium solani* interactions. *Plant Physiol.* (*Bethesda*) **66:**199–204.

Nishiguchi, M., Motoyoshi, F., and Oshima, N. 1978. Behavior of a temperature sensitive strain of tobacco mosaic virus in tomato leaves and protoplasts. *J. Gen. Virol.* **39:**53–61.

Nishimura, S., and Scheffer, R. P. 1965. Interactions between *Helminthosporium victoriae* spores and oat tissue. *Phytopathology* **55:**629–634.

Nishimura, S., Kohmoto, K., and Otani, H. 1974. Host specific toxins as an initiation factor for pathogenicity in *Alternaria kikichiana* and *A. mali*. *Rev. Plant Prot. Res.* **7:**21–32.

Nishimura, S., Kohmoto, K., Kuwata, M., and Watanabe, M. 1978. Production of a host-specific toxin by the pathogen causing black spot of strawberry. *Ann. Phytopathol. Soc. Jpn.* **44:**359 (Abstr.).

Nishimura, S., Kohmoto, K., and Otani, H. 1979. The role of host-specific toxins in saprophytic pathogens. In *Recognition and Specificity in Plant Host–Parasite Interactions* (J. M. Daly and I. Uritani, eds.), pp. 133–146. Japan Scientific Societies Press, Tokyo, and University Park Press, Baltimore.

Nishiyama, K., Sakai, R., Ezuka, A., Ichihara, A., Shiraishi, K., and Sakamura, S. 1977. Detection of coronatine in halo blight lesions of Italian ryegrass. *Ann. Phytopathol. Soc. Jpn.* **43:**219–220.

Northcote, D. H. 1963. The biology and chemistry of the cell walls of higher plants, algae, and fungi. *Int. Rev. Cytol.* **14:**223–265.

Northcote, D. H. 1972. Chemistry of the plant cell wall. *Ann. Rev. Plant Physiol.* **23:**113–132.

Novacky, A. 1972. Suppression of the bacterially induced hypersensitive reaction by cytokinins. *Physiol. Plant Pathol.* **2:**101–104.

Novacky, A., and Hanchey, P. 1974. Depolarization of membrane potentials in oat roots treated with victorin. *Physiol. Plant Pathol.* **4:**161–165.

Novacky, A., and Karr, A. I. 1977. Pathological alterations in cell membrane bioelectrical properties. In *Regulation of Cell Membrane Activities in Plants* (E. Marrè and O. Ciferri, eds.), pp. 137–144. Elsevier, Amsterdam.

Noyes, R. D., and Hancock, J. G. 1981. Role of oxalic acid in *Sclerotinia* wilt of sunflower. *Physiol. Plant Pathol.* **18:**123–132.

Nozue, M., Tomiyama, K., and Doke, N. 1977. Effect of blasticidin S on development of potential of potato tuber cells to react hypersensitively to infection by *Phytophthora infestans*. *Physiol. Plant Pathol.* **10:**181–189.

Nozue, M., Tomiyama, K., and Doke, N. 1978. Effect of adenosine 5′-triphosphate on hypersensitive death of potato tuber cells infected by *Phytophthora infestans*. *Phytopathology* **68:**873–876.

Nozue, M., Tomiyama, K., and Doke, N. 1979. Evidence of adherence of host plasmalemma to infecting hyphae of both compatible and incompatible races of *Phytophthora infestans*. *Physiol. Plant Pathol.* **15:**111–115.

Nozue, M., Tomiyama, K., and Doke, N. 1980. Effect of *N,N'*-diaretyl-D-chitobiose, the potato-lectin hapten, and other sugars on hypersensitive reaction of potato tuber cells infected by incompatible and compatible races of *Phytophthora infestans*. *Physiol. Plant Pathol.* **17:**221–227.

Oaks, A., and Johnson, F. J. 1972. Cyanide as an asparagine precursor in corn roots. *Phytochemistry (Oxf.)* **11:**3465–3471.

Obukowicz, M., and Kennedy, G. S. 1981. Phenolic ultracytochemistry of tobacco cells undergoing the hypersensitive reaction to *Pseudomonas solanacearum*. *Physiol. Plant Pathol.* **18:**339–344.

Odvody, G. N., Dunkle, L. D., and Edmunds, L. K. 1977. Characterization of the *Periconia circinata* population in a Milo disease nursery. *Phytopathology* **67:**1485–1489.

Ohashi, Y., and Shimomura, T. 1976. Leakage of cell constituents associated with local lesion formation on *Nicotiana glutinosa* leaf infected with tobacco mosaic virus. *Ann. Phytopathol. Soc. Jpn.* **42:**436–441.

Oku, H., Ouchi, S., Shiraishi, T., Utsumi, K., and Jeno, S. 1976a. A toxicity of a phytoalexin, pisatin, to mammalian cells. *Proc. Jpn. Acad. Ser. B Phys. Biol. Sci.* **52:**33–36.

Oku, H., Shiraishi, T., and Ouchi, S. 1976b. Effect of preliminary administration of pisatin to pea leaf tissues on the subsequent infection by *Erysiphe pisi* DC. *Ann. Phytopathol. Soc. Jpn.* **42:**597–600.

Oku, H., Shiraishi, T., and Ouchi, S. 1977. Suppression of induction of phytoalexin, pisatin by low-molecular weight substances from spore germination fluid of pea pathogen, *Mycosphaerella pinodes*. *Naturwiss. Med.* **64:**643–644.

Oku, H., Shiraishi, T., and Ouchi, S. 1979. The role of phytoalexins in host–parasite specificity. In *Recognition and Specificity in Plant Host–Parasite Interactions* (J. M. Daly and I. Uritani, eds.), pp. 317–333. Japan Scientific Societies Press, Tokyo, and University Park Press, Baltimore.

Okuno, T., Ishita, Y., Sawai, K., and Matsumoto, T. 1974. Characterization of alternariolide, a host-specific toxin produced by *Alternaria mali* Roberts. *Chem. Lett. (Chem. Soc. Japan)* **1974:**635–638.

Okuno, T., Furusawa, I., and Hiruki, C. 1977. Infection of barley protoplasts with brome mosaic virus. *Phytopathology* **67:**610–615.

Olien, C. R. 1957. Electrophoretic displacement of the necrotic area from the region of mycelial development in *Khapli emmer* infected with race 56 of *Puccinia graminis* var. *tritici*. *Phytopathology* **47:**26 (Abstr.).

Olsen, M. W., Misaghi, I. J., Goldstein, D., and Hine, R. B. 1982. Water relations in *Phymatotrichum*-infected cotton plants Phytopathology **72** (in press).

Otani, H., Nishimura, S., and Kohmoto, K. 1973. Nature of specific susceptibility to *Alternaria kikuchiana* in Nijisseiki cultivar among Japanese pears (part II). *J. Fac. Agric. Tottori Univ.* **8:**14–20.

Otsuki, Y., Shimomura, T., and Takebe, I. 1972. Tobacco mosaic virus multiplication and expression of the N-gene in necrotic responding tobacco varieties. *Virology* **50:**45–50.

Ouchi, S., Oku, H., and Hibino, C. 1976. Localization of induced resistance and susceptibility in barley leaves inoculated with the powdery mildew fungus. *Phytopathology* **66:**901–905.

Ouchi, S., Hibino, C., Oku, H., Fujiwara, M., and Nakabayashi, H. 1979. The induction of resistance or susceptibility. In *Recognition and Specificity in Plant Host–Parasite Interactions* (J. M. Daly and I. Uritani, eds.), pp. 49–65. Japan Scientific Societies Press, Tokyo, and University Park Press, Baltimore.

Overeem, J. C. 1976. Pre-existing antimicrobial substances in plants and their role in disease resistance. In *Biochemical Aspects of Plant–Parasite Relationships* (J. Friend and D. R. Threlfall, eds.), pp. 195–206. Academic Press, London and New York.

Owen, P. C. 1957a. The effect of infection with tobacco etch virus on the rates of respiration and photosynthesis of tobacco leaves. *Ann. Appl. Biol.* **45:**327–331.

Owen, P. C. 1957b. The effect of infection with tobacco mosaic virus on the photosynthesis of tobacco leaves. *Ann. Appl. Biol.* **45:**456–461.

Palmerley, R. A., and Callow, J. A. 1978. Common antigens in extracts of *Phytophthora infestans* and potatoes. *Physiol. Plant Pathol.* **12:**241–248.

Panopoulos, N. J., and Schroth, M. N. 1974. Role of flagellar motility in the invasion of bean leaves by *Pseudomonas phaseolicola. Phytopathology* **64:**1389–1397.

Panopoulos, N. J., and Staskawicz, B. J. 1981. Genetics of production. In *Toxins in Plant Disease* (R. D. Durbin, ed.), pp. 79–107. Academic Press, New York.

Panopoulos, N., Faccioli, G., and Gold, A. H. 1972. Translocation of phosphate in curly top virus-infected tomatoes. *Plant Physiol.* (*Bethesda*) **50:**266–270.

Papendick, R. I., and Campbell, G. S. 1975. Water potential in the rhizosphere and plant and methods of measurement and experimental control. In *Biology and Control of Soil-borne Plant Pathogens* (G. W. Bruehl, ed.), pp. 39–49. The American Phytopathological Society, St. Paul, MN.

Papendick, R. J., and Cook, R. J. 1974. Plant water stress and development of Fusarium root rot of wheat subjected to different cultural practices. *Phytopathology* **64:**358–363.

Park, P., Fukutomi, M., and Akai, S. 1976. Effect of the host-specific toxin from *Alternaria kikuchiana* on the ultrastructure of plasma membrane of cells in leaves of Japanese pear. *Physiol. Plant Pathol.* **9:**167–174.

Park, P., Nishimura, S., Kohomoto, K., Otani, H., and Tsujimoto, K. 1981. Two action sites of AM-toxin I. Produced by apple pathotype of *Alternaria alternata* in host cells: An ultrastructural study. *Can. J. Bot.* **59:**301–310.

Parsons, C. L., and Beardsley, R. E. 1968. Bacteriophage activity in homogenates of crown gall tissue. *J. Virol.* **2:**651.

Partridge, J. E., and Keen, N. T. 1977. Soybean phytoalexins: Rates of synthesis are not regulated by activation of initial enzymes in flavonoid biosynthesis. *Phytopathology* **67:**50–55.

Partridge, J. E., Shannon, L. M., Gumpf, D. J., and Colbaugh, P. 1974. Glycoprotein in the capsid of plant viruses as a possible determinant of seed transmissibility. *Nature* (*Lond.*) **247:**391–392.

Patil, S. S. 1974. Toxins produced by phytopathogenic bacteria. *Annu. Rev. Phytopathol.* **12:**259–279.

Patil, S. S. 1980. Defenses triggered by the invader: Detoxifying the toxins. In *Plant Disease: An Advanced Treatise* (J. G. Horsfall and E. B. Cowling, eds.), Vol. 5, pp. 269–277. Academic Press, New York.

Patil, S. S., and Dimond, A. E. 1967. Inhibition of *Verticillium* polygalacturonase by oxidation products of polyphenols. *Phytopathology* **57:**492–496.

Patil, S. S., and Dimond, A. E. 1968. Repression of polygalacturonase synthesis in *Fusarium oxysporum* f. sp. *lycopersici* by sugars and its effect on symptom reduction in infected tomato plants. *Phytopathology* **58:**676–682.

Patil, S. S., and Gnanamanickam, S. S. 1976. Suppression of bacterially induced hypersensitive reaction and phytoalexin accumulation in bean by phaseotoxin. *Nature* (*Lond.*) **259:**486–487.

Patil, S. S., Kolattukudy, P. E., and Dimond, A. E. 1970. Inhibition of ornithine carbamoyltransferase from bean plants by the toxin of *Pseudomonas phaseolicola*. I. Toxin specificity, chlorosis, and ornithine accumulation. *Plant Physiol.* (*Bethesda*) **49:**803–807.

Patil, S. S., Hayward, A. C., and Emmons, R. 1974. An ultraviolet-induced nontoxigenic mutant of *Pseudomonas phaseolicola* of altered pathogenicity. *Phytopathology* **64:**590–595.

Patil, S. S., Youngblood, P., Christiansen, P., and Moore, R. E. 1976. Phaseotoxin A: An antimetabolite from *Pseudomonas phaseolicola. Biochem. Biophys. Res. Commun.* **69:**1019–1027.

Patrick, T. W., Hall, R., and Fletcher, R. A. 1977. Cytokinin levels in healthy and *Verticillium*-infected tomato plants. *Can. J. Bot.* **55:**377–382.

Paulson, R. D., and Webster, J. M. 1970. Giant cell formation in tomato roots caused by *Meloidogyne incognita* and *Meloidogyne hapla* (Nematoda) infection. A light and electron microscope study. *Can. J. Bot.* **48:**271–276.

Paxton, J. D., and Chamberlain, D. W. 1967. Acquired local resistance of soybean plants to *Phytophthora* spp. *Phytopathology* **57:**352–353.

Payne, G. A., Knoche, H. W., Kono, Y., and Daly, J. M. 1980a. Biological activity of purified host-specific pathotoxin produced by *Bipolaris (Helminthosporium) maydis,* race T. *Physiol. Plant Pathol.* **16:**227–239.

Payne, G. A., Kono, Y., and Daly, J. M. 1980b. A comparison of purified host specific toxin from *Helminthosporium maydis,* race T, and its acetate derivative on oxidation by mitochondria from susceptible and resistant plants. *Plant Physiol. (Bethesda)* **65:**785–791.

Paynot, M., and Martin, C. 1977. Phenylalanine ammonia lyase activity and hypersensitivity. In *Current Topics in Plant Pathology* (Z. Király, ed.), pp. 375–383. Akadèmiai Kiadò, Budapest.

Pearce, R. B., and Ride, J. P. 1980. Specificity of induction of the lignification response in wounded wheat leaves. *Physiol. Plant Pathol.* **16:**197–204.

Pearce, R. B., and Ride, J. P. 1982. Chitin and related compounds as elicitors of the lignification response in wounded wheat leaves. *Physiol. Plant Pathol.* **20:**119–123.

Pegg, G. F. 1959. Role of indole-3-acetic acid in development of disease symptoms in *Verticillium* wilt of tomato. *Proc. IX Int. Botan. Cong., Montreal,* p. 297.

Pegg, G. F. 1976a. Endogenous auxins in healthy and diseased plants. In *Physiological Plant Pathology* (R. Heitefuss and P. H. Williams, eds.), pp. 560–581. Springer-Verlag, Berlin.

Pegg, G. F. 1976b. The involvement of ethylene in plant pathogenesis. In *Physiological Plant Pathology* (R. Heitefuss and P. H. Williams, eds.), Vol. 4, pp. 582–591. Springer-Verlag, Berlin.

Pegg, G. F. 1976c. Endogenous gibberellins in healthy and diseased plants. In *Physiological Plant Pathology* (R. Heitefuss and P. H. Williams, eds.), pp. 592–606. Springer-Verlag, Berlin.

Pegg, G. F. 1976d. Endogenous inhibitors in healthy and diseased plants. In *Physiological Plant Pathology* (R. Heitefuss and P. H. Williams, eds.), pp. 607–616. Springer-Verlag, Berlin.

Pegg, G. F. 1976e. The response of ethylene-treated tomato plants to infection by *Verticillium albo-atrum. Physiol. Plant Pathol.* **9:**215–226.

Pegg, G. F. 1976f. Ethylene production in tomato plants infected with *Verticillium albo-atrum. Physiol. Plant Pathol.* **8:**279–295.

Pegg, G. F. 1977. Glucanohydrolases of higher plants: A possible defense mechanism against parasitic fungi. In *Cell Wall Biochemistry Related to Specificity in Host-Plant Pathogen Interactions* (B. Solheim and J. Raa, eds.), pp. 305–345. Columbia University Press, New York.

Pegg, G. F. 1981a. The role of nonspecific toxins and hormone changes in disease severity. In *Plant Disease Control: Resistance and Susceptibility* (R. C. Staples and G. H. Toenniessen, eds.), pp. 13–31. John Wiley and Sons, New York.

Pegg, G. F. 1981b. The involvement of growth regulators in the diseased plant. In *Effects of Disease on the Physiology of the Growing Plant* (P. G. Ayres, ed.), pp. 149–177. Cambridge University Press, Cambridge.

Pegg, G. F., and Cronshaw, D. K. 1976. The relationship of *in vitro* and *in vivo* ethylene production in *Pseudomonas solanacearum* infection of tomato. *Physiol. Plant Pathol.* **9:**145–154.

Pegg, G. F., and Selman, I. W. 1959. An analysis of the growth response of young tomato plants to infection by *Verticillium albo-atrum.* II. The production of growth substances. *Ann. Appl. Biol.* **47:**222–231.

Pegg, G. F., and Sequeira, L. 1968. Stimulation of aromatic biosynthesis in tobacco plants infected with *Pseudomonas solanacearum. Phytopathology* **58:**476–483.

Pegg, G. F., and Vessey, J. C. 1973. Chitinase activity in *Lycopersicon esculentum* and its relationship to the *in vivo* lysis of *Verticillium albo-atrum* mycelium. *Physiol. Plant Pathol.* **3:**207–222.

Pellizzari, E. D., Kuć, J., and Williams, E. B. 1970. The hypersensitive reaction in *Malus* species: Changes in the leakage of electrolytes from apple leaves after inoculation with *Venturia inaequalis. Phytopathology* **60:**373–376.

Perchorowicz, J. T., Raynes, D. A., and Jensen, R. G. 1981. Light limitation of photosynthesis and activation of ribulose bisphosphate carboxylase in wheat seedlings. *Proc. Natl. Acad. Sci. U. S. A.* **78:**2985–2989.

Perrin, D. D., and Perrin, D. R. 1962. The N. M. R. spectrum of pisatin. *J. Am. Chem. Soc.* **84:**1922–1925.

Perrin, D. R., and Cruickshank, I. A. M. 1969. The antifungal activity of pterocarpans towards *Monilinia fructicola*. *Phytochemistry* **8:**971–978.

Petit, A., and Tourneur, J. 1972. Perte de virulence associée à la perte d'une activité enzymatique chez *Agrobacterium tumefaciens*. *C. R. Acad. Sci. Ser. D* **275:**137–139.

Petit, A., Delhaye, S., Tempé, J., and Morel, G. 1970. Recherches sur les guanidines des tissus de crown-gall. Mise en évidence d'une relation biochimique spécifique entre les souches d'*Agrobacterium tumefaciens* et les tumerus qu'elles induisent. *Physiol. Vég.* **8:**205–213.

Phelps, R. H., and Sequeira, L. 1968. Auxin biosynthesis in a host–parasite complex. In *Biochemistry and Physiology of Plant Growth Substances* (F. Wightman and G. Setterfield, eds.), pp. 197–212. Runge Press, Ottawa.

Pierpoint, W. S. 1971. Formation of o-quinones in some processes of agricultural importance. In *Report Rothamsted Exp. Station for 1970, part 2,* pp. 199–218.

Pilet, P. E. 1973. Effect of wall-degrading enzymes on protoplasts' transaminase activity. *Experientia (Basel)* **29:**1206–1207.

Pinkerton, F., and Strobel, G. A. 1976. Serinal as an activator of toxin production in attenuated cultures of *Helminthosporium sacchari*. *Proc. Natl. Acad. Sci. U. S. A.* **73:**4007–4011.

Pitt, D. 1970. Changes in hydrolase activity of *Solanum* tuber tissues during infection by *Phytophthora erythroseptica*. *Trans. Br. Mycol. Soc.* **55:**257–266.

Pitt, D., and Galpin, M. 1973. Role of lysosomal enzymes in pathogenicity. In *Fungal Pathogenicity and the Plant's Response* (R. J. W. Byrde and C. V. Cutting, eds.), pp. 449–467. Academic Press, London and New York.

Platt, S. G., Henriques, F., and Rand, L. 1979. Effects of virus infection on the chlorophyll content, photosynthetic rate and carbon metabolism of *Tolmiea menziesii*. *Physiol. Plant Pathol.* **15:**351–365.

Plotnikova, Y. M., Littlefield, L. J., and Miller, J. D. 1978. Scanning electron microscopy of the haustorium-host interface regions in wheat infected with *Puccinia graminis* f. sp. *tritici*. *Physiol. Plant Pathol.* **14:**37–39.

Plumbley, R. A., and Pitt, D. 1979. Purification and properties of a pectin methyl trans-eliminase, a phospholipase and two aminopeptidases produced in culture by *Phoma medicagenis* var. *pinodella*. *Physiol. Plant Pathol.* **14:**313–328.

Politis, D. J., and Goodman, R. N. 1978. Localized cell wall appositions: Incompatibility response of tobacco leaf cells to *Pseudomonas pisi*. *Phytopathology* **68:**309–316.

Politis, D. J., and Wheeler, H. 1973. Ultrastructural study of penetration of maize leaves by *Colletotrichum graminicola*. *Physiol. Plant Pathol.* **3:**465–471.

Porter, F. M. 1969. Protease, cellulase, and differential localization of endo- and exopolygalacturonase in conidia and conidial matrix of *Colletotrichum orbiculare*. *Phytopathology* **59:**1209–1213.

Powell, C. C., and Hildebrand, D. C. 1970. Fire blight resistance in *Pyrus:* Involvement of arbutin oxidation. *Phytopathology* **60:**337–340.

Powning, R. F., and Irzykiewicz, H. 1965. Studies on the chitinase system in bean and other seeds. *Comp. Biochem. Physiol.* **14:**127–133.

Pozsár, B. I., and Király, Z. 1966. Phloem transport in rust-affected plants and cytokinin-directed long distance movement of nutrients. *Phytopathol. Z.* **56:**297–309.

Preiss, J., and Kosuge, T. 1976. Regulation of enzyme activity in metabolic pathways. In *Plant Biochemistry* (J. Bonner and J. E. Varner, eds.), 3rd Ed., pp. 277–336. Academic Press, New York.

Price, K. R., Howard, B., and Coxon, D. T. 1976. Stress metabolite production in potato tubers infected by *Phytophthora infestans, Fusarium avenaceum* and *Phoma exigua*. *Physiol. Plant Pathol.* **9:**189–197.

Pridham, J. B. 1965. Low molecular weight phenols in higher plants. *Annu. Rev. Plant Physiol.* **16:**13–36.

Primrose, S. B. 1976. Formation of ethylene by *Escherichia coli. J. Gen. Microbiol.* **95:**159–165.

Pringle, R. B. 1970. Chemical constitution of the host–specific toxin of *Helminthosporium carbonum. Plant Physiol. (Bethesda)* **46:**45–49.

Pringle, R. B. 1971. Amino acid composition of the host-specific toxin of *Helminthosporium carbonum. Plant Physiol. (Bethesda)* **48:**756–759.

Pringle, R. B. 1972. Chemistry of host-specific phytotoxins. In *Phytotoxins in Plant Diseases* (R. K. S. Wood, A. Ballio and A. Graniti, eds.), pp. 139–155. Academic Press, London.

Pringle, R. B., and Braun, A. C. 1957. The isolation of the toxin of *Helminthosporium victoriae. Phytopathology* **47**:369–371.

Pringle, R. B., and Scheffer, R. P. 1963. Purification of the selective toxin of *Periconia circinata. Phytopathology* **53**:785–787.

Pringle, R. B., and Scheffer, R. P. 1964. Host-specific plant toxins. *Annu. Rev. Phytopathol.* **2**:133–156.

Pringle, R. B., and Scheffer, R. P. 1966. Amino acid composition of a crystaline host-specific toxin. *Phytopathology* **56**:1149–1151.

Pringle, R. B., and Scheffer, R. P. 1967. Multiple host-specific toxins from *Periconia circinata. Phytopathology* **57**:530–532.

Prusky, D., Dinoor, A., and Jakoby, B. 1980. The sequence of death of haustoria and host cells during the hypersensitive reaction of oat to crown rust. *Physiol. Plant Pathol.* **17**:33–40.

Pueppke, S. G. 1978. Some intriguing aspects of the putative role of isoflavonoid phytoalexins in plant disease. *Mycopathologia* **65**:115–119.

Pueppke, S. G., and Kluepfel, D. A. 1982. Interaction of *Agrobacterium* with potato lectin and concanavalin A and its effect on tumor induction in potato. *Physiol. Plant Pathol.* **20**:35–42.

Pueppke, S. G., and Van Etten, H. D. 1974. Pisatin accumulation and lesion development in peas infected with *Aphanomyces euteiches, Fusarium solani* f. sp. *pisi* or *Rhizoctonia solani. Phytopathology* **64**:1433–1440.

Pueppke, S. G., and Van Etten, H. D. 1976. The relation between pisatin and the development of *Aphanomyces euteiches* in diseased *Pisum sativum. Phytopathology* **66**:1174–1185.

Pueppke, S. G., Bauer, W. D., Keegstra, K., and Ferguson, A. I. 1978. Role of lectins in plant–microorganism interactions. *Plant Physiol. (Bethesda)* **61**:779–784.

Pupillo, P., Mazzucchi, U., and Pierini, G. 1976. Pectic lyase isozymes produced by *Erwinia chrysanthemi* Burkh. *et al.* in polypectate broth or in *Dieffenbachia* leaves. *Physiol. Plant Pathol.* **9**:113–120.

Purdy, R. E., and Kolattukudy, P. E. 1975a. Hydrolysis of plant cuticle by plant pathogens. Purification, amino acid composition and molecular weight of two isozymes of cutinase and a nonspecific esterase from *Fusarium solani* f. sp. *pisi. Biochemistry* **14**:2824–2831.

Purdy, R. E., Kolattukudy, P. E. 1975b. Hydrolysis of plant cuticle by plant pathogens. Properties of cutinase I, cutinase II and a nonspecific esterase isolated from *Fusarium solani* f. sp. *pisi. Biochemistry* **14**:2832–2840.

Pure, G. A., Chakravorty, A. K., and Scott, K. J. 1979. Cell-free translation of polysomal messenger RNA isolated from healthy and rust-infected wheat leaves. *Physiol. Plant Pathol.* **15**:201–209.

Pure, G. A., Chakravorty, A. K., and Scott, K. J. 1980. Changes in wheat leaf polysomal messenger RNA populations during the early stages of rust infection. *Plant Physiol. (Bethesda)* **66**: 520–524.

Rabenantoandro, Y., Auriol, P., and Touzé, A. 1976. Implication of β-(1→3) glucanase in melon anthracnose. *Physiol. Plant Pathol.* **8**:313–324.

Raggi, V. 1978. The $CO_2$ compensation point, photosynthesis and respiration in rust infected bean leaves. *Physiol. Plant Pathol.* **13**:135–139.

Raggi, V. 1980. Correlation of $CO_2$ compensation point (Γ) with photosynthesis and respiration and $CO_2$-sensitive Γ in rust-infected bean leaves. *Physiol. Plant Pathol.* **16**:19–24.

Rahe, J. E. 1973. Occurrence and levels of the phytoalexin phaseollin in relation to delimitation at sites of infection of *Phaseolus vulgaris* by *Colletotrichum lindemuthianum*. *Can. J. Bot.* **51**:2423–2430.

Rahe, J. E., and Arnold, R. M. 1975. Injury-related phaseollin accumulation in *Phaseolus vulgaris* and its implications with regard to host–parasite interaction. *Can. J. Bot.* **53**:921–928.

Rahe, J. E., Kuć, J., Chuang, C., and Williams, E. B. 1969a. Induced resistance in *Phaseolus vulgaris* to bean anthracnose. *Phytopathology* **59**:1641–1645.

Rahe, J. E., Kuć, J., Chuang, C., and Williams, E. B. 1969b. Correlation of phenolic metabolism with histological changes in *Phaseolus vulgaris* inoculated with fungi. *Neth. J. Plant Pathol.* **75**:58–71.

Rai, J. N., and Dhawan, S. 1976. Studies on purification and identification of toxin metabolites produced by *Sclerotinia sclerotiorum* causing white rot disease of crucifer. *Indian Phytopathol.* **29**:407–411.

Rai, P. V., and Strobel, G. A. 1969. Phytotoxic glycopeptides produced by *Corynebacterium michiganense*. I. Methods of preparation, physical and chemical characterization. *Phytopathology* **59**:47–52.

Rancillac, M., Kaur-Sawhney, R., Staskawicz, B., and Galston, A. W. 1976. Effects of cyclohexamide and kinitin pretreatment on responses of susceptible and resistant *Avena* leaf protoplasts to the phytotoxin victorin. *Plant Cell Physiol.* **17**:987–995.

Randazzo, G., Evidente, A., Capasso, R., Lasaponara, M., Tuttobello, L., and Ballio, A. 1978. On the biosynthesis of fusicoccin. *Gazz. Chim. Ital.* **108**:139–142.

Rathmell, W. G. 1973. Phenolic compounds and phenylalanine ammonia lyase activity in relation to phytoalexin biosynthesis in infected hypocotyls of *Phaseolus vulgaris*. *Physiol. Plant Pathol.* **3**:259–267.

Rathmell, W. G., and Sequeira, L. 1975. Induced resistance in tobacco leaves. The role of inhibitors of bacterial growth in the intercellular fluid. *Physiol. Plant Pathol.* **5**:65–73.

Rautela, G. S., and Payne, M. G. 1970. The relationship of peroxidase and ortho-diphenol oxidase to resistance of sugarbeets to *Cercospora* leaf spot. *Phytopathology* **60**:238–245.

Ravise, A., and Kirkiacharian, B. S. 1976. Influence de la structure de composés phenoliques sur l'inhibition du *Phytophthora parasitica* et d'enzymes participant aux processus parasitaires. I. Isoflavonoides et coumestanes. *Phytopathol. Z.* **85**:74–85.

Rawn, C. D. 1974. Victorin-induced changes in carbohydrate metabolism in oat leaves. Ph.D. Thesis. University of Kentucky, Lexington.

Rawn, C. D. 1977. Simultaneous changes in the rate and pathways of glucose oxidation in victorin-treated oat leaves. *Phytopathology* **67**:338–343.

Raymundo, A. K., and Ries, S. M. 1980. Chemotaxis of *Erwinia amylovora*. *Phytopathology* **70**:1066–1069.

Raymundo, A. K., and Ries, S. M. 1981. Motility of *Erwinia amylovora*. *Phytopathology* **71**:45–49.

Reddi, K. K. 1959. Studies on tobacco leaf ribonuclease. III. Its role in the synthesis of tobacco mosaic virus nucleic acid. *Biochim. Biophys. Acta* **33**:164–169.

Reddi, K. K. 1966. Ribonuclease induction in cells transformed by *Agrobacterium tumefaciens*. *Proc. Natl. Acad. Sci. U. S. A.* **56**:1207–1214.

Reese, E. T. (ed.) 1963. *Advances in Enzymatic Hydrolysis of Cellulose and Related Materials*. Pergamon Press, New York.

Reese, E. T., and Shibata, Y. 1965. β-Mannanases of fungi. *Can. J. Microbiol.* **11**:167–183.

Retig, N. 1974. Changes in perioxidase and polyphenoloxidase associated with natural and induced resistance of tomato to *Fusarium* wilt. *Physiol. Plant Pathol.* **4**:145–150.

Rhodes, J. M., and Wooltorton, L. S. C. 1978. The biochemistry of phenolic compounds in wounded plant storage tissues. In *Biochemistry of Wounded Plant Tissues* (G. Kahl, ed.), pp. 243–286. De Gruyter, Berlin and New York.

Rich, D. H., and Bhatnagar, P. K. 1978. Conformational studies of tentoxin by nuclear magnetic resonance spectroscopy. Evidence for a new conformation for a cyclic tetrapeptide. *J. Am. Chem. Soc.* **100**:2212–2218.

Rich, D. H., and Mathiaparanam, P. 1974. Synthesis of the cyclic tetrapeptide tentoxin. Effect of an N-methyl-dehydrophenylalanyl residue on conformation of linear tetrapeptides. *Tetrahedron Lett.* **46**:4037–4040.

Rich, J. R., Keen, N. T., and Thomason, I. J. 1977. Association of coumestans with the hypersensitivity of lima bean roots to *Pratylenchus scribneri*. *Physiol. Plant Pathol.* **10**:105–116.

Rich, S. 1963. The role of stomata in plant disease. In *Stomata and Water Relations in Plants* (I. Zelitch, ed.), pp. 102–116. *Conn. Agric. Exp. Stn. Bull. 664*.

Richmond, S., Kuć, J., and Elliston, J. E. 1979. Penetration of cucumber leaves by *Colletotrichum lagenarium* is reduced in plants systemically protected by previous infection with the pathogen. *Physiol. Plant Pathol.* **14**:329–338.

Ride, J. P. 1973. Rapid lignin formation as a response of wheat to fungal infection. *Proc. Soc. Gen. Microbiol.* **1**:28 (Abstr.).

Ride, J. P. 1975. Lignification in wounded wheat leaves in response to fungi and its possible role in resistance. *Physiol. Plant Pathol.* **5**:125–134.

Ride, J. P. 1978. The role of cell wall alterations in resistance to fungi. *Ann. Appl. Biol.* **89**:302–306.

Ride, J. P. 1980. The effect of induced lignification on the resistance of wheat cell walls to fungal degradation. *Physiol. Plant Pathol.* **16**:187–196.

Ride, J. P., and Pearce, R. B. 1979. Lignification and papilla formation at sites of attempted penetration of wheat leaves by nonpathogenic fungi. *Physiol. Plant Pathol.* **15**:79–92.

Ries, S. M., and Strobel, G. A. 1972a. A phytotoxic glycopeptide from cultures of *Corynebacterium insidiosum*. *Plant Physiol.* **49**:676–684.

Ries, S. M., and Strobel, G. A. 1972b. Biological properties and pathological role of phytotoxic glycopeptide from *Corynebacterium insidiosum*. *Physiol. Plant Pathol.* **2**:133–142.

Rissler, J. R., and Millar, R. L. 1977. Contribution of a cyanide-insensitive alternate respiratory system to increases in formamide hydro-lyase activity and to growth in *Stemphylium loti in vitro*. *Plant Physiol.* (*Bethesda*) **60**:857–861.

Robb, J., Busch, L., and Lu, B. C. 1975. Ultrastructure of wilt syndrome caused by *Verticillium dahliae*. I. In chrysanthemum leaves. *Can. J. Bot.* **53**:901–913.

Roberts, W. P., and Kerr, A. 1974. Crown gall induction: Serological reactons, isozyme patterns and sensitivity to mitomycin C and to bacteriocin of pathogenic and non-pathogenic strains of *Agrobacterium radiobacter*. *Physiol. Plant Pathol.* **4**:81–91.

Robertsen, B. K., Aman, P., Darvill, A. G., NcNeil, M., and Albersheim, P. 1981. Host–symbiont interactions. V. The structure of acidic extracellular polysaccharide secreted by *Rhizobium leguminosarum* and *Rhizobium trifolii*. *Plant Physiol.* (*Bethesda*) **67**:389–400.

Rodrigues, C. J., and Arny, D. C. 1966. Role of oxidative enzymes in the physiology of the leaf spot of coffee caused by *Mycena citricola*. *Phytopathol. Z.* **57**:375–384.

Rohde, R. A. 1972. Expression of resistance in plants to nematodes. *Annu. Rev. Phytopathol.* **10**:233–252.

Rohringer, R., and Samborski, D. J. 1967. Aromatic compounds in the host–parasite interaction. *Annu. Rev. Phytopathol.* **5**:77–86.

Rohringer, R., Fuchs, A., Lunderstadt, J., and Samborski, D. J. 1967. Metabolism of aromatic compounds in healthy and rust-infected primary leaves of wheat. I. Studies with $^{14}CO_2$, quinate-U-$^{14}C$, and shikimate-U-$^{14}C$ as precursors. *Can. J. Bot.* **45**:863–889.

Rombouts, F. M., and Pilnik, W. 1972. Research on pectin depolymerases in the sixties—a literature review. *C. R. C. Crit. Rev. Food Technol.* **3**:1–26.

Ross, A. F. 1961. Localized acquired resistance to plant virus infection in hypersensitive hosts. *Virology* **14**:329–339.

Ross, A. F. 1965. Systemic effects of local lesion formation. In *Viruses of Plants* (A. B. R. Beemster and J. Djikstra, eds.), pp. 127–150. North-Holland, Amsterdam.

Rossall, S., Mansfield, J. W., and Hutson, R. A. 1980. Death of *Botrytis cinerea* and *B. fabae* following exposure to wyerone derivatives *in vitro* and during infection development in broad bean leaves. *Physiol. Plant Pathol.* **16**:135–146.

Rottini, G., Dri, P., Romeo, D., and Patriarca, P. 1976. Influence of *E. coli* polysaccharide on the interaction of *E. coli* K+ and K− with polymorphonuclear leukocytes *in vitro*. *Zentralbl. Bakteriol. Parasitenkd Infektionskr. Hyg. Abt. 1: Orig., Reihe A. Med. Microbiol. Parasitol.* **234**:189–201.

Rovira, A. D., and Davey, C. B. 1974. Biology of the rizosphere. In *The Plant Root and its Environment* (E. W. Carson, ed.), pp. 153–204. University Press of Virginia, Charlottesville.

Royle, D. J. 1976. Structural features of resistance to plant diseases. In *Biochemical Aspects of Plant–Parasite Relationships* (J. Friend and D. R. Trelfall, eds.), pp. 161–193. Academic Press, New York.

Royle, D. J., and Thomas, G. G. 1973. Factors affecting zoospore responses towards stomata in hop downy mildew (*Pseudoperonospora humuli*) including some comparisons with grapevine downy mildew (*Plasmopara viticola*). *Physiol. Plant Pathol.* **3**:405–417.

Rubin, B. A., and Artsikhovskaya, E. V. 1963. *Biochemistry and Physiology of Plant Immunity*. Pergamon Press, Oxford.

Rubin, B. A., and Artsikhovaskaya, E. V. 1964. Biochemistry of pathological darkening of plant tissues. *Annu. Rev. Phytopathol.* **2**:157–178.

Rudolph, K. 1972. The halo-blight toxin of *Pseudomonas phaseolicola*: Influence on host parasite relationships and counter effect of metabolites. In *Phytotoxins in Plant Diseases* (R. K. S. Wood, A. Ballio, and A. Graniti, eds.), pp. 373–375. Academic Press, London and New York.

Rudolph, K. 1976. Forces by which the pathogen attacks the host plant: Non-specific toxins. In *Physiological Plant Pathology, Encyclopedia of Plant Physiology* (R. Heitefuss and P. H. Williams, eds.), Vol. 4, pp. 270–315. Springer-Verlag, Berlin, Heidelberg, and New York.

Russell, S. L., and Kimmins, W. C. 1971. Growth regulators and the effect of barley yellow dwarf virus on barley (*Hordeum vulgare* L.). *Ann. Bot. (Lond.) (N. S.)* **35**:1037–1043.

Rust, L. A., Fry, W. E., and Beer, S. V. 1980. Hydrogen cyanide sensitivity in bacterial pathogens of cyanogenic and non-cyanogenic plants. *Phytopathology* **70**:1005–1008.

Sachse, B., Wolf, G., and Fuchs, W. H. 1971. Nukleinsaure abbauende Enzyme in Blattern von *Triticum aestivum* nach Infektion mit *Puccinia graminis tritici*. (Enzymes decomposing nucleic acid in leaves of *Triticum aestivum* after infection with *Puccinia graminis tritici*). *Acta Phytopathol. Acad. Sci. Hung.* **6**:39–49.

Saftner, R. A., and Evans, M. L. 1974. Selective effects of victorin on growth and the auxin response in *Avena*. *Plant Physiol.* (*Bethesda*) **53**:382–387.

Saijo, R., and Kosuge, T. 1978. The conversion of 3-deoxyarabino-heptulosonate 7-phosphate to 3-dehydroquinate by sorghum seedling preparations. *Phytochemistry* (*Oxf.*) **17**:223–225.

Sakai, R., Nishiyama, K., Ichihara, A., Shiraishi, K., and Sakamura, S. 1979. The relation between bacterial toxin action and plant growth regulation. In *Recognition and Specificity in Plant Host–Parasite Interactions* (J. M. Daly and I. Uritani, eds.), pp. 165–179. Japan Scientific Societies Press, Tokyo, and University Park Press, Baltimore.

Salemink, C. A., Rebel, H., Kerling, L. C. P., and Tchernoff, V. 1965. Phytotoxin isolated from liquid cultures of *Ceratocystis ulmi*. *Science* **149**:202–203.

Samaddar, K. R., and Scheffer, R. P. 1968. Effect of the specific toxin in *Helminthosporium victoriae* on host cell membranes. *Plant Physiol.* (*Bethesda*) **43**:21–28.

Samaddar, K. R., and Scheffer, R. P. 1970. Effects of *Helminthosporium victoriae* toxin on germination and aleurone secretion by resistant and susceptible seeds. *Plant Physiol.* (*Bethesda*) **45**:586–590.

Samaddar, K. R., and Scheffer, R. P. 1971. Early effects of *Helminthosporium victoriae* toxin on plasma membranes and counteraction by chemical treatments. *Physiol. Plant Pathol.* **1**:319–328.

Samborski, D. J., and Shaw, M. 1956. The physiology of host–parasite relations II. The effect of *Puccinia graminis tritici* Eriks and Henn. on the respiration of resistant and susceptible species of wheat. *Can. J. Bot.* **34**:601–619.

Samborski, D. J., Kim, W. K., Rohringer, R., Howes, N. K., and Baker, R. J. 1977. Histological studies on host-cell necrosis conditioned by the Sr6 gene for resistance in wheat to stem rust. *Can. J. Bot.* **55**:1445–1452.

Samborski, D. J., Rohringer, R., and Kim, W. K. 1978. Translation and transcription in diseased plants. In *Plant Disease: An Advanced Treatise* (J. G. Horsfall and E. B. Cowling, eds.), Vol. 3, pp. 375–390. Academic Press, New York.

Sargent, J. A., Tommerup, I. C., and Ingram, D. S. 1973. The penetration of a susceptible lettuce variety by the downy mildew fungus *Bremia lactucae* Regel. *Physiol. Plant Pathol.* **3**:231–239.

Sasser, M. 1978. Involvement of bacterial protein synthesis in induction of the hypersensitive reaction in tobacco. *Phytopathology* **68**:361–363.

Sato, N., Kitazawa, K., and Tomiyama, K. 1971. The role of rishitin in localizing the invading hyphae of *Phytophthora infestans* in infection sites at the cut surfaces of potato tubers. *Physiol. Plant Pathol.* **1**:289–295.

Schafer, J. F. 1971. Tolerance to plant disease. *Annu. Rev. Phytopathol.* **9**:235–252.

Scharen, A. I., and Krupinsky, J. M. 1969. Effect of *Septoria nodorum* infection on $CO_2$ absorption and yield of wheat. *Phytopathology* **59**:1298–1301.

Scheffer, R. P. 1976. Host-specific toxins in relation to pathogenesis and disease resistance. In *Physiological Plant Pathology* (R. Heitefuss and P. H. Williams, eds.), Vol. 4, pp. 247–269. Springer-Verlag, Berlin, New York.

Scheffer, R. P., and Livingston, R. S. 1980. Sensitivity of sugarcane clones to toxin from *Helminthosporium sacchari* as determined by electrolyte leakage. *Phytopathology* **70**:400–404.

Scheffer, R. P., and Pringle, R. B. 1961. A selective toxin produced by *Periconia circinata*. *Nature* (*Lond.*) **191**:912–913.

Scheffer, R. P., and Pringle, R. B. 1967. Pathogen-produced determinants of disease and their effects on host plants. In *The Dynamic Role of Molecular Constituents in Plant–Parasite Interaction* (C. J. Mirocha and I. Uritani, eds.), pp. 217–236. Bruce Publishing Co., St. Paul, MN.

Scheffer, R. P., and Samaddar, K. R. 1970. Host-specific toxins as determinants of pathogenicity. *Rec. Adv. in Phytochem.* **3**:123–142.

Scheffer, R. P., and Ullstrup, A. J. 1965. A host-specific toxic metabolite from *Helminthosporium carbonum*. *Phytopathology* **55**:1037–1038.

Scheffer, R. P., and Walker, J. C. 1953. The physiology of *Fusarium* wilt of tomato. *Phytopathology* **43**:116–125.

Scheffer, R. P., and Yoder, O. C. 1972. Host-specific toxins and selective toxicity. In *Phytotoxins in Plant Diseases* (R. K. S. Wood, A. Ballio, and A. Graniti, eds.), pp. 251–272. Academic Press, New York.

Scheffer, R. P., Nelson, R. R., and Ullstrup, A. J. 1967. Inheritance of toxin production and pathogenicity in *Cochliobolus carbonum* and *Cochliobolus victoriae*. *Phytopathology* **57**:1288–1291.

Schell, J., Van Larebeke, N., Engler, G., Holsters, M., Van den Elsacker, S., Zaenen, I., and Schilperoort, R. A. 1974. Large plasmid in *Agrobacterium tumefaciens* essential for crown gall-inducing ability. *Nature (Lond.)* **252**:169–170.

Schertz, K. F., and Tai, Y. P. 1969. Inheritance of reaction of *Sorghum biocolor* to toxin produced by *Periconia circinata*. *Crop Sci.* **9**:621–624.

Schipper, A. I., Jr., and Mirocha, C. J. 1969. The mechanism of starch depletion in leaves of *Phaseolus vulgaris* infected with *Uromyces phaseoli*. *Phytopathology* **59**:1722–1727.

Schlösser, E. 1975. Role of saponins in antifungal resistance. III. Tomatin-dependent development of fruit rot organisms on tomato fruits. *Z. Pflanzenkr. Pflanzenschutz* **82**:476–484.

Schlösser, E. 1976. Role of saponins in antifungal resistance. VII. Significance of tomatin in species-specific resistance of tomato fruits against fruit rotting fungi. *Meded. Fac. Landbouwwet. Rijksuniv. Gent.* **41**:499–503.

Schlösser, E. 1977. Role of saponins in antifungal resistance. IV. Tomatin-dependent development of species of *Alternaria* on tomato fruits. *Acta Phytopathol.* **11**:77–87.

Schlösser, E. W. 1980. Preformed internal chemical defenses. In *Plant Disease: An Advanced Treatise* (J. G. Horsfall and E. B. Cowling, eds.), Vol. 5, pp. 161–177. Academic Press, New York.

Schnathorst, W. C., and DeVay, J. E. 1963. Common antigens in *Xanthomonas malvacearum* and *Gossypium hirsutum* and their possible relationship to host specificity and disease resistance. *Phytopathology* **53**:1142 (Abstr.).

Schneider, I. R. 1965. Introduction, translocation and distribution of viruses in plants. *Adv. Virus Res.* **11**:163–222.

Scholander, P. E., Hammel, H. T., Brandstreet, E. D., and Hemmingsen, E. A. 1965. Sap pressure in vascular plants. *Science* **148**:339–346.

Schönbeck, F., and Schlösser, E. 1976. Preformed substances as potential protectants. In *Physiological Plant Pathology* (R. Heitefuss and P. H. Williams, eds.), Vol. 4, pp. 653–678. Springer-Verlag, Heidelberg, Berlin, and New York.

Schönbeck, F., and Schröeder, C. 1972. Role of antimicrobial substances (Tuliposides) in tulips attacked by *Botrytis* spp. *Physiol. Plant Pathol.* **2**:91–99.

Schramm, R. J., Jr., and Wolf, F. T. 1954. The transpiration of black shank-infected tobacco. *J. Elisha Mitchell Sci. Soc.* **70**:255–261.

Schroeder, C. 1972. Die Bedeutung der γ-Hydroxysauren fur das Wirt-Parasit-Verhaltnis von Tulpe und *Botrytis* spp. *Phytopathol. Z.* **74**:175–181.

Scott, K. J., Chakravorty, A. K., and Flynn, J. G. 1976. Biochemical aspects of gene expression in wheat and barley after infection with rusts and mildew fungi. *Aust. Plant Pathol. Soc. Newsl.* **5**:Suppl. 9.

Scott, K. J., and Smillie, R. M. 1966. Metabolic regulation in diseased leaves. I. The respiratory rise in barley leaves infected with powdery mildew. *Plant Physiol. (Bethesda)* **41**:289–297.

Scrubb, L. A., Chakravorty, A. K., and Shaw, M. 1972. Changes in the ribonuclease activity of flax cotyledons following inoculation with flax rust. *Plant Physiol. (Bethesda)* **50**:73–79.

Seevers, P. M., and Daly, J. M. 1970a. Studies on wheat stem rust resistance controlled at the Sr6 locus. I. The role of phenolic compounds. *Phytopathology* **60**:1322–1328.

Seevers, P. M., and Daly, J. M. 1970b. Studies on wheat stem rust resistance controlled at the Sr6 locus. II. Peroxidase activities. *Phytopathology* **60**:1642–1647.

Seevers, P. M., Catedral, F. F., and Daly, J. M. 1971. The role of peroxidase isozymes in resistance to wheat stem rust disease. *Plant Physiol. (Bethesda)* **48**:353–360.

Sela, I., Harpaz, I., and Birk, Y. 1966. Identification of the active component of an antiviral factor isolated from virus-infected plants. *Virology* **28**:71–78.

Sempio, C. 1950. Metabolic resistance to plant diseases. *Phytopathology* **40**:799–819.

Sequeira, L. 1964. Inhibition of indoleacetic acid oxidase in tobacco plants infected by *Pseudomonas solanacearum*. *Phytopathology* **54**:1078–1083.

Sequeira, L. 1965. Origin of indoleacetic acid in tobacco plants infected by *Pseudomonas solanacearum*. *Phytopathology* **55**:1232–1236.

Sequeira, L. 1973. Hormone metabolism in diseased plants. *Annu. Rev. Plant Physiol.* **24**: 353–380.

Sequeira, L. 1978. Lectins and their role in host–pathogen specificity. *Annu. Rev. Phytopathol.* **16**:453–481.

Sequeira, L. 1979. The acquisition of systemic resistance by prior inoculation. In *Recognition and Specificity in Plant Host–Parasite Interactions* (J. M. Daly and I. Uritani, eds.), pp. 231–251. Japan Scientific Societies Press, Tokyo, and University Park Press, Baltimore.

Sequeira, L. 1980. Defenses triggered by the invader: Recognition and compatibility phenomena. In *Plant Disease: An Advanced Treatise* (J. G. Horsfall and E. B. Cowling, eds.), Vol. 5, pp. 179–200. Academic Press, New York.

Sequeira, L., and Graham, T. L. 1977. Agglutination of avirulent strains of *Pseudomonas solanacearum* by potato lectin. *Physiol. Plant Pathol.* **11**:43–54.

Sequeira, L., and Hill, L. M. 1974. Induced resistance in tobacco leaves: The growth of *Pseudomonas solanacearum* in protected leaves. *Physiol. Plant Pathol.* **4**:447–455.

Sequeira, L., and Kelman, A. 1962. The accumulation of growth substances in plants infected by *Pseudomonas solanacearum*. *Phytopathology* **52**:439–448.

Sequeira, L., and Seevers, T. A. 1954. Auxin inactivation and its relation to leaf drop caused by the fungus *Omphalia flavida*. *Plant Physiol.* (*Bethesda*) **29**:11–16.

Sequeira, L., Gaard, G., and DeZoeten, G. A. 1977. Attachment of bacteria to host cell walls: Its relation to mechanisms of induced resistance. *Physiol. Plant Pathol.* **10**:43–50.

Shalla, T. A., and Petersen, L. J. 1973. Infection of isolated plant protoplasts with potato virus X. *Phytopathology* **63**:1125–1130.

Shaw, M. 1967. Cell biological aspects of host–parasite relations of obligate fungal parasites. *Can. J. Bot.* **45**:1205–1220.

Shaw, M., and Hawkins, A. R. 1958. The physiology of host–parasite relations. I. A preliminary examination of the level of free endogenous indoleacetic acid in rusted and mildewed cereal leaves and their ability to decarboxylate exogenously supplied radioactive indoleacetic acid. *Can. J. Bot.* **36**:1–16.

Shaw, M., and Samborski, D. J. 1957. The physiology of host–parasite relations III. The pattern of respiration in rusted and mildewed cereal leaves. *Can. J. Bot.* **35**:389–407.

Shaykh, M., Soliday, C., and Kolattukudy, P. E. 1977. Proof for the production of cutinase by *Fusarium solani* f. *pisi* during penetration into its host, *Pisum sativum*. *Plant Physiol.* (*Bethesda*) **60**:170–172.

Shepard, D. V., and Pitt, D. 1976. Purification of phospholipase from *Botrytis* and its effects on plant tissues. *Phytochemistry* **15**:1465–1470.

Sherwood, R. T., and Vance, C. P. 1976. Histochemistry of papillae formed in reed canarygrass leaves in response to noninfecting pathogenic fungi. *Phytopathology* **66**:503–510.

Sherwood, R. T., and Vance, C. P. 1980. Resistance to fungal penetration in Gramineae. *Phytopathology* **70**:273–279.

Shimony, C., and Friend, J. 1975. Ultrastructure of the interaction between *Phytophthora infestans* and leaves of two cultivars of potato (*Solanum tuberosum* L.). Orion and Majestic. *New Phytol.* **74**:59–65.

Shiraishi, T., Oku, H., Isono, M., and Ouchi, S. 1975. The injurious effect of pisatin on the plasma membrane of pea. *Plant Cell Physiol.* **16**:939–942.

Shiraishi, T., Oku, H., Ouchi, S., and Isono, M. 1976. Pisatin production prior to the cell necrosis demonstrated in powdery mildew of pea. *Ann. Phytopathol. Soc. Jpn.* **42**:609–612.

Shiraishi, T., Oku, H., Ouchi, S., and Tsuji, Y. 1977. Local accumulation of pisatin in tissues of pea seedlings infected by powdery mildew fungi. *Phytopathol. Z.* **88**:131–135.

Siegel, A. 1979. Recognition and specificity in plant virus infection. In *Recognition and Specificity in Plant Host–Parasite Interactions* (J. M. Daly and I. Uritani, eds.), pp. 253–270. Japan Scientific Societies Press, Toyko, and University Park Press, Baltimore.

Silverman, W. 1960. A toxin extract from Marquis wheat infected by race 38 of the stem rust fungus. *Phytopathology* **50**:130–136.

Simons, T. J., and Ross, A. F. 1970. Enhanced peroxidase activity associated with induction of resistance to tobacco mosaic virus in hypersensitive tobacco. *Phytopathology* **60**:383–384.

Sinden, S. L., and Durbin, R. D. 1970. A comparison of the chlorosis-inducing toxin from *Pseudomonas coronafaciens* and wildfire toxin from *Pseudomonas tobaci*. *Phytopathology* **60**:360–364.

Sinden, S. L., DeVay, J. E., and Backman, P. A. 1971. Properties of syringomycin, a wide spectrum antibiotic and phytotoxin produced by *Pseudomonas syringae*, and its role in the bacterial canker disease of peach trees. *Physiol. Plant Pathol.* **1**:199–213.

Sing, V. O., and Schroth, M. N. 1977. Bacteria–plant cell surface interactions: Active immobilization of saprophytic bacteria in plant leaves. *Science* **197**:759–761.

Singer, S. J. 1974. The molecular organization of membranes. *Annu. Rev. Biochem.* **43**:805–833.

Singer, S. J., and Nicolson, G. L. 1972. The fluid mosaic model of the structure of cell membranes. *Science* **175**:720–731.

Sinha, A. K., and Trivedi, N. 1969. Immunization of rice plants against *Helminthosporium* infection. *Nature* (*Lond.*) **223**:963–964.

Sjulin, T. M., and Beer, S. V. 1978. Mechanism of wilt induction by *amylovorin* in cotoneaster shoots and its relation to wilting of shoots infected by *Erwinia amylovora*. *Phytopathology* **68**:89–94.

Skipp, R. A., and Bailey, J. A. 1976. The effect of phaseollin on the growth of *Colletotrichum lindemuthianum* in bioassays designed to measure fungitoxicity. *Physiol. Plant Pathol.* **9**:253–263.

Skipp, R. A., and Bailey, J. A. 1977. The fungitoxicity of isoflavonoid phytoalexins measured using different types of bioassay. *Physiol. Plant Pathol.* **11**:101–112.

Skipp, R. A., and Deverall, B. J. 1973. Studies on cross-protection in the anthracnose disease of bean. *Physiol. Plant Pathol.* **3**:299–314.

Skipp, R. A., and Samborski, D. J. 1974. The effect of Sr6 gene for host resistance on histological events during the development of stem rust in near-isogenic wheat lines. *Can. J. Bot.* **52**:107–115.

Skipp, R. A., Selby, C., and Bailey, J. A. 1978. Toxic effect of phaseollin on plant cells. *Physiol. Plant Pathol.* **10**:221–227.

Slatyer, R. O. 1967. Plant–water relations. In *Water Deficits and Plant Growth* (T. T. Kozlowski, ed.), Vol. 1, pp. 49–76. Academic Press, New York.

Slykhuis, J. T. 1969. Mites as vectors of plant viruses. In *Viruses, Vectors and Vegetation* (K. Maramorosch, ed.), pp. 121–141. Interscience Publishers, New York.

Smidt, M., and Kosuge, T. 1978. The role of indole-3-acetic acid accumulation by alpha methyl tryptophan-resistant mutants of *Pseudomonas savastanoi* in gall formation on oleanders. *Physiol. Plant Pathol.* **13**:203–214.

Smith, A. M. 1973. Ethylene as a cause of soil fungistasis. *Nature* (*Lond.*) **246**:311–313.

Smith, A. M., and Cook, R. J. 1974. Implications of ethylene production by bacteria for biological balance of soil. *Nature* (*Lond.*) **252**:703–705.

Smith, D. A. 1978. Observations on the fungitoxicity of the phytoalexin, kievitone. *Phytopathology* **68**:81–87.

Smith, D. A., Van Etten, H. D., and Bateman, D. F. 1975. Accumulation of phytoalexins in *Phaseolus vulgaris* hypocotyls following infection by *Rhizoctonia solani*. *Physiol. Plant Pathol.* **5**:51–64.

Smith, D. A., Kuhn, P. J., Bailey, J. A., and Burden, R. S. 1980. Detoxification of phaseollidin by *Fusarium solani* f. sp. *phaseoli*. *Phytochemistry* **19**:1673–1675.

Smith, D. A., Harrer J. M., and Cleveland, T. E. 1981. Simultaneous detoxification of phytoalexins by *Fusarium solani* f. sp. *phaseoli*. *Phytopathology* **71**:1212–1215.

Smith, E. F., Brown, N. A., and Townsend, C. O. 1911. Crown gall of plants: Its cause and remedy. *U. S. Dept. Ag. Bur. Plant Indus. Bull. 213.*

Smith, J. L., and Doetsch, R. N. 1968. Motility in *Pseudomonas fluorescens* with special reference to survival and negative chemotaxis. Life Sci. Part II. *Physiol. Pharmacol.* **7**:875–886.

Smith, P. R., and Neales, T. F. 1977. Analysis of the effects of virus infection on the photosynthetic properties of peach leaves. *Aust. J. Plant Physiol.* **4**:723–732.

Smith, P. R., and Neales, T. F. 1980. The effect of peach rosette and decline disease on the translocation of assimilates in peach trees. *Physiol. Plant Pathol.* **16**:383–389.

Smith, S. J., McCall, S. R., and Harris, J. H. 1968. Auxin transport in curly top virus-infected tomato. *Phytopathology* **58**:1669–1671.

Snyder, E. B. 1950. Inheritance and associations of hydrocyanic acid potential, disease reactions, and other characters in sudan grass, *Sorghum vulgare* var. *sudanensis*. Ph.D. Thesis. University of Wisconsin, Madison.

Squire, G. R., and Mansfield, T. A. 1972. Studies of the mechanism of action of fusicoccin, the fungal toxin that induces wilting, and its interaction with abscisic acid. *Planta (Berl.)* **105**:71–78.

Stack, J. P., Mount, M. S., Berman, P. M., and Hubbard, J. P. 1980. Pectic enzyme complex from *Erwinia carotovora*: A model for degradation and assimilation of host pectic fractions. *Phytopathology* **70**:267–272.

Stahmann, M. A. 1965. The biochemistry of proteins of the host and parasite in some plant diseases. (Biochemische Probleme der Kranken Pflanze). *Tagungsberichte* **74**:9–40.

Stahmann, M. A., and Demorest, D. M. 1973. Changes in enzymes of host and pathogen with special reference to peroxidase interaction. In *Fungal Pathogenicity and the Plant's Response* (R. J. W. Byrde and C. V. Cutting, eds.), pp. 405–422. Academic Press, London and New York.

Stahmann, M. A., Clare, B. G., and Woodbury, W. 1966. Increased disease resistance and enzyme activity induced by ethylene and ethylene production by black rot-infected sweet potato tissue. *Plant Physiol. (Bethesda)* **41**:1505–1512.

Stahmann, M. A., Spencer, A. K., and Honold, G. R. 1977. Crosslinking of proteins *in vitro* by peroxidase. *Biopolymers* **16**:1307–1318.

Stakman, E. C., and Piemeisel, F. J. 1917. A new strain of *Puccinica graminis*. *Phytopathology* **7**:73.

Stall, R. E., and Cook, A. A. 1979. Evidence that bacterial contact with the plant cell is necessary for the hypersensitive reaction but not the susceptible reaction. *Physiol. Plant Pathol.* **14**:77–84.

Stanbridge, B., Gay, J. L., and Wood, R. K. S. 1971. Gross and fine structural changes in *Erysiphe graminis* and barley before and during infection. In *Ecology of Leaf Surface Microorganisms* (T. F. Preece and C. H. Dickinson, eds.), pp. 367–379. Academic Press, New York.

Stanghellini, M. E., and Aragaki, M. 1966. Relation of periderm formation and callose deposition to anthracnose resistance in papaya fruit. *Phytopathology* **56**:444–450.

Staskawicz, B. J., and Panopoulos, N. J. 1979. A rapid and sensitive microbiological assay for phaseolotoxin. *Phytopathology* **69**:663–666.

Staub, T., Dahmen, H., and Schwinn, F. J. 1974. Light- and scanning electron microscopy of cucumber and barley powdery mildew on host and nonhost plants. *Phytopathology* **64**:364–372.

Steadman, J. R., and Sequeira, L. 1970. Abscisic acid in tobacco plants. Tentative identification and its relation to stunting induced by *Pseudomonas solanacearum*. *Plant Physiol. (Bethesda)* **45**:691–697.

Steele, J. A., Uchytil, T. F., Durbin, R. D., Bhatnagar, P., and Rich, D. H. 1976. Chloroplast coupling Factor 1: A species-specific receptor for tentoxin. *Proc. Natl. Acad. Sci. U. S. A.* **73**:2245–2248.

Steele, J. A., Uchytil, T. F., and Durbin, R. D. 1978. The stimulation of coupling factor 1 ATPase by tentoxin. *Biochim. Biophys. Acta.* **504**:136–141.

Stein, D. B. 1962. The developmental morphology of *Nicotiana tabacum* ''White Burley'' as influenced by virus infection and gibberellic acid. *Am. J. Bot.* **49**:437.

Steiner, G. W., and Byther, R. S. 1971. Partial characterization and use of a host-specific toxin from *Helminthosporium sacchari* on sugarcane. *Phytopathology* **61**:691–695.

Steiner, G. W., and Strobel, G. A. 1971. Helminthosporoside, a host-specific toxin from *Helminthosporium sacchari. J. Biol. Chem.* **246**:4350–4357.

Stekoll, M., and West, C. A. 1978. Purification and properties of an elicitor of caster bean phytoalexin from culture filtrates of the fungus *Rhizopus stolonifer. Plant Physiol.* (*Bethesda*) **61**:38–45.

Stephens, S. J., and Wood, R. K. S. 1975. Killing of protoplasts by soft-rot bacteria. *Physiol. Plant Pathol.* **5**:165–181.

Sterne, R. E., Kaufmann, M. R., and Zentmeyer, G. A. 1978. Effect of Phytophthora root rot on water relations of avocado; interpretation with a water transport model. *Phytopathology* **68**:595–602.

Stewart, W. W. 1971. Isolation and proof of structure of wildfire toxin. *Nature* (*Lond.*) **229**:172–178.

Stoessl, A. 1981. Structure and biogenetic relations: Fungal non-host specific. In *Toxins in Plant Disease* (R. D. Durbin, ed.), pp. 109–219, Academic Press, New York.

Stoessl, A., Unwin, C. H., and Ward, E. W. B. 1973. Post-infectional fungus inhibitors from plants: Fungal oxidation of capsidiol in pepper fruit. *Phytopathology* **63**:1225–1230.

Stoessl, A., Stothers, J., and Ward, E. W. B. 1976. Sesquiterpenoid stress compounds of the *Solanacae. Phytochemistry* **15**:855–872.

Stoessl, A., Ward, E. W. B., and Stothers, J. B. 1977. Biosynthetic relationships of sesquiterpenoidal stress compounds from *Solanacae*. In *Host Plant Resistance to Pests* (P. A. Hedin, ed.), pp. 61–77. ACS Symp. Ser. No. 62. American Chemical Society, Washington, D. C.

Stonier, T. 1969. Studies on auxin protectors VII. Association of auxin protectors with crown gall development in sunflower stems. *Plant Physiol.* (*Bethesda*) **44**:1169–1174.

Stonier, T. 1971. The role of auxin protectors in autonomous growth. *Colloq. Int. Cent. Natl. Rech. Sci.* No. 193: 423–435.

Stout, R. G., and Cleland, R. E. 1980. Partial characterization of fusicoccin binding to receptor sites on oat root membranes. *Plant Physiol.* (*Bethesda*) **66**:353–359.

Strand, L. L., Rechtoris, C., and Mussell, H. 1976. Polygalacturonases release cell-well-bound proteins. *Plant Physiol.* (*Bethesda*) **58**:722–725.

Strobel, G. A. 1963. A xylanase system produced by *Diplodia viticola. Phytopathology* **53**:592–596.

Strobel, G. A. 1967. Purification and properties of a phytotoxic polysaccharide produced by *Corynebacterium sepedonicum. Plant Physiol.* (*Bethesda*) **42**:1433–1441.

Strobel, G. A. 1973. The helminthosporoside-binding protein of sugarcane: its properties and relationship to susceptibility to eyespot disease. *J. Biol. Chem.* **248**:1321–1328.

Strobel, G. A. 1974a. The toxin-binding protein of sugarcane, its role in the plant and in disease development. *Proc. Natl. Acad. Sci. U. S. A.* **71**:4232–4236.

Strobel, G. A. 1974b. Phytotoxins produced by plant parasites. *Annu. Rev. Plant Physiol.* **25**:541–566.

Strobel, G. A. 1976. Toxins of plant pathogenic bacteria and fungi. In *Biochemical Aspects of Plant–Parasite Relationships* (J. Friend and D. R. Threlfall, eds.), pp. 135–159. Academic Press, London, New York.

Strobel, G. A. 1977. Bacterial phytotoxins. *Annu. Rev. Microbiol.* **31**:205–224.

Strobel, G. A., and Hapner, K. D. 1975. Transfer of toxin susceptibility to plant protoplasts via the helminthosporoside-binding protein of sugarcane. *Biochem. Biophys. Res. Commun.* **63**:1151–1156.

Strobel, G. A., and Hess, W. M. 1974. Evidence for the presence of the toxin-binding protein on the plasma membrane of sugarcane cells. *Proc. Natl. Acad. Sci. U. S. A.* **71**:1413–1417.

Strobel, G. A., and Steiner, G. W. 1972. Runner lesion formation in relation to helminthosporoside in sugar cane leaves infected by *Helminthosporium sacchari*. *Physiol. Plant Pathol.* **2**:129–132.

Strobel, G. A., Talmadge, K. W., and Albersheim, P. 1972. Structure of the phytotoxic glycopeptide of *Corynebacterium sepedonicum*. *Biochim. Biophys. Acta* **261**:365–374.

Strobel, G. A., Van Alfen, N., Hapner, K. D., McNeil, N., and Albershiem, P. 1978. Some phytotoxic glycopeptides from *Ceratocystis ulmi*, the Dutch elm disease pathogen. *Biochim. Biophys. Acta* **538**:60–75.

Suhayda, C. G., and Goodman, R. N. 1981. Early proliferation and migration and subsequent xylem occlusion by *Erwinia amylovora* and the fate of its extracellular polysaccharide (EPS) in apple shoots. *Phytopathology* **71**:697–707.

Sukanya, N. K., and Vaidyanathan, C. S. 1964. Aminotranferases of *Agrobacterium tumefaciens*. Transamination between tryptophan and phenylpyruvate. *Biochem. J.* **92**:594–598.

Suzuki, H. 1980. Defenses triggered by previous invaders: Fungi. In *Plant Disease: An Advanced Treatise* (J. G. Horsfall and E. B. Cowling, eds.), Vol. 5, pp. 319–332. Academic Press, New York.

Swain, T. 1977. Secondary compounds as protective agents. *Annu. Rev. Plant Physiol.* **28**:479–501.

Sziráki, I., Balázs, E., and Király, Z. 1975. Increased levels of cytokinin and inoleacetic acid in peach leaves infected with *Taphrina deformans*. *Physiol. Plant Pathol.* **5**:45–50.

Sziráki, I., Balázs, E., and Király, Z. 1980. Role of different stresses in inducing systemic acquired resistance to TMV and increasing cytokinin level in tobacco. *Physiol. Plant Pathol.* **16**:277–284.

Tabachnik, M., and DeVay, J. E. 1980. Black root rot development in cotton roots caused by *Thielaviopsis basicola* and the possible role of methyl acetate in pathogenesis. *Physiol. Plant Pathol.* **16**:109–117.

Takebe, I. 1975. The use of protoplasts in plant virology. *Annu. Rev. Phytopathol.* **13**:105–125.

Takebe, I., and Otsuki, Y. 1969. Infection of tobacco mesophyll protoplasts by tobacco mosaic virus. *Proc. Natl. Acad. Sci. U. S. A.* **74**:843–848.

Talboys, P. W. 1968. Water deficits in vascular disease. In *Water Deficits and Plant Growth* (T. T. Kozlowski, ed.), Vol. 3, pp. 255–311. Academic Press, New York.

Talboys, P. W. 1972. Resistance to vascular wilt fungi. *Proc. R. Soc. Lond. B. Biol. Sci.* **181**:319–332.

Talboys, P. W. 1978. Dysfunction of the water system. In *Plant Disease: An Advanced Treatise* (J. G. Horsfall and E. B. Cowling, eds.), Vol. 3, pp. 141–181. Academic Press, New York.

Talmadge, K. W., Keegstra, K., Bauer, W. D., and Albersheim, P. 1973. The structure of plant cell walls. I. The macromolecular components of the walls of suspension cultured sycamore cells with a detailed analysis of the pectic polysaccharides. *Plant Physiol. (Bethesda)* **51**:158–173.

Tanaka, H., and Akai, S. 1960. On the mechanism of starch accumulation in tissues surrounding spots in leaves of rice plants due to the attack of *Cochliobolus miyabeanus*. II. On the activities of beta-amylase and invertase in tissues surrounding spots. *Ann. Phytopathol. Soc. Jpn.* **25**:80–84.

Tanaka, S. 1933. Studies on black spot disease of the Japanese pear (*Pyrus serotina*). *Mem. Coll. Agric. Kyoto Univ.* **28**:1–31.

Tani, T., and Yamamoto, H. 1978. Nucleic acid and protein synthesis in association with the resistance of oat leaves to crown rust. *Physiol. Plant Pathol.* **12**:113–121.

Tani, T., and Yamamoto, H. 1979. RNA and protein synthesis and enzyme changes during infection. In *Recognition and Specificity in Plant Host–Parasite Interactions* (J. M. Daly and I. Uritani, eds.), pp. 273–287. Japan Scientific Societies Press, Tokyo, and University Park Press, Baltimore.

Tani, T., Yoshikawa, M., and Naito, N. 1971. Changes in $^{32}$P-ribonucleic acids in oat leaves associated with susceptible and resistant reactions to *Puccinia coronata*. *Ann. Phytopathol. Soc. Jpn.* **37**:43–51.

Tani, T., Yoshikawa, M., and Naito, N. 1973. Template activity of ribonucleic acid extracted from oat leaves infected by *Puccinia coronata*. *Ann. Phytopathol. Soc. Jpn.* **39**:7–13.

Tani, T., Yamamoto, H., Onoe, T., and Naito, N. 1975. Initiation of resistance and host cell collapse in the hypersensitive reaction of oat leaves against *Puccinia coronata avenae*. *Physiol. Plant Pathol.* **7**:231–242.

Tani, T., Yamashita, Y., and Yamamoto, H. 1980. Initiation of induced nonhost resistance of oat leaves to rust infection. *Phytopathology* **70**:39–42.

Taylor, P. A., Schnoes, H. K., and Durbin, R. D. 1972. Characterization of chlorosis-inducing toxins from a plant pathogenic *Pseudomonas* sp. *Biochim. Biophys. Acta* **286**:107–117.

TeBeest, D. O., Durbin, R. D., and Kuntz, J. E. 1976. Stomatal resistance of red oak seedlings infected by *Ceratocystis fagacearum*. *Phytopathology* **66**:1295–1297.

Tegtmeier, K. J., and Van Etten, H. D. 1979. Genetic analysis of sexuality, phytoalexin sensitivity and virulence of *Nectria haematococca* MP VI (*Fusarium solani*). *Phytopathology* **69**:1047 (Abstr.).

Tempé, J., Petit, A., Holtsers, M., Van Montagu, M., and Schell, J. 1977. Thermosensitive step associated with transfer of the Ti plasmid during conjugation: Possible relation to transformations in crown gall. *Proc. Natl. Acad. Sci. U. S. A.* **74**:2848–2849.

Templeton, G. E. 1972. *Alternaria* toxins related to pathogenesis in plants. In *Microbial Toxins* (S. Kadis, A. Ciegler, and S. J. Ajl, eds.), Vol. 8, pp. 169–192. Academic Press, New York and London.

Templeton, G. E., Meyer, W. L., Garble, C. I., and Sigel, C. W. 1967. The chlorosis toxin from *Alternaria tenuis* is a cyclic tetrapeptide. *Phytopathology* **57**:833 (Abstr.).

Thatcher, F. S. 1939. Osmotic and permeability relations in the nutrition of fungus parasites. *Am. J. Bot.* **26**:449–458.

Thatcher, F. S. 1942. Further studies of osmotic and permeability relations in parasitism. *Can. J. Res. Sect. C. Bot. Sci.* **20**:283–311.

Thatcher, F. S. 1943. Cellular changes in relation to rust resistance. *Can. J. Res. Sect. C. Bot. Sci.* **21**:151–172.

Thimann, K. V., Tetley, R. R., and Van Thanh, T. 1974. The metabolism of oat leaves during senesence. *Plant Physiol.* (*Bethesda*) **54**:859–862.

Thomas, H. E. 1934. Studies on *Armillaria mellea* (Vahl.) Quel., infection, parasitism, and host resistance. *J. Agric. Res.* **48**:187–218.

Thomas, P. E., and Fulton, R. W. 1968. Correlation of ectodesmata number with non-specific resistance to initial virus infection. *Virology* **34**:459–469.

Thomashow, M. F., Nutter, R., Postle, K., Chilton, M. D., Blattner, F. R., Powell, A., Gordon, M. P., and Nester, E. W. 1980a. Recombination between plant DNA and Ti plasmid of *Agrobacterium tumefaciens*. *Proc. Natl. Acad. Sci. U. S. A.* **77**:6448–6457.

Thomashow, M. F., Panagopoulos, C. G., Gordon, M. P., and Nester, E. W. 1980b. Host range of *Agrobacterium tumefaciens* is determined by the Ti plasmid. *Nature* (*Lond.*) **283**:794–796.

Thomashow, M. F., Nutter, R., Montoya, A. l., Gordon, M. P., and Nester, E. W. 1980c. Integration and organization of Ti plasmid sequences in crown gall tumors. *Cell* **19**:729–739.

Threlfall, R. J. 1959. Physiological studies on the *Verticillium* wilt disease of tomato. *Ann. Appl. Biol.* **47**:57–77.

Tipton, C. L., Mondal, M. H., and Uhlig, J. 1973. Inhibition of the $K^+$ stimulated ATPase of maize root microsomes by *Helminthosporium maydis* race T pathotoxin. *Biochem. Biophys. Res. Commun.* **51**:525–528.

Tomiyama, K. 1967. Further observation on the time requirement for hypersensitive cell death of potatoes infected by *Phytophthora infestans* and its relation to metabolic activity. *Phytopathol. Z.* **58**:367–378.

Tomiyama, K., and Fukaya, M. 1975. Accumulation of rishitin in dead potato-tuber tissue following treatment with $HgCl_2$. *Ann. Phytopathol. Soc. Jpn.* **41**:418–420.

Tomiyama, K., Takakuwa, M., Takase, N., and Sakai, R. 1959. Alteration of oxidative metabolism in a potato tuber cell invaded by *Phytophthora infestans* and in neighboring tissues. *Phytopathol. Z.* **37**:113–144.

Tomiyama, K., Sakuma, T., Ishizaka, N., Sato, N., Katsui, N., Takasugi, M., and Masamune, T. 1968a. A new antifungal substance isolated from potato tuber tissue infected by pathogens. *Phytopathology* **58**:115–116.

Tomiyama, K., Ishizaka, N., Sato, N., Masamune, T., and Katsui, N. 1968b. "Rishitin" a phytoalexin-like substance. Its role in the defense reaction of potato tubers to infection. In *Biochemical Regulation in Diseased Plants or Injury* (T. Hirari, ed.), pp. 287–292. Kyoritsu Printing Co., Tokyo, Japan.

Tomiyama, K., Doke, N., and Lee, H. S. 1976. Mechanism of hypersensitive cell death in host–parasite interaction. In *Biochemistry and Cytology of Plant–Parasite Interaction* (K. Tomiyama, J. M. Daly, I. Uritani, H. Oku, and S. Ouchi, eds.), pp. 136–142. Kodansha Ltd., Tokyo.

Tomlinson, J. A., and Webb, M. J. W. 1978. Ultrastructural changes in chloroplasts of lettuce infected with beet western yellows virus. *Physiol. Plant Pathol.* **12**:13–18.

Toothman, P. 1982. Octopine accumulation early in crown gall development is progressive. *Plant Physiol.* **69**:214–219.

Touze, A., and Rossingnol, M. 1977. Lignification and the onset of premunition in muskmelon plants. In *Cell Wall Biochemistry Related to Specificity in Host–Plant Pathogen Interactions* (B. Solheim and J. Raa, eds.), pp. 289–293. Universitetsforlaget, Oslo, Norway.

Tribe, H. T. 1955. Studies in the physiology of parasitism. XIX. On the killing of plant cells by enzymes from *Botrytis cinerea* and *Bacterium avoideae*. *Ann. Bot. (N. S.)* **19**:351–368.

Trione, E. J. 1960. The HCN content of flax in relation to flax wilt resistance. *Phytopathology* **50**:482–486.

Troshin, A. S. 1966. *Problems of Cell Permeability* (English translation by M. G. Hell and W. F. Widdas). Pergamon Press, Oxford.

Tschesche, R., Kammerer, F. J., Wulff, G., and Schonbeck, F. 1968. Uber die antibiotisch wirksamen Substanzen der Tulpe (*Tulipa gesneriana*). *Tetrahedron Lett.* **6**:701–706.

Tseng, T. C., and Bateman, D. F. 1969. A phosphatidase produced by *Sclerotium rolfsii*. *Phytopathology* **59**:359–363.

Tseng, T. C., and Chang, L. H. 1970. Phosphatidases produced by some fungal rice pathogens *in vitro*. *Bot. Bull. Acad. Sin. (Taipei)* **11**:120–122.

Tseng, T. C., and Mount, M. S. 1974. Toxicity of endopolygalacturonate *trans*-eliminase, phosphotidase, and protease to potato and cucumber tissue. *Phytopathology* **64**:229–236.

Tu, J. C. 1977. Cyclic AMP in clover and its possible role in clover yellow mosaic virus-infected tissue. *Physiol. Plant Pathol.* **10**:117–123.

Tu, J. C., and Ford, R. E. 1968. Effect of maize dwarf virus infection on respiration and photosynthesis of corn. *Phytopathology* **58**:282–284.

Tukey, H. B., Jr. 1970. The leaching of substances from plants. *Annu. Rev. Plant Physiol.* **21**:305–324.

Turgeon, R., Wood, H. N., and Braun, A. C. 1976. Studies on the recovery of crown gall tumor cells. *Proc. Natl. Acad. Sci. U. S. A.* **73**:3562–3564.

Turner, J. G. 1976. The nonelectrolyte permeability of tobacco cell membranes in tissues inoculated with *Pseudomonas pisi* and *Pseudomonas tabaci*. Ph.D. Thesis. University of Missouri, Columbia.

Turner, N. C., and Dimond, A. E. 1972. Water balance and membrane damage studies with the pressure chamber technique. *Phytopathology* **62**:501. (Abstr.).

Turner, N. C., and Graniti, A. 1969. Fusicoccin: A fungal toxin that opens stomata. *Nature (Lond.)* **223**:1070–1071.

Turner, N. C., and Graniti, A. 1976. Stomatal response of two almond cultivars to Fusicoccin. *Physiol. Plant Pathol.* **9**:175–182.

Tyihák, E., Balla, J., Gáborjányi, R., and Balázs, E. 1978. Increased free formaldehyde level in crude extract of virus infected hypersensitive tobaccos. *Acta Phytopathol. Acad. Sci. Hung.* **13**:29–31.

Uchytil, T. F., and Durbin, R. D. 1980. Hydrolysis of tabtoxins by plant and bacterial enzymes. *Experientia (Basel)* **36**:301–302.

Ueno, T., Nakashima, T., Hayashi, Y., and Fukami, H. 1975a. Structure of AM-toxin I and II. Host-specific phytotoxic metabolites produced by *Alternaria mali. Agric. Biol. Chem.* **39**:1115–1122.

Ueno, T., Nakashima, T., Hayashi, Y., and Fukami, H. 1975b. Isolation and structure of AM toxin III, a host-specific phytotoxic metabolite produced by *Alternaria mali. Agric. Biol. Chem.* **39**:2081–2082.

Upper, C. D., Helgeson, J. P., Kemp, J. D., and Schmidt, C. J. 1970. Gas–liquid chromatographic isolation of cytokinins from natural sources. *Plant Physiol. (Bethesda)* **45**:543–547.

Urbanek, H., and Yirdaw, G. 1978. Acid proteases produced by *Fusarium* species in cultures and in infected seedlings. *Physiol. Plant Pathol.* **13**:81–87.

Uritani, I. 1971. Protein changes in diseased plants. *Annu. Rev. Phytopathol.* **9**:211–234.

Uritani, I. 1976. Protein metabolism. In *Physiological Plant Pathology, Encyclopedia of Plant Physiology* (R. Heitefuss and P. H. Williams, eds.), Vol. 4, pp. 509–525. Springer-Verlag, Berlin and New York.

Uritani, I., and Akazawa, T. 1959. Alteration of the respiratory pattern in infected plants. In *Plant Pathology: An Advanced Treatise I* (J. G. Horsfall and A. E. Dimond, eds.). Vol. 1, pp. 349–391. Academic Press, New York.

Uritani, I., and Kojima, M. 1979. Spore agglutinating factor and germ tube growth inhibiting factor in host plant. In *Recognition and Specificity in Plant Host–Parasite Interactions* (J. M. Daly and I. Uritani, eds.), pp. 181–191. Japan Scientific Societies Press, Tokyo, and University Park Press, Baltimore.

Uritani, I., and Stahmann, M. A. 1961. Pectolytic enzymes of *Ceratocystis fimbriata. Phytopathology* **51**:277–285.

Valent, B. S., and Albersheim, P. 1977. Role of elicitors of phytoalexin accumulation in plant disease resistance. In *Host Plant Resistance to Pests* (P. A. Hedin, ed.), pp. 27–34. American Chemical Society, Washington, D. C.

Van Alfen, N. K., and Allard–Turner, V. 1979. Susceptibility of plants to vascular disruption by macromolecules. *Plant Physiol. (Bethesda)* **63**:1072–1075.

Van Alfen, N. K., and McMillan, B. D. 1982. Macromolecular plant-wilting toxins: Artifacts of the bioassay methods? *Phytopathology* **72**:132–135.

Van Alfen, N. K., and Turner, N. C. 1975a. Changes in alfalfa stem conductance induced by *Corynebacterium insidiosum* toxin. *Plant Physiol. (Bethesda)* **55**:559–561.

Van Alfen, N. K., and Turner, N. C. 1975b. Influence of a *Ceratocystis ulmi* toxin on water relations of elm (*Ulmus americana*). *Plant Physiol. (Bethesda)* **55**:312–316.

Van Andel, O. M., and Fuchs, A. 1972. Interference with plant growth regulation by microbial metabolites. In *Phytotoxins in Plant Diseases* (R. K. S. Wood, A. Ballio, and A. Graniti, eds.), pp. 227–249. Academic Press, London and New York.

Vance, C. P., and Sherwood, R. T. 1976. Cycloheximide treatments implicate papilla formation in resistance of reed canarygrass to fungi. *Phytopathology* **66**:498–502.

Vance, C. P., Anderson, J. O., and Sherwood, R. T. 1976. Soluble and cell wall peroxidases in reed canarygrass in relation to disease resistance and localized lignin formation. *Plant Physiol. (Bethesda)* **57**:920–922.

Vance, C. P., Kirk, T. K., and Sherwood, R. T. 1980. Lignification as a mechanism of disease resistance. *Annu. Rev. Phytopathol.* **18**:259–288.

Van den Ende, G., and Linskens, H. F. 1974. Cutinolytic enzymes in relation to pathogenesis. *Annu. Rev. Phytopathol.* **12**:247–258.

Van den Heuvel, J., and Glazener, J. A. 1975. Comparative abilities of fungi pathogenic and nonpathogenic to bean (*Phaseolus vulgaris*) to metabolize phaseollin. *Neth. J. Plant Pathol.* **81**:125–137.

Vander Molen, G. E., Beckman, C. H., and Rodehorst, E. 1977. Vascular gelation: A general response phenomenon following infection. *Physiol. Plant Pathol.* **11**:95–100.

Van der Wal, A. F., Smeitink, H., and Mann, G. C. 1975. An ecophysiological approach to crop losses exemplified in the system of wheat, leaf rust and glume blotch. *Neth. J. Plant Pathol.* **81**:1–13.

Van Etten, H. D. 1973. Differential sensitivity of Fungi to pisatin and phaseollin. *Phytopathology* **63**:1477–1482.

Van Etten, H. D. 1976. Antifungal activity of pterocarpans and other selected isoflavonoids. *Phytochemistry* **15**:655–659.

Van Etten, H. D. 1979. Relationship between tolerance to isoflavonoid phytoalexins and pathogenicity. In *Recognition and Specificity in Plant Host–Parasite Interactions* (J. M. Daly and I. Uritani, eds.), pp. 301–316. Japan Scientific Societies Press, Tokyo, and University Park Press, Baltimore.

Van Etten, H. D., and Bateman, D. F. 1965. Proteolytic activity in extracts of *Rhizoctonia*-infected bean hypocotyls. *Phytopathology* **55**:1285 (Abstr.).

Van Etten, H. D., and Bateman, D. F. 1969. Enzymatic degradation of galactan, galactomannan, and xylan by *Sclerotium rolfsii*. *Phytopathology* **59**:968–972.

Van Etten, H. D., and Bateman, D. F. 1971. Studies on the mode of action of the phytoalexin phaseollin. *Phytopathology* **61**:1363–1372.

Van Etten, H. D., and Pueppke, S. G. 1976. Isoflavonoid phytoalexins. In *Biochemical Aspects of Plant–Parasite Relationships* (J. Friend and D. R. Threlfall, eds.), pp. 239–289. Academic Press, New York.

Van Etten, H. D., and Smith, D. A. 1975. Accumulation of antifungal isoflavonoids and 1α-hydroxyphaseollone, a phaseollin metabolite, in bean tissue infected with *Fusarium solani* f. sp. *phaseoli*. *Physiol. Plant Pathol.* **5**:225–237.

Van Etten, H. D., and Stein, J. I. 1978. Differential response of *Fusarium solani* to pisatin and phaseollin. *Phytopathology* **68**:1276–1283.

Van Etten, H. D., Matthews, P. S., Tegtmeier, K. J., Dietert, M. F., and Stein, J. I. 1980. The association of pisatin tolerance and demethylation with virulence on pea in *Nectria haematococca*. *Physiol. Plant Pathol.* **16**:257–268.

Van Larebeke, N., Engler, G., Holsters, M., Van den Elsacker, S., Zaenen, I., Schilperoort, R. A., and Schell, J. 1974. Large plasmid in *Agrobacterium tumefaciens* essential for crown gall-inducing ability. *Nature* (*Lond.*) **252**:169–170.

Van Larebeke, N., Genetello, C., Schell, J., Schilperoort, R. A., Hermans, A. K. Hernalsteens, J. P., and Van Montagu, M. 1975. Acquisition of tumor-inducing ability by non-oncogenic agrobacteria as a result of plasmid transfer. *Nature* (*Lond.*) **255**:742–743.

Van Montagu, M., and Schell, J. 1979. The plasmids of *Agrobacterium tumefaciens*. In *Plasmids of Medical, Environmental and Commercial Importance* (K. N. Timmis and A. Puhler, eds.), pp. 71–95. Elsevier/North-Holland Biomedical Press, Amsterdam.

Varner, J. E., and Ho, D. T. H. 1976. Hormones. In *Plant Biochemistry* (J. Bonner and J. E. Varner, eds.), 3rd ed., pp. 713–770. Academic Press, New York.

Varns, J. L., and Kuć, J. 1971. Suppression of rishitin and phytuberin accumulation and hypersensitive response in potato by compatible races of *Phytophthora infestans*. *Phytopathology* **61**:178–181.

Varns, J. L., Currier, W. W., and Kuć, J. 1971a. Specificity of rishitin and phytuberin accumulation by potato. *Phytopathology* **61**:968–971.

Varns, J. L., Kuć, J., and Williams, E. B. 1971b. Terpenoid accumulation as biochemical response of the potato tuber to *Phytophthora infestans*. *Phytopathology* **61**:174–177.

Veech, J. A. 1976. Localization of Peroxidase in *Rhizoctonia solani*-infected cotton seedlings. *Phytopathology* **66**:1072–1076.

Veech, J. A. 1978. An apparent relationship between methoxy-substituted terpenoid aldehydes and the resistance of cotton to *Meloidogyne incognita*. *Nematologica* **24**:81–87.

Veech, J. A., and McClure, M. A. 1977. Terpenoid aldehydes in cotton roots susceptible and resistant to the root-knot nematode, *Meloidogyne incognita*. *J. Nematol.* **9**:225–229.

Vidhyasekaran, P. 1974. Carbohydrate metabolism of ragi plants infected with *Helminthosporium nodulosum*. *Phytopathol. Z.* **79**:130–141.

Von Broembsen, S. L., and Hadwiger, L. A. 1972. Characterization of disease resistance responses in certain gene-for-gene interactions between flax and *Melampsora lini*. *Physiol. Plant Pathol.* **2**:207–215.

Wacek, T. J., and Sequeira, L. 1973. The peptidoglycan of *Pseudomonas solanacearum*: Chemical composition and biological activity in relation to the hypersensitive reaction in tobacco. *Physiol. Plant Pathol.* **3**:363–369.

Wade, M., Cline, K., Hahn, M., Valent, B. S., and Albersheim, P. 1977. The stimulation of phytoalexin accumulation in higher plants by fungal glucans. *Abstracts of the 69th American Phytopathological Society Annual Meeting. East Lansing, Michigan,* p. 89. American Phytopathological Society, St. Paul.

Wagoner, W., Loschke, D. C., and Hadwiger, L. A. 1982. Two-dimensional electrophoretic analysis of *in vivo* and *in vitro* synthesis of proteins in peas inoculated with compatible and incompatible *Fusarium solani*. *Physiol. Plant Pathol.* **20**:99–107.

Walker, J. C., and Stahmann, M. A. 1955. Chemical nature of disease resistance in plants. *Annu. Rev. Plant Physiol.* **6**:351–366.

Wallis, F. M. 1977. Ultrastructural histopathology of tomato plants infected with *Corynebacterium michiganense*. *Physiol. Plant Pathol.* **11**:333–342.

Wallis, F. M., and Truter, J. 1978. Histopathology of tomato plants infected with *Pseudomonas solanacearum,* with emphasis on ultrastructure. *Physiol. Plant Pathol.* **13**:307–317.

Walton, J. D., Earle, E. D., Yoder, O. C., and Spanswick, R. M. 1979. Reduction of adenosine triphosphate levels in susceptible maize mesophyll protoplasts by *Helminthosporium maydis* race T toxin. *Plant Physiol.* (*Bethesda*) **63**:806–810.

Ward, E. W. B., and Stoessl, A. 1976a. Phytoalexins from potatoes: Evidence for the conversion of lubimin to 1,5-dihydrolubimin by fungi. *Phytopathology* **66**:468–471.

Ward, E. W. B., and Stoessl, A. 1976b. On the question of "elicitors" or "inducers" in incompatible interactions between plants and fungal pathogens. *Phytopathology* **66**:940–941.

Wardlaw, I. F., and Passioura, J. B., eds. 1976. *Transport and Transfer Processes in Plants*. Academic Press, New York.

Waterman, M. A., Aist, J. R., and Israel, H. W. 1978. Centrifugation studies help to clarify the role of papilla formation in compatible barley–powdery mildew interactions. *Phytopathology* **68**:797–802.

Watson, B., Currier, T. C., Gordon, M. P., Chilton, M. D., and Nester, E. W. 1975. Plasmid required for virulence of *Agrobacterium tumefaciens*. *J. Bacteriol.* **123**:255–264.

Wattenbarger, D. W., Gray, E., Rice, J. S., and Reynolds, J. H. 1968. Effects of frost and freezing on hydrocyanic acid potential of sorghum plants. *Crop Sci.* **8**:526–528.

Webb, J. L. 1966. *Quinones, Enzymes and Metabolic Inhibitors,* Vol. 3. Academic Press, New York.

Webster, J. M. 1969. The host–parasite relationships of plant-parasitic nematodes. *Adv. Parsitol.* **7**:1–40.

Weiler, E. W., and Spanier, K. 1981. Phytohormones in the formation of crown gall tumors. *Planta* **153**:326–337.

Weinhold, A. R., and Hancock, J. G. 1980. Defense at the perimeter: Extruded chemicals. In *Plant Disease: An Advanced Treatise* (J. G. Horsfall and E. B. Cowling, eds.), Vol. 5, pp. 121–138. Academic Press, New York.

Weinstein, L. I., Hahn, M. G., and Albersheim, P. 1981. Isolation and biological activity of glycinol, a pterocarpan phytoalexin synthesized by soybeans. *Plant Physiol.* (*Bethesda*) **68**:358–363.

Wenkert, W., Lemon, E. R., and Sinclair, T. R. 1978. Changes in water potential during pressure bomb measurement. *Agronomy J.* **70:**353–355.

Weststeijn, E. A. 1978. Permeabiity changes in the hypersensitive reaction of *Nicotiana tobacum* cv. Xanthi mc after infection with tobacco mosaic virus. *Physiol. Plant Pathol.* **13**:253–258.

Whatley, M. H. 1977. Studies on the adherence step essential for tumor induction by *Agrobacterium*. Ph.D. Thesis. Northwestern University, Evanston, IL.

Whatley, M. H., and Sequeira, L. 1981. Bacterial attachment to plant cell walls. In *Recent Advances in Phytochemistry: The Phytochemistry of Cell Recognition and Cell Surface Interactions* (F. A. Loewus and C. A. Ryan, eds.), Vol. 15, pp. 213–240. Plenum Press, New York.

Whatley, M. H., Bodwin, J. S., Lippincott, B. B., and Lippincott, J. A. 1976. Role for *Agrobacterium* cell envelope lipopolysaccharide in infection site attachment. *Infect. Immun.* **13**:1080–1083.

Whatley, M. H., Margot, J. B., Schell, J., Lippincott, B. B., and Lippincott, J. A. 1978. Plasmid and chromosomal determination of *Agrobacterium* adherence specificity. *J. Gen. Microbiol.* **107**:395–398.

Whatley, M. H., Hunter, N., Cantrell, M. A., Keegstra, K., and Sequeira, L. 1979. Bacterial lipopolysaccharide structure and the induction of the hypersensitive response in tobacco. *Plant Physiol.* (*Bethesda*) **63** (Suppl.):134 (Abstr.).

Whatley, M. H., Hunter, N., Cantrell, M. A., Hendrick, C., Keegstra, K., and Sequeira, L. 1980. Lipopolysaccharide composition of the wilt pathogen, *Pseudomonas solanacearum* correlation with the hypersensitive response in tobacco. *Plant Physiol.* (*Bethesda*) **65**:557–559.

Wheeler, H. 1969. The fate of victorine in susceptible and resistant oat coleoptiles. *Phytopathology* **59**:1093–1097.

Wheeler, H. 1975. *Plant Pathogenesis*. Springer-Verlag, Berlin and New York.

Wheeler, H. 1976a. Permeability alterations in diseased plants. In *Physiological Plant Pathology* (R. Heitefuss and P. H. Williams, eds.), Vol. 4, pp. 413–429. Springer-Verlag, New York.

Wheeler, H. 1976b. The role of phytotoxins in specificity. In *Specificity in Plant Diseases* (R. K. S. Wood and A. Graniti, eds.), pp. 217–230. Plenum Press, New York.

Wheeler, H. 1977. Increase with age in sensitivity of oat leaves to victorin. *Phytopathology* **67**:859–861.

Wheeler, H. 1978. Disease alterations in permeability and membranes. In *Plant Disease: An Advanced Treatise* (J. G. Horsfall and E. B. Cowling, eds.), Vol. 3, pp. 327–347. Academic Press, New York.

Wheeler, H. 1981. Role in pathogenesis. In *Toxins in Plant Disease* (R. D. Durbin, ed.), pp. 477–494. Academic Press, New York.

Wheeler, H., and Ammon, V. D. 1977. Effects of *Helminthosporium maydis* T-toxin on the uptake of uranyl salts in corn roots. *Phytopathology* **67**: 325–330.

Wheeler, H., and Black, H. S. 1963. Effects of *Helminthosporium victoriae* and victorin upon permeability. *Am. J. Bot.* **50**:686–693.

Wheeler, H., and Luke, H. H. 1963. Microbial toxins in plant disease. *Annu. Rev. Microbiol.* **17**:223–242.

Whenham, R. J., and Fraser, R. S. S. 1981. Effect of systemic and local-lesion-forming strains of tobacco mosaic virus on abscisic acid concentration in tobacco leaves: Consequences for the control of leaf growth. *Physiol. Plant Pathol.* **18**:267–278.

White, J. A., Calvert, D. H., and Brown, M. F. 1973. Ultrastructural changes in corn leaves after inoculation with *Heminthosporium maydis,* race T. *Phytopathology* **63**:296–300.

Wickner, W. 1980. Assembly of proteins into membranes. *Science* **210**:861–868.

Wieringa-Brants, D. H. 1981. The role of the epidermis in virus-induced local lesions on cowpea and tobacco leaves. *J. Gen. Virol.* **54**:209–212.

Wiese, M. V., and DeVay, J. E. 1970. Growth regulator changes in cotton associated with defoliation caused by *Verticillium albo-atrum. Plant Physiol.* (*Bethesda*) **45**:304–309.

Williams, E. B., and Kuć, J. 1969. Resistance in *Malus* to *Venturia inaequalis. Annu. Rev. Phytopathol.* **7**:223–246.

Williams, P. H. 1979. How fungi induce disease. In *Plant Disease: An Advanced Treatise* (J. G. Horsfall and E. B. Cowling, eds.), Vol. 4, pp. 163–179. Academic Press, New York.

Williams, P. H., Reddy, M. N., and Strandburg, J. O. 1969. Growth of noninfected and *Plasmodiophora brassicae*-infected cabbage callus in culture. *Can. J. Bot.* **47**:1217–1221.

Williams, P. H., Aist, S. J., and Bhattacharya, P. K. 1973. Host–parasite relations in cabbage clubroot. In *Fungal Pathogenicity and the Plant's Response* (R. J. W. Byrde and C. V. Cutting, eds.), pp. 141–158. Academic Press, New York.

Willmitzer, L., DeBeuckeleer, M., Lemmers, M., Van Montagu, M., and Schell, J. 1980. DNA from Ti plasmid present in nucleus and absent from plastids of crown gall plant cells. *Nature (Lond.)* **287**:359–361.

Wilson, C. L. 1973. A lysosomal concept for plant pathology. *Annu. Rev. Phytopathol.* **11**: 247–272.

Wilson, C. L. 1978. Plant teratomas—who's in control of them? In *Plant Disease: An Advanced Treatise* (J. G. Horsfall and E. B. Cowling, eds.), Vol. 3, pp. 215–230. Academic Press, New York.

Wimalajeewa, D. L. S., and DeVay, J. E. 1971. The occurrence and characterization of a common antigen relationship between *Ustilago maydis* and *Zea mays. Physiol. Plant Pathol.* **1**:523–535.

Wolpert, J. S., and Albersheim, P. 1976. Host–symbiont interactions. I. The lectins of legumes interact with O-antigen containing lipopolysaccharides of their symbiont rhizobia. *Biochem. Biophys. Res. Commun.* **70**:729–737.

Wolpert, T. J., and Dunkle, L. D. 1980. Purification and partial characterization of host-specific toxins produced by *Periconia circinata. Phytopathology* **70**:872–876.

Wong, W. C., and Preece, T. F. 1978. *Erwinia salicis* in cricket bat willows: Peroxidase, polyphenoloxidase, β-glucosidase, pectinolytic and cellulolytic enzyme activity in diseased wood. *Physiol. Plant Pathol.* **12**:333–347.

Wood, H. N. 1970. Revised identification of the chromophore of a cell division factor from crown gall tumor cells of *Vinca rosea* L. *Proc. Natl. Acad. Sci. U. S. A.* **67**:1283–1287.

Wood, H. N. 1972. The development of a capacity for autonomous growth of the crown gall tumor cell. *Prog. Exp. Tumor Res.* **15**:78–92.

Wood, K. R., and Barbara, D. J. 1971. Virus multiplication and peroxidase activity in leaves of cucumber (*Cucumis sativus* L.) cultivars systemically infected with the W strain of cucumber mosaic virus. *Physiol. Plant Pathol.* **1**:73–81.

Wood, R. K. S. 1967. *Physiological Plant Pathology*. Blackwell Scientific, Oxford and Edinburgh.

Wood, R. K. S. 1972. The killing of plant cells by soft-rot parasites. In *Phytotoxins in Plant Diseases* (R. K. S. Wood, A. Ballio, and A. Graniti, eds.), pp. 273–288. Academic Press, New York.

Wood, R. K. S. 1976a. Killing of protoplasts. In *Biochemical Aspects of Plant–Parasite Relationships* (J. Friend and D. R. Threlfall, eds.), pp. 105–116. Academic Press, New York.

Wood, R. K. S. 1976b. Specificity—an assessment. In *Specificity in Plant Diseases* (R. K. S. Wood, A. Ballio, and A. Graniti, eds.), pp. 327–338. Plenum Press, New York.

Woodward, J. R., Keane, P. J., and Stone, B. A. 1980. Structure and properties of wilt-inducing polysaccharides from *Phytophthora* species. *Physiol. Plant Pathol.* **16**:439–454.

Wright, S. T. C. 1969. An increase in the 'inhibitor-β' content of detached wheat leaves following a period of wilting. *Planta* (*Berl.*) **86**:10–20.

Wright, S. T. C. 1977. The relation between leaf water potential and the levels of abscisic acid and ethylene in excised wheat leaves. *Planta* **134**:183–189.

Wright, S. T. C., and Hiron, R. W. P. 1969. (+)-abscisic acid, the growth inhibitor induced in detached wheat leaves following a period of wilting. *Nature* (*Lond.*) **224**:719.

Wullems, G. J., Molendijk, L., Ooms, G., and Schilperoort, R. A. 1981a. Retention of tumor markers in $F_1$ progeny plants from *in vitro*-induced optopine and nopaline tumor tissues. *Cell* **24**:719–727.

Wullems, G. J., Molendijk, L., Ooms, G., and Schilperoort, R. A. 1981b. Differential expression of crown gall tumor markers in transformant, obtained after *in vitro Agrobacterium tumefaciens*-induced transformation of cell wall regenerating protoplasts derived from *Nicotiana tabacum*. *Proc. Natl. Acad. Sci. U. S. A.* **78**:4344–4348.

Wyen, N. V., Udvardy, J., and Farkas, G. L. 1971. Changes in nucleolytic enzymes in virus-infected tobacco leaf tissues. *Acta Phytopathol. Acad. Sci. Hung.* **6**:37–38.

Wyman, J. G., and Van Etten, H. D. 1978. Antibacterial activity of selected isoflavonoids. *Phytopathology* **68**:583–589.

Wynn, W. K. 1963. Photosynthetic phosphorylation by chloroplasts isolated from rust-infected oats. *Phytopathology* **53**:1376–1377.

Yabuta, T., and Hayashi, T. 1939. Biochemical studies on ''bakanae'' fungus of rice. II. Isolation of gibberellin, the active principle which produces slender rice seedlings. *J. Agric. Chem. Soc. Jpn.* **15**:257–266.

Yadav, N. S., Postle, K., Saiki, R. K., Thomashow, M. F., and Chilton, M. D. 1980. T-DNA of a grown gall teratoma is covalently joined to host plant DNA. *Nature* (*Lond.*) **287:**458–461.

Yamamoto, H., Tani, T., and Hokin, H. 1976. Protein synthesis linked with resistance of oat leaves to crown rust fungus. *Ann. Phytopathol. Soc. Jpn.* **42**:583–590.

Yang, F., and Simpson, R. B. 1981. Revertant seedlings from crown gall tumors retain a portion of the bacterial Ti plasmid DNA sequences. *Proc. Natl. Acad. Sci. U. S. A.* **78**:4151–4155.

Yang, F., Montoya, A. I., Merlo, D. J., Drummond, M. H., Chilton, M.-D., Nester, E. W., and Gordon, M. P. 1980. Foreign DNA sequences in crown gall teratoma and their fate during the loss of the tumorous traits. *Mol. Gen. Genet.* **177**:707–714.

Yang, S. F., and Pratt, H. K. 1978. The physiology of ethylene in wounded plant tissue. In *Biochemistry of Wounded Plant Storage Tissues* (G. Kahl, ed.), pp. 595–622. de Gruyter, Berlin.

Yang, S. F., Ku, H. S., and Pratt, H. K. 1967. Photochemical production of ethylene from methionine and its analogues in the presence of FMN. *J. Biol. Chem.* **242**(22):5274–5280.

Yarwood, C. E. 1954. Mechanism of acquired immunity to a plant rust. *Proc. Natl. Acad. Sci. U. S. A.* **40**:374–377.

Yarwood, C. E. 1967. Responses to parasites. *Annu. Rev. Plant Physiol.* **18**:419–438.

Yarwood, C. E. 1976. Modification of the host response—Predisposition. In *Physiological Plant Pathology: Encyclopedia of Plant Physiology* (R. Heitefuss and P. H. Williams, eds.), Vol. 4. pp. 703–718 Springer-Verlag, Berlin.

Yoder, O. C. 1973. A selective toxin produced by *Phyllosticta maydis*. *Phytopathology* **63**:1361–1366.

Yoder, O. C. 1976. Evaluation of the role of *Helminthosporium maydis,* race T toxin in Southern Corn Leaf Blight. In *Biochemistry and Cytology of Plant–Parasite Interaction* (K. Tomiyama, J. M. Daly, I. Uritani, H. Oku, and S. Ouchi, eds.), pp. 16–24. Kodansha, Ltd., Tokyo.

Yoder, O. C. 1980. Toxins in pathogenesis. *Annu. Rev. Phytopathol.* **18**:103–129.

Yoder, O. C., and Gracen, V. E. 1975. Segregation and pathogenicity types and host-specific toxin production in progenies of crosses between races T and O of *Helminthosporium maydis* (*Cochliobolus heterostrophus*). *Phytopathology* **65**:273–276.

Yoder, O. C., and Scheffer, R. P. 1969. Role of toxin in early interactions of *Helminthosporium victoriae* with susceptible and resistant oat tissue. *Phytopathology* **59**:1954–1959.

Yoder, O. C., and Scheffer, R. P. 1973. Effects of *Helminthosporium carbonum* toxin on absorption of solutes by corn roots. *Plant Physiol.* (*Bethesda*) **52**:518–523.

York, D. W. 1978. *Ultrastructural and Biochemical Aspects of Southern Corn Leaf Blight Disease*. Ph.D. Thesis. Cornell University, Ithaca, New York.

Yoshikawa, M. 1978. Diverse modes of action of biotic and abiotic phytoalexin elicitors. *Nature (Lond.)* **275**:546–547.

Yoshikawa, M., Masago, H., and Keen, N. T. 1977a. Activated synthesis of poly(A)-containing messenger RNA in soybean hypocotyls inoculated with *Phytophthora megasperma* var. *sojae*. *Physiol. Plant Pathol.* **10**:125–138.

Yoshikawa, M., Tsukadaira, T., Masago, H., and Minoura, S. 1977b. A nonpectolytic protein from *Phytophthora capsici* that macerates plant tissue. *Physiol. Plant Pathol.* **11**:61–70.

Yoshikawa, M., Yamauchi, K., and Massago, H. 1978a. Glyceollin: Its role in restricting fungal growth in resistant soybean hypocotyls infected with *Phytophthora megasperma* var. *sojae*. *Physiol. Plant Pathol.* **12**:73–82.

Yoshikawa, M., Yamauchi, K., and Massago, H. 1978b. *De novo* messenger RNA and protein synthesis are required for phytoalexin-mediated disease resistance in soybean hypocotyls. *Plant Physiol.* (*Bethesda*) **61**:314–317.

Yoshikawa, M., Yamauchi, K., and Massago, H. 1979. Biosynthesis and biodegradation of glyceollin by soybean hypocotyls infected with *Phytophthora megasperma* var. *sojae*. *Physiol. Plant Pathol.* **14**:157–169.

Yoshikawa, M., Matama, M., and Massago, H. 1981. Release of a soluble phytoalexin elicitor from mycelial walls of *Phytophthora megasperma* var. *sojae* by soybean tissues. *Plant Physiol.* (*Bethesda*) **67**:1032–1035.

Young, K. 1978. A simple psychrometer for continuous field use. *J. Agric. Eng. Res.* **23**:339–341.

Yu, P. H. 1971. Untersuchungen über die Beeinflussung des Wuchsstoff- und Nukleinsaure-Haushaltes von *Sinapis alba* L. durch Infektion mit *Albugo candida* (Lev.) Kunze. Ph.D. Thesis. Georg–August University, Göttingen.

Zaenen, I., Van Larebeke, N., Teuchy, H., Van Montagu, M., and Schell, J. 1974. Supercoiled circular DNA in crown-gall inducing *Agrobacterium* strains. *J. Mol. Biol.* **86**:109–127.

Zaitlin, M. 1976. Virus cross protection: More understanding is needed. *Phytopathology* **66:** 382–383.

Zaitlin, M., and Hesketh, J. D. 1965. The short term effects of infection by tobacco mosaic virus on apparent photosynthesis of tobacco leaves. *Ann. Appl. Biol.* **55**:239–263.

Zaitlin, M., Leonard, D. A., Shalla, T. A., and Petersen, L. J. 1981. A possible function for the 30,000 MW protein coded by TMV RNA. Fifth International Congress of Virology, Abstract, p. 254.

Zaki, A. I., Keen, N. T., and Erwin, D. C. 1972a. Implications of vergosin and hemigossypol in the resistance of cotton to *Verticillium albo-atrum*. *Phytopathology* **62**:1402–1406.

Zaki, A. I., Keen, N. T., Sims, J. J., and Erwin, D. C. 1972b. Vergosin and hemigossypol antifungal compounds produced in cotton plants inoculated with *Verticillium albo-atrum*. *Phytopathology* **62**:1398–1401.

Zaki, A. I., Zentmyer, G. A., Pettus, J., Sims, J. J., Keen, N. T., and Sing, V. O. 1980. Borbonol from *Persea* spp.—chemical properties and antifungal activity against *Phytophthora cinnamoni*. *Physiol. Plant Pathol.* **16**:205–212.

Zaki, A. I., and Durbin, R. D. 1965. Effect of bean rust on the translocation of photosynthetic products from diseased leaves. *Phytopathology* **55**:528–529.

Zeigler, R. S., Powell, L. E., and Thurston, H. D. 1980. Gibberellin $A_4$ production by *Sphaceloma manihoticola,* causal agent of cassava superelongation disease. *Phytopathology* **70**:589–593.

Zimmerman, M. H., and McDonough, J. 1978. Dysfunction in the flow of food. In *Plant Disease: An Advanced Treatise* (J. G. Horsfall and E. B. Cowling, eds.), Vol. 3, pp. 117–140. Academic Press, New York.

Zucker, M. 1968. Sequential induction of phenyalanine ammonia-lyase and a lyase-inactivating system in potato tuber disks. *Plant Physiol.* (*Bethesda*) **43**:365–374.

Zucker, M. L., and Hankin, L. 1970. Regulation of pectate lyase synthesis in *Pseudomonas fluorescence* and *Erwinia carotovora*. *J. Bact.* **104**:13–18.

# Index